SPRINGER HANDBOOK OF AUDITORY RESEARCH

Series Editors: Richard R. Fay and Arthur N. Popper

Springer Handbook of Auditory Research

Volume 1: The Mammalian Auditory Pathway: Neuroanatomy
Edited by Douglas B. Webster, Arthur N. Popper, and Richard R. Fay

Volume 2: The Mammalian Auditory Pathway: Neurophysiology
Edited by Arthur N. Popper and Richard R. Fay

Volume 3: Human Psychophysics
Edited by William Yost, Arthur N. Popper, and Richard R. Fay

Volume 4: Comparative Hearing: Mammals
Edited by Richard R. Fay and Arthur N. Popper

Volume 5: Hearing by Bats
Edited by Arthur N. Popper and Richard R. Fay

Volume 6: Auditory Computation
Edited by Harold L. Hawkins, Teresa A. McMullen, Arthur N. Popper, and Richard R. Fay

Volume 7: Clinical Aspects of Hearing
Edited by Thomas R. Van De Water, Arthur N. Popper, and Richard R. Fay

Volume 8: The Cochlea
Edited by Peter Dallos, Arthur N. Popper, and Richard R. Fay

Volume 9: Development of the Auditory System
Edited by Edwin W Rubel, Arthur N. Popper, and Richard R. Fay

Volume 10: Comparative Hearing: Insects
Edited by Ronald Hoy, Arthur N. Popper, and Richard R. Fay

Volume 11: Comparative Hearing: Fish and Amphibians
Edited by Richard R. Fay and Arthur N. Popper

Volume 12: Hearing by Whales and Dolphins
Edited by Whitlow W.L. Au, Arthur N. Popper, and Richard R. Fay

Volume 13: Comparative Hearing: Birds and Reptiles
Edited by Robert Dooling, Arthur N. Popper, and Richard R. Fay

Volume 14: Genetics and Auditory Disorders
Edited by Bronya J.B. Keats, Arthur N. Popper, and Richard R. Fay

Volume 15: Integrative Functions in the Mammalian Auditory Pathway
Edited by Donata Oertel, Richard R. Fay, and Arthur N. Popper

Volume 16: Acoustic Communication
Edited by Andrea Simmons, Arthur N. Popper, and Richard R. Fay

Volume 17: Compression: From Cochlea to Cochlear Implants
Edited by Sid P. Bacon, Richard R. Fay, and Arthur N. Popper

Volume 18: Speech Processing in the Auditory System
Edited by Steven Greenberg, William Ainsworth, Arthur N. Popper, and Richard R. Fay

Volume 19: The Vestibular System
Edited by Stephen M. Highstein, Richard R. Fay, and Arthur N. Popper

Volume 20: Cochlear Implants: Auditory Prostheses and Electric Hearing
Edited by Fan-Gang Zeng, Arthur N. Popper, and Richard R. Fay

Volume 21: Electroreception
Edited by Theodore H. Bullock, Carl D. Hopkins, Arthur N. Popper, and Richard R. Fay

Continued after index

Jacqueline F. Webb
Richard R. Fay
Arthur N. Popper
Editors

Fish Bioacoustics

With 81 Illustrations

Springer

Editors
Jacqueline F. Webb
Department of Biological Sciences
University of Rhode Island
100 Flagg Road
Kingston, RI 02881
USA
jacqueline_webb@mail.uri.edu

Arthur N. Popper
Department of Biology
University of Maryland
College Park, MD 20742
USA
apopper@umd.edu

Richard R. Fay
Parmly Sensory Sciences Institute
6525 North Sheridan Road
Loyola University Chicago
Chicago, IL 60626
USA
rfay@luc.edu

Series Editors
Richard R. Fay
Parmly Sensory Sciences Institute
6525 North Sheridan Road
Loyola University Chicago
Chicago, IL 60626
USA

Arthur N. Popper
Department of Biology
University of Maryland
College Park, MD 20742
USA

ISBN: 978-0-387-73028-8 e-ISBN: 978-0-387-73029-5

Library of Congress Control Number: 2007927886

Cover illustration: The image includes parts of Figures 2.2 and 8.1 appearing in the text.

Printed on acid-free paper

9 8 7 6 5 4 3 2 1

springer.com

Dedication

This volume is dedicated to the memory of Professor Arthur A. Myrberg, Jr (1933–2005). Art was a true pioneer in the field of animal bioacoustics. His insightful experimental studies and his creative thinking moved the field forward, and stand as benchmarks against which all subsequent work must be evaluated. And, in addition to being a great scholar, Art was a marvelous colleague, teacher, and mentor. But most importantly, Art was a great and valued friend.[1]

[1]The editors are grateful that they were able to let Art know of their intent to dedicate this book to him prior to his passing away. He was very touched by this, and provided two pictures we might use. Art told us in a letter that he could not decide which picture would be best—a recent photo that shows him with his beloved books, or a somewhat earlier photo showing him, slate in hand, returning from observing animals underwater. He also told us that he asked everyone who visited him which photo to use and the vote was evenly split between the two. His request to us was that we use both photos if at all possible. We are pleased to honor Art's request.

Contents

Contributors

ANDREW H. BASS
Department of Neurobiology and Behavior, Cornell University, Ithaca, NY 14853-3901, USA, Email: ahb3@cornell.edu

HORST BLECKMANN
Institut für Zoologie der Rheinischen, Friedrich-Wilhelms-Universität Bonn, D-53115 Bonn, Germany, Email: Bleckmann@uni-bonn.de

CHRISTOPHER B. BRAUN
Department of Psychology, Hunter College of The City University of New York (CUNY), New York, NY 10021, USA, Email: cbraun@hunter.cuny.edu

PEGGY L. EDDS-WALTON
Parmly Sensory Sciences Institute, Loyola University Chicago, Chicago, IL 60626, USA, Email: pewalton@luc.edu

RICHARD R. FAY
Parmly Sensory Sciences Institute, Loyola University Chicago, Chicago, IL 60626, USA, Email: rfay@luc.edu

TERRY GRANDE
Department of Biology, Loyola University Chicago, Chicago, IL 60626, USA, Email: tgrande@luc.edu

ANTHONY D. HAWKINS
Loughine Ltd., Kincraig, Blairs Aberdeen, AB12 5YT, UK, Email: a.hawkins@btconnect.com

J. MICHAEL JECH
NOAA/NMFS Northeast Fisheries Science Center, Woods Hole, MA 02543, USA, Email: michael.jech@noaa.gov

FRIEDRICH LADICH
Department of Neurobiology and Behavior, University of Vienna, 1090 Vienna, Austria, Email: friedrich.ladich@univie.ac.at

DAVID A. MANN
University of South Florida, College of Marine Science, St. Petersburg, FL 33701, USA, Email: dmann@marine.usf.edu

JOACHIM MOGDANS
Institut für Zoologie der Rheinischen, Friedrich-Wilhelms-Universität Bonn, D-53115 Bonn, Germany, Email: mogdans@uni-bonn.de

JOHN C. MONTGOMERY
Leigh Marine Laboratory and School of Biological Sciences, University of Auckland, Auckland 1142, New Zealand, Email: j.montgomery@auckland.ac.nz

ARTHUR N. POPPER
Department of Biology, University of Maryland, College Park, MD 20742, USA, Email: apopper@umd.edu

PETER H. ROGERS
The George W. Woodruff School of Mechanical Engineering, Georgia Institute of Technology, Atlanta, GA 30332, USA, Email: peter.rogers@gatech.edu

OLAV SAND
Department of Molecular Biosciences, The University of Oslo, N-0316 Oslo, Norway, Email: olav.sand@bio.uio.no

CARL R. SCHILT
LGL Ltd. Environmental Research Associates, North Bonneville, WA 98639, USA, Email: cschilt@lgl.com

JACQUELINE F. WEBB
Department of Biological Sciences, University of Rhode Island, Kingston, RI 02881, USA, Email: Jacqueline_webb@mail.uri.edu

DAVID G. ZEDDIES
Department of Biology, Center for the Comparative and Evolutionary Biology of Hearing, University of Maryland, College Park, MD 20742, USA, Email: dzeddies@umd.edu

Series Preface

The Springer Handbook of Auditory Research presents a series of comprehensive and synthetic reviews of the fundamental topics in modern auditory research. The volumes are aimed at all individuals with interests in hearing research including advanced graduate students, postdoctoral researchers, and clinical investigators. The volumes are intended to introduce new investigators to important aspects of hearing science and to help established investigators to better understand the fundamental theories and data in fields of hearing that they may not normally follow closely.

Each volume presents a particular topic comprehensively, and each serves as a synthetic overview and guide to the literature. As such, the chapters present neither exhaustive data reviews nor original research that has not yet appeared in peer-reviewed journals. The volumes focus on topics that have developed a solid data and conceptual foundation rather than on those for which a literature is only beginning to develop. New research areas will be covered on a timely basis in the series as they begin to mature.

Each volume in the series consists of a few substantial chapters on a particular topic. In some cases, the topics will be ones of traditional interest for which there is a substantial body of data and theory, such as auditory neuroanatomy (Vol. 1) and neurophysiology (Vol. 2). Other volumes in the series deal with topics that have begun to mature more recently, such as development, plasticity, and computational models of neural processing. In many cases, the series editors are joined by a co-editor having special expertise in the topic of the volume.

RICHARD R. FAY, Chicago, IL
ARTHUR N. POPPER, College Park, MD

Volume Preface

Fishes comprise the largest group of vertebrates by far. Indeed, there are more extant species of fishes than there are of all other vertebrate species combined. And, with this diversity in species, fishes show remarkable diversity and adaptations in the ways in which they deal with the aquatic environment. The diversity of structure and function in sensory systems is exceptional, and suggests that, as fishes have evolved, they have found "new" ways to gather information about their highly diverse environments. This diversity is particularly evident in the octavolateralis system of fishes, the inner ear and the lateral line: the senses that detect water motion and sound.

This volume provides an overview of the octavolateralis system of fishes, but unlike earlier volumes on the topic, it takes an approach that explores fish bioacoustics both from a basic perspective of understanding how fishes detect and process signals and from an applied perspective that explores how bioacoustics is used to understand and affect fish behavior.

In Chapter 1, Fay, Popper, and Webb provide an historical perspective on the topic of fish bioacoustics and also give a brief introduction to "who" fishes are. This is followed in Chapter 2 by Popper and Schilt, in which the authors explore hearing capabilities and mechanisms of fishes, and put these findings into the context of several applied approaches to fish bioacoustics, including a discussion of attempts that have been made to use sound to "control" fish behavior. In Chapter 3, Fay and Edds-Walton continue the discussion of fish hearing, but examine the topic from the perspective of the physiology of the ear and the central nervous system. They emphasize the strong similarities between fishes and terrestrial vertebrates in the organization and function of the auditory brain. Finally, the issue of evolutionary adaptations of the auditory system for the detection and processing of the sound pressure waveform is examined by Braun and Grande in Chapter 4.

Despite being aware of the presence of the lateral line for centuries, it has only been relatively recently that investigators have started to really understand its critical function in the lives of fishes. The role of the lateral line is discussed from the viewpoint of morphology, physiology, and function in Chapter 5 by Webb, Montgomery, and Mogdans. They also discuss the interaction of input to the lateral line and inner ear which is expanded upon in Chapter 6 by Sand and Bleckmann, who discuss one of the most fascinating of all issues in fish bioacoustics: the orientation and localization to sound by fish. Sound source localization is also treated in Chapter 7 by Rogers and Zeddies, who present a new and important model of the mechanism by which fish are likely to localize sound, and by Fay and Edds-Walton (Chapter 3), who discuss the central neural circuits that may underlie sound source localization.

Fishes use sound in a wide range of behavioral contexts, and this is explored by Bass and Ladich in Chapter 8. The theme of acoustic communication is continued in Chapter 9, where Mann, Hawkins, and Jech consider the use of sounds produced by fishes in applied approaches to fisheries biology.

As with other volumes in the Springer Handbook of Auditory Research series, the chapters in this volume are complemented by chapters in earlier volumes. Volume 9 in the series, *Comparative Hearing: Fish and Amphibians* (edited by Fay and Popper) has several chapters relevant to this volume including the structure of the ear (Popper and Fay), hearing capabilities (Fay and Megela Simmons), anatomy of the auditory CNS (McCormick), the lateral line (Coombs and Montgomery), and acoustic communication (Zelick, Mann, and Popper). The physical acoustics of underwater communication are discussed by Bass and Clark in Vol. 16, *Acoustic Communication* (edited by Megela Simmons, Popper, and Fay). Volume 22 in the series, *Evolution of the Vertebrate Auditory System* (edited by Manley, Popper, and Fay), has several chapters on the evolution of the octravolateralis system in fish including an examination of the evolution of the ear (Ladich and Popper), sensory hair cells (Coffin, Kelley, Manley, and Popper), and on environmental constraints on hearing and the concept of auditory scene analysis (Lewis and Fay). In the same volume, Clack and Allin discuss the transition from fish to land vertebrates in terms of changes in the ear. In Vol. 25, *Sound Source Localization* (edited by Popper and Fay), Fay discusses fish sound localization capabilities.

Two other volumes in the SHAR series are relevant and related to this one. *Electroreception*, Vol. 21 (edited by Bullock, Hopkins, Popper, and Fay) discusses another major sensory system of fishes that is related, in an evolutionary sense, to the octavolateralis system. Volume 28, *Hearing and Sound Communication in Amphibians* (edited by Narins, Feng, Fay, and Popper), considers many of the same topics that are considered in this volume, in a group of vertebrates that may be very instructive to help us further understand fish bioacoustics.

Jacqueline F. Webb, Kingston, RI
Richard R. Fay, Chicago, IL
Arthur N. Popper, College Park, MD

1
Introduction to Fish Bioacoustics

RICHARD R. FAY, ARTHUR N. POPPER, AND JACQUELINE F. WEBB

1. Introduction

The field of fish bioacoustics was essentially inaugurated in the 1960s with two meetings and their subsequent published proceedings, which were organized and edited by Professor William N. Tavolga. These two volumes, *Marine Bio-Acoustics* (Tavolga 1964) and *Marine Bio-Acoustics II* (Tavolga 1967), define the scope and content of the field of marine bioacoustics to this day. Hearing and sound production, underwater acoustics, and a plethora of other topics are discussed in the volumes. Authors of chapters in the volumes emphasized that investigators must examine all organisms from invertebrates to marine mammals when considering the underwater sound environment.

Although fish bioacoustics was an important component of the two Tavolga volumes, interest in the bioacoustics of fish started far earlier. Some of the earliest discussion of sound and fish was by Pliny the Elder more than 2,000 years ago when he wrote in "The History of the World" that:

> FISHES verily have no eares, ne yet any holes to serve for hearing: and yet plaine it is that they doe heare. Which we may daily see in certaine fish-ponds and stewes where fishes bee kept: for when those that have the charge of them make a noise with clapping of their hands: as wild as they bee otherwise, they shall have them come in great flockes to take their meat that is throwne into them: and this are they wont to doe daily . . . . Hereupon it is, that the Mullet, sea-Pike, Stockfish, and Chronius, are thought to heare best of all others, and therfore live very ebbe among the shelves and shallowes.

Other well-known writers suggested that they were aware that fish could detect sounds. Indeed, the famous English fisherman Issac Walton cautioned that one should walk very slowly near a fishing site so that the fish would not detect the sounds of the walker and be frightened away.

In modern times, as reviewed by Moulton (1963) and Tavolga (1971), perhaps the earliest studies that tested fish hearing were by G.H. Parker (e.g., 1903), who was the first to demonstrate that fishes are able to detect sounds. Later, Karl von Frisch (who went on to win the Nobel Prize for his studies on the language of bees) and his students (e.g., von Frisch 1923; von Frisch and Dijkgraaf 1935) did a set of monumental studies that demonstrated that fishes use their ears for hearing and also provided the first quantitative measures of hearing sensitivity and signal discrimination in fishes. Although it is beyond the scope of this chapter or volume to write a history of the field of fish bioacoustics, many of the formative articles were republished (in English) in two important volumes by Tavolga (1976, 1977).

The field of fish bioacoustics has expanded greatly since the publication of Tavolga's 1964 and 1967 volumes. The present volume is an update on the field as it now is defined, covering the topics of central and peripheral mechanisms in the ear and the lateral line system, sound production and communication, the evolution of sensory specializations, acoustics and its application to the biomechanics of the ear, acoustic orientation, and the use of acoustics to locate and assess populations.

Fish bioacoustics was, and continues to be, interdisciplinary, and an understanding of the field requires contributions from psychology, biology, evolution, population biology, biomechanics, physical acoustics, and mathematical modeling. Although the field has expanded in the last four decades with many new observations and concerns, some of the fundamental questions posed in the 1960s have remained only partially answered and new questions have arisen. These questions include a full understanding of the myriad of peripheral and central hearing mechanisms and their biomechanics among species, the diversity of sound production mechanisms, the ubiquity of sound communication behaviors among fishes, the relationships between the auditory and lateral line systems in sound detection and source perception, and the mechanisms for acoustic orientation and source localization behaviors of fishes using both the ears and lateral line systems. To these old and persistent questions may be added the effects of anthropogenic (human-generated) sound on fishes, a topic of current importance (e.g., Popper 2003). This volume addresses these and other questions now arising in the field of fish bioacoustics.

There are currently more than 30,000 named species of living fishes (see www.fishbase.org), but only a mere fraction have been investigated with reference to their ability to detect acoustic stimuli (via the ears and lateral line) and to produce sound. Nevertheless, approaches from comparative biology (e.g., informed taxonomic sampling and phylogenetic inference) can be used to estimate the number of fish species that are likely to be hearing specialists based on anatomical features shared among members of particular fish taxa. For instance, all approximately 360 species of clupeiform fishes (e.g., herrings, shads, and alewives) have an air bubble associated with the inner ear, and the approximately 7,800 species of otophysans (e.g., goldfish, catfish, and carp), which represent more than two-thirds of freshwater fish species and more than 25 % of all fish species, have a series of bones, the Weberian ossicles, that mechanically connect the swim bladder to the inner ear. In addition, there is a demonstration of or evidence for hearing specializations in representatives of 26 other families across the spectrum of teleost fishes (see Braun and Grande, Chapter 4). Thus, it is now known that more than one in five teleost families have members that have, or are likely to have, specializations that enhance hearing sensitivity and frequency range of hearing (e.g., are probably "hearing specialists"), thus demonstrating the importance of hearing in the lives of a great diversity of fishes.

2. Peripheral and Central Hearing Mechanisms

It was understood decades ago (e.g., von Frisch and Dijkgraaf 1935) that some fish species were apparently specialized for hearing and would respond in proportion to sound pressure (e.g., the Otophysi with their Weberian ossicles), whereas other species had no such specializations. In recent years, fishes in the former group have been referred to as "hearing specialists" based on the presence of structures linking the swim bladder and ears, whereas fishes without specializations have been referred to as "hearing generalists." But more recently, this dichotomy has been called into question (e.g., Popper and Schilt, Chapter 2). What do we really mean by "specialists" and "generalists," and how is a species thus characterized? Braun and Grande (Chapter 4) suggest that specializations may be more varied and numerous than previously recognized, with possible hearing specializations having evolved independently 20 or more times (also see Ladich and Popper 2004). But the functions and biological roles of these putative hearing specializations have not been studied in most fish taxa. Thus, except for the Otophysi, clupeids, and mormyrids (elephantfishes), the term "specialist" can be applied only uncertainly. In addition, it has been shown that the marine catfish *Arius felis* (an otophysan) hears best at the lower frequencies characteristic of the generalists, although with the great sensitivity characteristic of other otophysans (Popper and Tavolga 1981). Most recently, one group of "hearing specialists," the alosine clupeids (including the American shad, *Alosa sapidissima*), has been shown to hear ultrasound (up to 200 kHz) but to have rather insensitive hearing in the "normal" range of other hearing specialists (Popper and Schilt, Chapter 2). Other fishes (butterflyfishes in the genus (*chaetodon*) have an intimate connection between the swim bladder and the lateral line canal system (laterophysic connection; Webb 1998; Webb, Montgomery, and Mogdans Chapter 5), which may turn out to be a sensory specialization. In what way is hearing or lateral line function enhanced by this relationship of the swim bladder to the lateral line canal system?

It is now understood that species lacking a swim bladder (e.g., flatfishes, some tunas, and all sharks) most likely do not respond to sound pressure but do respond directly to hydrodynamic water motions in the acoustic near field and to acoustic particle motion in the acoustic far field (Sand and Bleckmann, Chapter 6; Rogers and Zeddies, Chapter 7). This has been shown in two flatfishes, the plaice and dab (Chapman and Sand 1974), for which particle motion audiograms have been determined. Most other species that have swim bladders but lack known specializations linking the swim bladder and the ears fall into an unknown category in which there is uncertainty and controversy regarding their pressure and particle motion sensitivity. In at least one case, an unspecialized fish has been shown to respond to sound pressure over the higher-frequency portion of their hearing range (i.e., a damselfish, Myrberg and Spires 1980), and Tavolga and Wodinsky (1963) in their classic paper on fish psychoacoustics noted that some species may actually "switch" between detecting pressure and particle motion. Are "hearing generalists" likely to respond to pressure, particle motion, or both [as is required by Schuijf's (1975) phase model of directional hearing (see Sand and Bleckmann, Chapter 6; Rogers and Zeddies Chapter 7)]? This is an important question because in an attempt to determine a species' sensitivity to sound and to assess the effects of anthropogenic noise on fish auditory systems (Popper and Schilt, Chapter 2), it is important to know what acoustical quantity to manipulate and measure. For example, a sound pressure audiogram for a species tested in the laboratory is meaningless if the species responds primarily to acoustic particle motion. This problem is greatest in the acoustic near field or where sound does not propagate (the conditions under which all laboratory investigations of hearing sensitivity have taken place) because the ratio of pressure to particle motion varies with distance from the source, frequency, proximity to the water surface, and other factors; the acoustic particle motion amplitude must be measured, but this has been done only rarely (e.g., Myrberg and Spires 1980). So the assumptions that all fishes with a swim bladder respond in proportion to sound pressure (an implicit assumption of many studies on fish hearing) or, alternatively, that all hearing generalists respond only in proportion to acoustic particle motion, may not be useful. Again, relevant sensory anatomy and the relative contributions of acoustic particle motion and sound pressure must be determined empirically among a diversity of species until it is possible to correctly infer physiological function from anatomical structure.

The demonstration that at least some unspecialized fishes respond to both pressure and particle motion (Myrberg and Spires 1980) but in different ratios in particular frequency ranges makes understanding hearing in fishes all the more uncertain and difficult. Hearing sensitivity and frequency range may be a function of various aspects of anatomy as well as of water depth, fish depth, source distance, and other aspects of the underwater environment that determine the actual ratio of pressure to particle motion (effective impedance of the medium). On the other hand, sensitivity to both pressure and particle motion means that a fish is capable of determining more about acoustic sources and environments using sound than animals sensitive to only one or the other acoustic quantity (e.g., terrestrial animals with only sound pressure sensitivity).

When Tavolga's original volumes on marine bioacoustics (1964, 1967) were published, very little was understood about the central auditory systems of anamniotes, including fish, beyond generalizations from the classical work of Herrick (1948) on the tiger salamander. Thanks in large part to the persistent focus of Northcutt and of McCormick on the auditory brains of fishes (e.g., Northcutt 1980, 1981; McCormick and Hernandez 1996; McCormick 1999), it is now clear that the organization of the auditory central nervous system (CNS) in fishes is consistent with that understood for most other vertebrates at levels from the lower hindbrain to the telencephalon. At most levels, auditory nuclei of the amniotic vertebrates have functional analogies among the fishes investigated so far. At the same time, however, it has not been possible to identify homologies among nuclei across vertebrate taxa, and the highly analogous pathways and functions that we see must still be attributed to parallel or convergent evolution (Grose et al. 2004).

As discussed by Fay and Edds-Walton (Chapter 3), in most fishes investigated to date, there are five octaval nuclei of the medulla ("octaval column"; Northcutt 1980). Recently, McCormick and Hernandez (1996) and McCormick (1999) have described the secondary octaval nuclei in fishes (possibly analogous to the superior olivary complex of terrestrial vertebrates) and what was called the secondary octaval population (SOP) by Fay and Edds-Walton (Chapter 3). The SOP of fishes may be composed of one to three subdivisions, with the dorsally positioned SOdor population being the most consistently present. Axons from auditory sites in the medulla in fishes travel via the lateral lemniscus to the torus semicircularis in the midbrain as in other vertebrates. In nonelectric fishes (e.g., species other than mormyrids and gymnotids), the midbrain includes the auditory nucleus centralis (NC) and the lateral line nucleus ventrolateralis (NV). Reciprocal connections exist between the auditory nucleus centralis of the torus semicircularis in the midbrain and the central posterior nucleus of the dorsal thalamus. Other potential auditory sites based on projections from NC include the ventromedial nucleus of the ventral thalamus, the preglomerular complex, and the anterior tuberal nucleus of the hypothalamus in both otophysines ("auditory specialists") and auditory generalists. There are multiple nuclei of the telencephalon identified anatomically (Streidter 1991), but little is known about their function.

The physiology of auditory brain cells has been studied in selected species primarily at the levels of the primary afferents of the auditory nerve, the medulla (primarily in the descending octaval nucleus), and the midbrain (nucleus centralis of the torus semicircularis). There is one report on the response properties of thalamic cells in the goldfish (central posterior nucleus; Lu and Fay 1996).

Fay and Edds-Walton (Chapter 3) conclude that anatomical, physiological, and behavioral (psychophysical) studies have revealed that the general flow of auditory information from the periphery to the midbrain in teleost fishes is similar to that in most other vertebrates investigated. The response properties revealed by single-unit and multiunit studies indicate that basic acoustic features are encoded by the auditory afferents of teleost fishes and have much in common with terrestrial vertebrates, including frequency selectivity at the periphery, highly selective and discontinuous tuning curves in the brain that are not seen in the periphery, a gradual loss of phase locking as the auditory system is ascended, and many of the same temporal response properties of single cells that are known for tetrapods. Units of the central nervous system encode temporal patterns and frequency via phase locking, with frequency selectivity and directionality encoded at all levels as in all other vertebrate brains so far investigated. Limited evidence indicates that additional frequency selectivity and directional sharpening occur in the midbrain, probably through excitatory–inhibitory interactions within auditory nuclei, as occur in other vertebrates.

The data Fay and Edds-Walton (Chapter 3) review are consistent with the hypothesis that many of the basic functions of auditory processing in terrestrial vertebrates also are found in fishes. The most important of these functions must be the common fundamental auditory capacity of all vertebrates, i.e., the capacity to determine and perceptually segregate sources of sound so that appropriate behavior may occur with respect to them (Lewis and Fay 2004).

3. Sound Production Mechanisms and Behaviors

As Bass and Ladich (Chapter 8) discuss, it appears that the number of species known to produce sounds and to communicate acoustically has steadily grown over the years. They propose a classification scheme for the sound production mechanisms in fishes that is based on anatomical structures adapted exclusively for sound production and communication. The main group of sound production mechanisms includes sonic swim bladder mechanisms with their numerous morphological variations. Intrinsic drumming muscles attach solely to the swim bladder walls, whereas extrinsic muscles originate on other structures such as the skull, ribs, and vertebrae. The second major group of adaptations for sound production includes movements of the pectoral girdle, pectoral fin rays, or fin tendons. It has also been proposed that the grating of pharyngeal teeth results in the production of sounds in many species. In addition, bubbles emitted from the cloaca in herring produce a stereotyped series of high-frequency pulses, which might have some communicative value, but this still needs to be demonstrated.

There are several exciting new studies of the neural and behavioral mechanisms of acoustic communication among teleost fishes, including studies of neuroendocrine vocal and auditory mechanisms (e.g., Sisneros and Bass 2003; Bass and Ladich, Chapter 8). There is tremendous diversity of reproductive and acoustic behaviors among teleosts. These studies also show that such mechanisms are likely common to all vertebrates given the conserved pattern of the organization of the neuroendocrine, vocal, and auditory systems. Many of these traits are also likely to be shared with other vertebrates. For example, the ascending auditory system of fishes that communicate acoustically largely resembles that of teleosts and of vertebrates in general, including those that are not known to produce sounds (Bass and Ladich, Chapter 8). Thus, the central mechanisms responsible for processing communication sounds, at least initially, are likely to be shared among all fishes (and possibly among all vertebrates; Fay and Edds-Walton, Chapter 3). Most communication sounds used by fishes are temporally patterned and their interpretation would require neural circuits adapted for the analysis of such patterns. Fay and Edds-Walton (Chapter 3) emphasize what we understand of these temporal pattern analyzers in the brains of several species. There is a need for more behavioral and neural studies on the sensitivity of fishes to the temporal parameters of acoustic signals (e.g., Fay 1985; Crawford 1997;

Bodnar and Bass 1999), particularly at forebrain levels where species-specific sound communication processing strategies (if they exist) are likely to be found.

A better understanding of sound communication among fishes has also aided the assessment and localization of fish populations using acoustical survey methods (Mann, Hawkins, and Jech, Chapter 9). Passive acoustics uses listening to communication sounds produced by fishes to understand the distribution of fish populations and, because most sounds are produced in to the context of courtship and spawning, to understand the dynamics of spawning. In passive acoustics, it is usually easy to identify which species are being studied because most the communication sounds are species specific, but it has been more difficult to quantify fish abundance from the sounds they produce. The future of these fields depends on developing algorithms to process large data sets and to classify automatically the species under study.

Active acoustics uses a pulse of sound generated by a transducer, and one or more receivers are used to "listen" for echoes. Because fish scatter sounds, especially from their swim bladders, active acoustics can be used to map and quantify fish abundance. The challenge of the use of active acoustics has been the development of models of fish sound scattering and the ability to quantify numbers and identify fish species based on the characteristics of scattered sound. There is great potential for combining passive and active acoustic systems to study fish populations. Many of the issues related to understanding fish populations and their distributions that are difficult to study with passive acoustics could be answered with active acoustic systems. At the same time, there has been considerable discussion among fisheries biologists, as discussed by Popper and Schilt (Chapter 2), that the very sounds used to find fish stocks may also have an effect on fishing if the frequency range of the echosounder or fishing vessel overlaps with the hearing range of the fish that are ensonified.

4. Relationship Between Auditory and Lateral Line Systems

At the time of Tavolga's volumes (1964, 1967) and Cahn's volume entitled *Lateral Line Detectors* (1967), the lateral line system was generally thought to be an accessory hearing organ that responded to the frequency range below the normal range of the ear. In fact, reviewers required that any study of hearing in any fish species had to demonstrate that the lateral line system was not involved and vice versa. Tavolga and Wodinsky (1963) determined sound pressure audiograms for nine species of fishes that were considered to be hearing generalists and observed "double audiograms" for some. These alternate audiograms appeared after extensive avoidance training at the lowest frequencies and were hypothesized to arise from responses by the lateral line system. The origin of these alternative audiograms is still not clear and they have not been reported in subsequent studies of hearing in fishes (Fay 1988). But, at the time, the findings of Tavolga and Wodinsky (1963) did raise awareness of the idea that results from hearing studies could possibly be explained by contributions of both the ear and the lateral line system. This notion was supported by van Bergeijk (1967) in his belief that the lateral line system was required for directional hearing (Sand and Bleckmann, Chapter 6), and some contemporary investigators still cite the lateral line system as responsible for low-frequency hearing in fishes.

It is now understood that the biomechanics of the lateral line system is fundamentally different from that of the ears. The ears are stimulated inertially by motion of the fish's body (whole body motion engaged by the motion of the surrounding medium) and, in some cases, by sound pressure. In contrast, the lateral line system, which is composed of neuromast receptor organs in pored lateral line canals and on the skin's surface, is activated by relative motions between the water medium and the fish's body. This relative motion can cause pressure gradients at adjacent lateral line canal pores, resulting in displacements of the fluid in the canals, which stimulates canal neuromasts. Superficial neuromasts on the skin surface are activated by near-field hydrodynamic motions of the medium relative to the body surface (Webb, Montgomery, and Mogdans Chapter 5). In general, these relative motions occur only in the acoustic near field where there are steep amplitude gradients of hydrodynamic motions (Sand and Bleckmann, Chapter 6). Thus, stimulation of the lateral line occurs only very close to the acoustic source (within one or two body lengths). The lateral line cannot respond in the far field because although the fish's body may move with water particles, relative motion between the body and the medium does not significantly occur. Of course, it is possible that stimulation of the lateral line system may occur in laboratory studies of hearing where the primary stimulus is generally near-field particle motion (e.g., Tavolga and Wodinsky 1963). However, if the lateral line canal system is linked to the swim bladder (as in butterflyfishes of the genus *Chaetodon*), the lateral line system could be made sensitive to sound pressure.

Webb, Montgomery, and Mogdans (Chapter 5) discuss the diverse nature and sources of lateral line stimuli, the functional roles of the lateral line system, the functional attributes of different components of the lateral line system (e.g., ability of canal vs. superficial neuromasts to detect vibratory stimuli against a background of flowing waters), and the multimodal integration of hydrodynamic (lateral line) and acoustic (ear) stimuli by the CNS. It is clear that the functional evolution of the ear and lateral line system has occurred in response to the complexity of underwater acoustics but in ways that are still not fully understood.

5. Orientation and Sound Source Localization

The issue of determination of sound source direction and orientation is longstanding and complex (e.g., see van Bergeijk 1967; chapters in Tavolga 1976). Sand and Bleckman (Chapter 6) discuss the theories and data on directional hearing and lateral line source localization, whereas Rogers and Zeddies (Chapter 7) propose a new way of thinking about sound source localization that involves the ear. With regard to sound source localization, it is pointed out that nearly all published studies have left open the question as to whether fishes are able to determine the actual location of the source as opposed to being able to discriminate differences in location. If one generalizes from what is understood about source location in humans and presumably other terrestrial animals, the many behavioral studies on source location discrimination and directional masking support the idea that at least some fish species are able to determine where the discriminated sources are located. But an unequivocal demonstration of source localization in fishes, as in humans, has yet to be done. In the few demonstrations of fishes orienting to or approaching sound sources (e.g., Popper et al. 1973; McKibben and Bass 1999), it is not clear whether the fish were using near-field mechanisms (possibly including the lateral line system) or far-field hearing.

In any case, Sand and Bleckmann (Chapter 6) point out that all aspects of directional hearing mediated by the ears must rely on direct, inertial stimulation of the otolith organs and a sort of "vectorial weighting" resulting from the diversity of hair cell orientations documented for the otolithic organs in all species investigated (Popper and Schilt, Chapter 2). Fay and Edds-Walton (Chapter 3) discuss the fate of this directionally encoded information in the brain of the oyster toadfish (*Opsanus tau*). It is remarkable that the vast majority of brain stem cells investigated reflect, and even enhance, the directionality that is set up at the periphery. It seems unlikely that this directionality operates only to discriminate differences in location (or to aid in some other undefined hearing function) but not to somehow represent the actual location of sound sources. At the same time, all the mechanisms by which fishes may unambiguously determine the location of sound sources remain a mystery in terms of experimental evidence. As pointed out by Sand and Bleckmann (Chapter 6), Schuijf's phase model (1975), the only substantial theory of sound source localization that we have had (but see Rogers and Zeddies, Chapter 7), may contain several untenable assumptions. This state of affairs has led to Kalmijn's (1989, 1997) notion of a "guidance" procedure that explains how a fish could be successful in approaching a continuous sound source using only a vectorial weighting scheme. Although attractive, this ethological account of sound source localization leaves the question of whether fishes can determine the location of sources, as suggested by the large location discrimination literature that Sand and Bleckmann (Chapter 6) discuss, unexplained.

Sand and Bleckmann (Chapter 6) also discuss another mystery about orientation and sound source localization concerning the directionality of the Mauthner cell (M-cell) response that mediates reflex or fast responses to nearby sources. There is a fairly complete account of the sensory information used by the M-cells in making this rapid decision and a way to view a solution of the 180° ambiguity problem. Moulton and Dixon (1967) first presented behavioral data with respect to this problem in Tavolga's (1967) *Marine Bio-Acoustics II* volume. They demonstrated that binaural hearing was necessary for this reflexive response and that rather high-frequency hearing (not involving the lateral line system) was sufficient for directed behaviors in goldfish. A question that lingers since

Moulton (1967) concerns the relationships, if any, between this reflex directional response and sound source localization (or source location discrimination). This was subsequently studied behaviorally at various times by investigators including Hawkins, Sand, Johnstone, Schuijf, Buwalda, and others. Although the problem may have been solved at a reticulospinal level, this probably cannot simultaneously explain how fishes can remember and discriminate sound source location or possibly "know" about source location for action at a later time. Perhaps fishes have solved the directional hearing problem twice, in parallel, using entirely different neural elements and circuits. This seems inefficient, but perhaps the mechanisms (and theory) for one solution have served as a model for the other.

Experiments by Schuijf et al. (1977) on the ide (*Leuciscus idus*), an otophysan, showed that an otophysan species with Weberian ossicles displays directional hearing and is able to discriminate between sounds from sources 180° apart. Because the saccular hair cell orientation patterns in otophysans are all aligned on an essentially vertical (dorsoventral) axis, an account of sound source localization making use of the vectorial weighting hypothesis is difficult to formulate for the saccule alone. As pointed out by Sand and Bleckmann (Chapter 6), however, localization in azimuth may be accomplished using binaural processing (Sand 1974; Fay and Edds-Walton 1997) and only localization in elevation requires a diversity of hair cell orientations (but see Rogers and Zeddies, Chapter 7). In the case of azimuth, the monaural vectorial-weighting mechanisms need not operate because a difference in the overall activation of the right and left ears are all that may be required. If so, then the observations of Ma and Fay (2002) that all the cells of the goldfish midbrain respond to sounds that are oriented essentially vertically (unlike the unspecialized toadfish, *Opsanus tau*) may not be inconsistent with good localization in azimuth. The auditory torus semicircularis of the midbrain appears to represent the neurally coded output of the saccule in both specialized and unspecialized fishes. This leaves the question of elevation localization for Otophysi open and unaccounted for without the use of the lagena and the assumption that the sampling of midbrain cells by Ma and Fay (2002) may have missed input influences from the lagena or utricle. Perhaps the Otophysi have no need for localization abilities in elevation due to the shallow-water environments in which they live and the likelihood that birds are an important predator and always attack from above.

Finally, it is worth considering, as both Sand and Bleckman (Chapter 6) and Fay (2005) have pointed out, that the focus on the 180° ambiguity problem in fish audition and the efforts to develop a single, unifying model for its solution may have been exaggerated, and this idea is reinforced by Rogers and Zeddies (Chapter 7), who propose a new model for localization for which these 180° ambiguities do not arise. Humans and all terrestrial vertebrates also encounter auditory ambiguities (i.e., the "cones of confusion" problems). Because all sources located on these cones produce the same (including zero) interaural differences in timing, phase, and intensity, this problem is a general one. All terrestrial animals must handle these ambiguities in both the up–down (dorsoventral) and front–back (rostrocaudal) axes. Such ambiguities are solved by various means, including movements of the head and pinnae, directional filtering by the ears (the head-related transfer functions), visual and other sensory cues, and estimation of the most likely source location based on experience. Considering the various and familiar solutions to the auditory ambiguity problems in terrestrial vertebrates, it is reasonable to suggest that fishes may also employ a variety of mechanisms to resolve the 180° ambiguity. After all, sound sources are real objects having many physical and chemical attributes, and perhaps it is asking too much of the auditory system to solve every problem associated with them.

Finally, the issue of the 180° ambiguity may be resolvable if the biomechanics of the fish ear is reconsidered, as has been done by Rogers and Zeddies (Chapter 7). In this chapter, Rogers and Zeddies argue that if one assumes a quadrupole model rather than a monopole or dipole model for sound detection, it is possible for fishes, even those without a swim bladder (e.g., sharks and flatfishes), to determine sound source direction and to do so without the 180° ambiguity of earlier models. This quadrupole model, while needing experimental testing, is important in that it provides new thinking that solves the localization problem for all species and for all kinds of sounds as opposed to earlier models that required a swim bladder (or other air bubble) and that were primarily designed to localize pure tones (totally nonbiological signals).

6. Effects of Anthropogenic Noise

Over the past decade, it has become apparent that there are actually two groups of investigators interested in fish bioacoustics. One group has primarily been asking basic questions about fish hearing and sound detection, including questions of form and function, behavior, and physiology. The second group has taken a far more applied approach to fish bioacoustics and has been asking questions related to how sound can be used to understand the behavior of fish and to control the behavior of fish. Interestingly, it was rare that these two groups of investigators interacted to share knowledge and ideas. Although there has been more communication in recent years (e.g., Popper and Carlson 1998), there is still a need for further interactions. Developing such interactions in order to inform the two groups of one another's interests and concerns is one of the foci of Popper and Schilt (Chapter 2).

The other focus of this chapter is to bring to the field of fish bioacoustics issues that are increasingly becoming more important as humans add sound to the aquatic environment. Although it is impossible to know how much noisier the marine environment is now as compared to the days prior to steam shipping (e.g., early 1800s), it is clear that with the advent of larger and louder noise sources (e.g., shipping, oil exploration, and sonar), the oceans have become noisier. However, it was not only recently (e.g., National Research Council 1994; Wartzog et al. 2004) that investigators, regulators, and industries started to develop a concern that the added noise budget in the environment might have a deleterious effect on marine organisms. This issue looms larger and larger now, and although there are still few data to show how increased human-generated (anthropogenic) sound impacts fishes (e.g., Popper et al. 2003; Popper et al. 2004), there is growing international concern about this issue.

7. Summary

The field of fish bioacoustics has grown considerably since the two pioneering volumes by Tavolga (1964, 1967) and the equally important volume on lateral line by Cahn (1967). Indeed, ever since two more recent volumes that covered fish bioacoustics (Tavolga et al. 1981; Fay and Popper 1999) and one on the lateral line (Coombs et al. 1989), the field has grown. Yet a large number of questions remain.

Although many of these remaining questions have already been discussed in this chapter, one that has only been "hinted" at is that of patterns and processes in the evolution of both hearing and sound production mechanisms. There are on the order of 30,000 species of living fishes. As demonstrated by Retzius in 1881, there is extraordinary diversity in the anatomy of fish ears and particularly in those parts of the ears associated with hearing (Popper et al. 2003; Ladich and Popper 2004). Similarly, there is remarkable diversity in peripheral structures associated with enhancing sound detection (e.g., the swim bladder and vertebral elements; see Braun and Grande, Chapter 4) and in the lateral line system (Webb 1989; Webb, Montgomery, and Mogdans, Chapter 5). What this means for sound production, sound detection, and acoustic behavior remains a mystery. Indeed, as investigators start to have the capabilities of studying a greater diversity of species, including those living at great depths (e.g., Popper 1980), it becomes apparent that the diversity in structure and, presumably, in function of the ear and lateral line system is even greater than previously appreciated.

References

Bodnar DA, Bass AH (1999) Midbrain combinatorial code for temporal and spectral information in concurrent acoustic signals. J Neurophysiol 81:552–563.

Cahn PH, ed (1967) Lateral Line Detectors. Bloomington, IN: Indiana University Press.

Chapman CJ, Sand O (1974) Field studies of hearing in two species of flatfish, *Pleuronectes platessa* (L.) and *Limanda limanda* (L.) (family Pleuronectidae). Comp Biochem Physiol 47A:371–385.

Coombs S, Görner P, Münz H, eds (1989) The Mechanosensory Lateral Line: Neurobiology and Evolution. New York: Springer-Verlag.

Crawford JD (1997) Feature-detecting auditory neurons in the brain of a sound-producing fish. J Comp Physiol 180:439–450.

Fay RR (1985) The goldfish ear codes the axis of acoustic particle motion in three dimensions. Science 225:951–954.

Fay RR (1988) Hearing in Vertebrates: A Psychophysics Databook. Winnetka, IL: Hill-Fay Associates.

Fay RR (2005) Sound source localization by fishes. In: Popper AN, Fay RR (eds) Sound Source Localization. New York: Springer Science+Business Media, pp. 36–66.

Fay RR, Edds-Walton PL (1997) Directional response properties of saccular afferents of the toadfish, *Opsanus tau*. Hear Res 111:1–21.

Fay RR, Popper AN, eds (1999). Comparative Hearing: Fishes and Amphibians. New York: Springer-Verlag.

Grose B, Carr CE, Casseday JH, Fritzsch B, Koppl C (2004) The evolution of central pathways and their neural processing patterns. In: Manley GA, Popper AN, Fay RR (eds) Evolution of the Vertebrate Auditory System. New York: Springer-Verlag, pp. 289–359.

Herrick CJ (1948) The Brain of the Tiger Salamander. Chicago, IL: The University of Chicago Press.

Kalmijn AJ (1989) Functional evolution of lateral line and inner ear systems. In: Coombs S, Görner P, Münz H (eds) The Mechanosensory Lateral Line: Neurobiology and Evolution. New York: Springer-Verlag, pp. 187–216.

Kalmijn AJ (1997) Electric and near-field acoustic detection, a comparative study. Acta Physiol Scand 638:25–38.

Ladich F, Popper AN (2004) Parallel evolution in fish hearing organs. In: Manley G, Popper A, Fay R (eds) Evolution of the Vertebrate Auditory System. New York: Springer-Verlag, pp. 95–127.

Lewis ER, Fay RR (2004) Environmental variables and the fundamental nature of hearing. In: Manley G, Popper A, Fay R (eds) Evolution of the Vertebrate Auditory System. New York: Springer-Verlag, pp. 27–54.

Lu Z, Fay RR (1996) Acoustic response properties of single neurons in the central posterior nucleus of the thalamus of the goldfish (*Carassiuis auratus*). J Comp Physiol 176:747–760.

Ma W-L, Fay RR (2002) Neural representations of the axis of acoustic particle motion in nucleus centralis of the torus semicircularis of the goldfish, *Carassius auratus*. J Comp Physiol 188:301–313.

McCormick CA (1999) Anatomy of the central auditory pathways of fish and amphibians. In: Fay RR, Popper AN (eds) Comparative Hearing: Fish and Amphibians. New York: Springer-Verlag, pp. 155–217.

McCormick CA, Hernandez DV (1996) Connections of octaval and lateral line nuclei of the medulla in the goldfish, including the cytoarchitecture of the secondary octaval population in goldfish and catfish. Brain Behav Evol 47:113–137.

McKibben JR, Bass AH (1999) Peripheral encoding of behaviorally relevant acoustic signals in a vocal fish: single tones. J Comp Physiol A 184:563–576.

Moulton JM (1963) Acoustic behaviour of fishes. In: Busnel R-G (ed) Acoustic Behaviour of Animals. Amsterdam: Elsevier, pp. 655–693.

Moulton JM, Dixon RH (1967) Directional hearing in fishes. In: Tavolga WN (ed) Marine Bio-Acoustics II. Oxford: Pergamon Press, pp. 187–228.

Myrberg AA Jr, Spires JY (1980) Hearing in damselfishes: an analysis of signal detection among closely related species. J Comp Physiol 140:135–144.

National Research Council (1994) Low-Frequency Sound and Marine Mammals: Current Knowledge and Research Needs. National Research Council, National Academy Press, Washington, DC.

Northcutt RG (1980). Central auditory pathways in anamniotic vertebrates. In: Popper AN, Fay RR (eds) Comparative Studies of Hearing in Vertebrates. New York: Springer-Verlag, pp. 79–118.
Northcutt RG (1981) Audition in the central nervous system of fishes. In: Tavolga WN, Popper AN, Fay RR (eds) Hearing and Sound Communication in Fishes. New York: Springer-Verlag, pp. 331–355.
Parker GH (1903) The sense of hearing in fishes. Am Nat 37:185–203.
Popper AN (1980) Scanning electron microscopic studies of the sacculus and lagena in several deep-sea fishes. Am J Anat 157:115–136.
Popper AN (2003) Effects of anthropogenic sound on fishes. Fisheries 28:24–31.
Popper AN, Carlson TJ (1998) Application of sound and other stimuli to control fish behavior. Trans Am Fish Soc 127:673–707.
Popper AN, Tavolga WN (1981) Structure and function of the ear of the marine catfish, *Arius felis.* J Comp Physiol 144:27–34.
Popper AN, Salmon M, Parvulescu A (1973) Sound localization by two species of Hawaiian squirrelfish, *Myripristis berndti* and *M. argyromus.* Anim Behav 21:86–97.
Popper AN, Fay RR, Platt C, Sand O (2003) Sound detection mechanisms and capabilities of teleost fishes. In: Collin SP, Marshall NJ (eds) Sensory Processing in Aquatic Environments. New York: Springer-Verlag, pp. 3–38.
Popper AN, Fewtrell J, Smith ME, McCauley RD (2004) Anthropogenic sound: effects on the behavior and physiology of fishes. Mar Technol Soc J 37:35–40.
Retzius G (1881) Das Gehörorgan der Wirbelthiere, Vol. I. Stockholm: Samson and Wallin.
Sand O (1974) Directional sensitivity of microphonic potentials from the perch ear. J Exp Biol 60:881–899.
Schuijf A (1975) Directional hearing of cod (*Gadus morhua*) under approximate free field conditions. J Comp Physiol 98:307–332.
Schuijf A, Visser C, Willers AFM, Buwalda RJA (1977) Acoustic localization in an ostariophysan fish. Experientia 33:1062–1063.
Sisneros JA, Bass AH (2003) Seasonal plasticity of peripheral auditory frequency selectivity. J Neurosci 23:1049–1058.
Striedter GF (1991) Auditory, electrosensory, and mechanosensory lateral line pathways through the forebrain in channel catfishes. J Comp Neurol 312:311–331.
Tavolga WN, ed (1964) Marine Bio-Acoustics. Oxford: Pergamon Press.
Tavolga WN, ed (1967) Marine Bio-Acoustics II. Oxford: Pergamon Press.
Tavolga WN (1971) Sound production and detection. In: Hoar WS, Randall DJ (eds) Fish Physiology, Vol. V. New York: Academic Press, pp. 135–205.
Tavolga WN, ed (1976) Sound Reception in Fishes—Benchmark Papers in Animal Behavior, Vol. 7. Stroudsburg PA: Dowden, Hutchinson & Ross.
Tavolga WN, ed (1977) Sound Production in Fishes—Benchmark Papers in Animal Behavior, Vol. 9. Stroudsburg PA: Dowden, Hutchinson & Ross.
Tavolga WN, Wodinsky J (1963) Auditory capacities in fishes. Pure tone thresholds in nine species of marine teleosts. Bull Am Mus Nat Hist 126:177–240.
Tavolga WN, Popper AN, Fay RR, eds (1981) Hearing and Sound Communication in Fishes. New York: Springer-Verlag.
van Bergeijk WA (1967) The evolution of vertebrate hearing. In: Neff WD (ed) Contributions to Sensory Physiology. New York: Academic Press, pp. 1–49.
von Frisch K (1923) Ein Zwergwels der kommt, wenn man ihm pfeift. Biol Zentralbl Leipzig 43:439–446.

von Frisch K, Dijkgraaf S (1935). Können Fische die Schallrichtung wahrnehmen? Z Vergl Physiol 22:641–655.

Wartzog D, Popper AN, Gordon J, Merrill J (2004) Factors affecting the responses of marine mammals to acoustic disturbance. Mar Technol Soc J 37:6–15.

Webb JF (1989) Gross morphology and evolution of the mechanoreceptive lateral line system in teleost fishes. Brain Behav Evol 33:34–53.

Webb JF (1998) Laterophysic connection: a unique link between the swim bladder and the lateral-line system in *Chaetodon* (Perciformes: Chaetodontidae). Copeia 1998: 1032–1036.

2
Hearing and Acoustic Behavior: Basic and Applied Considerations

Arthur N. Popper and Carl R. Schilt

1. Introduction

Over the past several decades, two different groups of investigators have been involved with fish bioacoustics but with only marginal interaction and cross-fertilization of findings and ideas between them. One group has been trying to understand the basic biology of fish hearing and vestibular system and lateral line function as well as orientation, sound production, acoustic communication, and the acoustic ecology of fishes (see Fay and Edds-Walton, Chapter 3; Bass and Ladich, Chapter 8; Braun and Grande, Chapter 4; Rogers and Zeddies, Chapter 7). The other group, with more applied needs and interests, has sought to use sound and hydrodynamic phenomena for applications in fisheries science (see Mann, Hawkins, and Jech, Chapter 9). Besides the ubiquitous use of various kinds of sonar in fisheries, a topic that is not considered here (but see Mann, Hawkins, and Jech, Chapter 9), a frequent goal of these applications has been to use sound and other hydromechanical stimuli to influence or control fish behavior. Often the objective is to restrict or otherwise alter the local distributions of the fish in a given industry-influenced environment. Although there have been some attempts to bring the ideas and findings of the two separate groups together (e.g., Popper and Carlson 1998), this has not been done extensively.

The purpose of this chapter is to provide a broad overview of the findings and issues of these two research communities and to provide a context for sharing ideas and efforts. The intent is to provide some insights that may facilitate the work of both groups of investigators and to encourage collaboration between them.

The chapter is divided into three parts. The first considers some basic aspects of fish hearing that are most germane to applied issues that are discussed later in the chapter. For a more detailed discussion of the fish auditory system, readers are referred to other chapters in this volume as well as to recent reviews (e.g., Popper and Fay 1999; Popper et al. 2003; Ladich and Popper 2004). Detailed discussions of fish hearing capabilities are presented in Fay and Megela Simmons (1999) and of fish sound localization in Fay (2005). The second part of the chapter considers the efforts that have been made to use sound and other hydromechanical sensory stimuli, including flows and turbulence, to control local fish distributions, primarily to reduce the harm of human activities for fishes. This material has been reviewed in greater detail by Popper and Carlson (1998). Finally, the chapter addresses the ways in which anthropogenic sounds from human activities ranging from shipping and construction noise to mineral exploration and seismic geology studies, offshore wind farms, and sonar may affect fish.

2. Basic Mechanosensory Systems and Capabilities of Fishes

Fishes have evolved a wide array of sensory systems and behavioral responses with which they perceive and respond to their environments (see discussions of the aquatic sensory environment in chapters in Atema et al. 1988 and in Collin and Marshall 2003). The mechanosensory systems of fishes include (1) the hearing of sound pressure oscillations through the inner ears; (2) orientation and body motion sensation (the vestibular system), which is also mediated by the inner ears; and (3) detection of hydromechanical stimulation near the fish that is mediated by the lateral line. The lateral line system consists of an array of neuromasts composed of hair cells and found within pored, bony canals and on the epithelium of the head, trunk, and tail (Coombs et al. 1988; Coombs and Montgomery 1999). It senses local water motions and differential pressures, which are induced by water flows (referred to as "svenning" by Platt et al. 1989 in honor of the extensive and insightful work on lateral line structure and function done by Professor Sven Dijkgraaf [e.g., Dijkgraaf 1963, 1989]).

The evolutionary and functional relationships that relate the auditory, vestibular, and lateral line systems are beyond the scope of this chapter (but see Popper et al. 2003; Ladich and Popper 2004). These sensory capabilities enable a wide variety of life functions including prey and predator location, group cohesion and coordination, mate attraction and courtship, and, perhaps most fundamentally, a general awareness of the environment and things in it (Fay and Popper 2000; Fay 2008).

2.1 Origin of Hearing Capabilities in Fish

Hearing has been studied in a number of fishes and has been reviewed extensively (e.g., Fay 1988; Popper and Fay 1993; Fay and Megela Simmons 1999; Fay and Popper 1999; Popper et al. 2003; Ladich and Popper 2004). One of the fundamental questions to ask with regard to hearing in fish (as in all other vertebrates) is why hearing has evolved. Clearly, hearing is used by many species for interspecific communication (e.g., Myrberg and Spires 1980; Zelick et al. 1999). However, more recent analysis leads to the suggestion that rather than having evolved for acoustic communication per se, hearing evolved to provide fish (and other vertebrates and, perhaps, invertebrates) with a "sense" of their environment that extends a considerable distance from the animal. In effect, because visual signals are only effective in adequate light and in directions in which the eye is looking and chemical signals do not carry for great distances with any speed or directional characteristics, sound has the potential to provide fish with information about the environment from considerable distances, at high rates of speed, and with significant directional information (Fay and Popper 2000; Fay 2008). In water, turbidity presents an additional problem for light sensing and signaling, and it may be suggested that the selective pressures that resulted in the evolution of hearing were for the detection of distant predators and prey as well as for detection of objects in the environment, the location of coral reefs, and numerous other things (Fay and Popper 2000; Fay 2008).

This overview of the acoustic environment has been called the "auditory scene" (Bregman 1990). The auditory scene provides the animal with a perceptual "world" that extends far beyond other senses, thereby increasing survival chances. Loss of hearing sensitivity, as might occur in a noisy environment (e.g., from human-generated masking sounds), can potentially have a significant effect on the survival of fish and their populations because they would lose the broader perspective of the environment.

In considering the evolution of vertebrate hearing, Fay and Popper (2000) argued that in order for any animal to make use of its "auditory scene," it must also be able to do "stream segregation," which is the ability to discriminate between sounds that are and are not of biological relevance (see also Fay 2008). To do stream segregation, all vertebrates must have certain basic auditory functions including the ability to discriminate between frequency and intensity of sounds, determine the direction of a sound source, and detect signals in the presence of other sounds that might otherwise interfere with detection (e.g., "masking" sounds).

2.2 Hearing Capabilities

Fish demonstrate all of the capabilities needed for use of the auditory scene including the ability to discriminate between signals and determine sound source direction (see reviews by Fay and Megela Simmons 1999; Fay 2005). Measures of hearing sensitivity (see Fig. 2.1) have demonstrated that fish of most species hear over a relatively narrow range of frequencies. Generally, this ranges from 50 Hz or below to 1,000 or 1,500 Hz. Sensitivity at these frequencies is often not very good, and there is considerable variation in hearing sensitivity in different species. As pointed out by Ladich and Popper (2004), there is no known clear correlation between the taxonomic position of species and hearing capabilities, and too little is known about the hearing capabilities in different species to be able to correlate hearing capabilities in different environments or ecological niches. Moreover, there is considerable variation in ear structure and hearing capabilities within some taxonomic groups. For example, Coombs and Popper (1979) showed that two different genera of squirrelfish have very different ear structures and hearing capabilities despite the two species living sympatrically and using similar sounds for communication. As a consequence, without sufficient data, it is

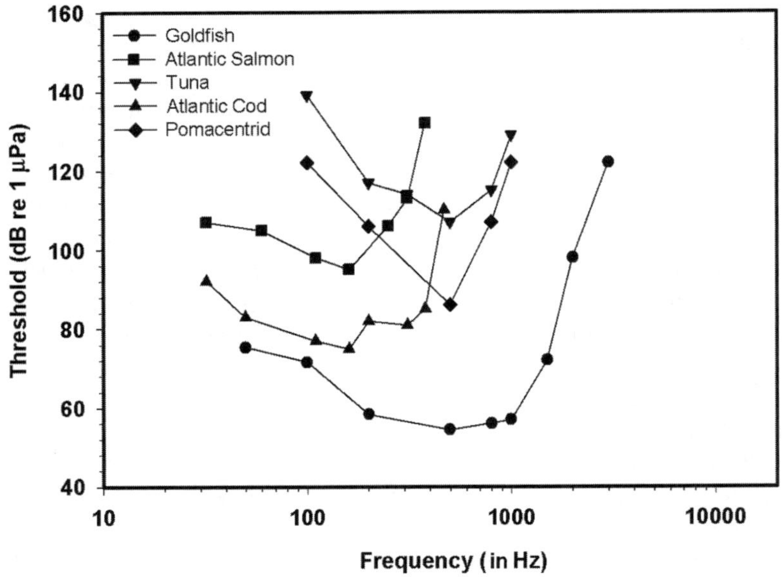

FIGURE 2.1. Auditory thresholds from a select group of teleost fishes. (All data from Fay 1988.)

often not realistic to generalize about hearing capabilities even between closely related taxa.

At the same time, as shown in Fig. 2.1, there are some species, referred to as hearing "specialists" as opposed to the aforementioned hearing "generalists," that are able to detect sounds to greater than 3,000 Hz. Moreover, even at the lower frequencies that both types of fish can hear, the specialists can detect lower intensity sounds than the generalists so that the specialists hear better in the frequency range that they share with the generalists and also hear over a wider frequency range.

2.2.1 Hearing Specialists versus Generalists

The hearing specialists, which include species as diverse as otophysans (goldfish, carp, catfish), mormyrids (elephantfishes), and possibly myctophids (deep-sea lantern fishes), all have specializations peripheral to the ear that mechanically couple the motion of the swim bladder (or other air bubble), which vibrates in response to pressure stimulation, directly to the inner ear. Because a gas bubble expands and contracts in response to pressure signals much more than does water or fish tissue, the air bubble converts pressure to motion and thereby stimulates the auditory end organs of the inner ears. In hearing specialists, this motion is coupled directly to the ear with minimal loss of energy. In contrast, hearing generalists often have a swim bladder, but they do not have a coupling between the gas bubble and the ear. Thus, much less of the pressure-generated motion of the swim bladder gets to the ear than is the case in the specialists. How much, if any, pressure-generated motion gets to the ear in the generalists is not known, although there is reason to believe that there may be some swim bladder contributions to hearing in at least some generalists.

Specializations to enhance hearing vary widely among different hearing specialist species. The best-known specializations are the Weberian ossicles in the otophysan fishes (e.g., goldfish, catfish, carp, and relatives). This series of bones, derived from parts of vertebrae, directly connects the swim bladder to the fluids of the inner ear, thereby coupling swim bladder motions to the ear. Other specialist species have anterior projections from the swim bladder that terminate near or are directly in contact with the inner ear, thereby bringing pressure-generated motions to the ear without intervening structures. Finally, there are fishes such as mormyrids (elephantfishes) and clupeids (herrings, anchovies, shads, and relatives) in which there is an ancillary bubble of air near or in contact with the ear.

2.2.2 Infrasound and Ultrasound

Although hearing specialists generally hear to no more than 3–5 kHz, recent studies show that fish in one clupeid subfamily (the Alosinae or the anadromous herrings and menhadens) can detect sounds well into the ultrasonic range (Kynard and O'Leary 1990; Mann et al. 1997, 2001). As discussed in Section 3.8, there is evidence suggesting that the evolutionary origin of ultrasound detection may have enabled these animals to detect and avoid dolphin predators.

Finally, a number of species are able to detect sounds into the infrasonic range (below the human lower range of about 20 Hz; e.g., Sand and Karlsen 1986; Sand and Karlsen 2000; Sand et al. 2000, 2001; Popper et al. 2003). Although there has not been an extensive analysis of infrasound detection, this has been demonstrated in species as diverse as Atlantic salmon (*Salmo salar*), Atlantic cod (*Gadus morhua*), and European silver eels (*Anguilla anguilla*) (Sand et al. 2000). In all cases, however, infrasound detection seems to primarily take place when the fish is relatively near the sound source.

2.3 Structure and Function of the Inner Ear

The inner ear of sharks and bony fishes (Fig. 2.2) consists of three semicircular canals, three otolith organs (saccule, lagena, and utricle), and, in some species, a relatively diminutive macula (or papilla) neglecta (see Popper et al. 2003 for a detailed description of fish ears).

The transducing elements of the ear, or the cells that convert mechanical energy into a signal that can stimulate the nervous system, are the sensory hair cells. Each sensory hair cell has a typical cell body as well as an apically located ciliary bundle made up of a single kinocilium and many stereocilia (or stereovilli; see Fig. 2.3). Bending of the ciliary bundle by mechanical energy results in a cascade of intracellular events that leads to the release of a neurotransmitter and the stimulation of the innervating eighth cranial nerve (e.g., Hudspeth 1985, 1997).

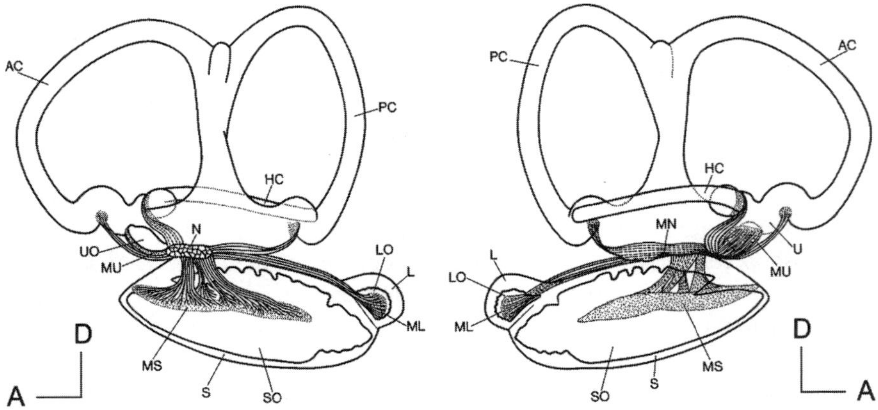

FIGURE 2.2. Inner ear of a perch (from Ladich and Popper 2004). Medial view on the left and lateral view on the right. AC, HC, PC, anterior, horizontal, and posterior semicircular canals; L, lagena; LO, lagena otolith; MN, macula (papilla) neglecta; MU, utricular epithelium; MS, saccular epithelium; N, eighth cranial nerve; S, saccule; SO, saccular otolith; UO, utricular otolith.

FIGURE 2.3. Ciliary bundles from a teleost fish. The apical surface of each sensory cell has a group of cilia, the longest of which is the kinocilium. The longest of the graded stereocilia (or stereovilli) is closet to the kinocilium. Each of the sensory cells is surrounded by support cells that have apical microvilli. Note that all of the ciliary bundles are oriented so that the kinocilium is to the upper right in the figure. That is, they are all oriented in the same direction.

The three semicircular canals are oriented in three mutually perpendicular planes and each has a sensory region or ampulla at its base. Movement of fluids in the canal, resulting from angular acceleration of the head, produces bending of a gelatinous cupula in which are embedded ciliary bundles from sensory hair cells. Cupula motion results in bending of the ciliary bundles that results in neural activity and the detection of angular acceleration (Platt 1983; Popper et al. 2003).

The otolithic end organs each have an epithelium composed of sensory and nonsensory cells (Fig. 2.3). The sensory cells number in the thousands to hundreds of thousands depending on the species and the size of the fish (Lombarte and Popper 1994). The sensory epithelium lies close to a dense calcareous otolith and is separated from the otolith by a thin otolithic membrane that mechanically couples them together (Popper et al. 2003, 2005a).

Hair cell stimulation results from the relative motion between the sensory epithelium and the otolith. In effect, the epithelium and otolith move at different amplitudes and phases because of their different densities.

Fish otolithic end organs are likely to have two functions. One is to determine head position relative to gravity as in terrestrial vertebrates (see Platt 1983 for a review). The saccule, lagena, and, very likely, the utricle are also involved in sound detection (e.g., Popper et al. 2003). The precise role of each end organ is not known, and the relative contributions of each to sound detection may vary in different species. For example, in the otophysan fishes, the connection between the swim bladder and saccule may result in that end organ being the primary detector of sound pressure (Rogers and Zeddies, Chapter 7), whereas in clupeiform fishes, the utricle may be the major sound detection end organ, at least for higher frequency sounds (e.g., Mann et al. 2001; Higgs et al. 2004; Plachta et al. 2004).

2.3.1 Sensory Cell Organization on the Otolithic End Organs

A significant feature of the otolithic end organs is that the sensory cells are organized into "orientation groups" based on the position of the eccentrically placed kinocilium (Fig. 2.3). All ciliary bundles in each region on the epithelium are oriented with the kinocilium in the same direction. The morphological polarization is accompanied by a physiological polarization whereby bending of the bundle results in hair cell responses that are graded and proportional to the vector component in the axis of best physiological sensitivity (Hudspeth 1985; Lu and Popper 2001). Thus, each sensory cell is potentially capable of measuring the direction of the particle motion of a sound source.

On discovery of this orientation pattern (e.g., Dale 1976; Popper 1976), it was suggested that this grouping of like-oriented hair cells may provide fishes with an ability to determine the direction of the particle motion of a sound source and thus provide information about sound source direction (Popper et al. 2003; Fay 2005; Rogers and Zeddies, Chapter 7).

Recent physiological data support this hypothesis (e.g., Lu et al. 1996; Fay and Edds-Walton 1997; Edds-Walton 1998; Lu and Popper 2001). The assumption is

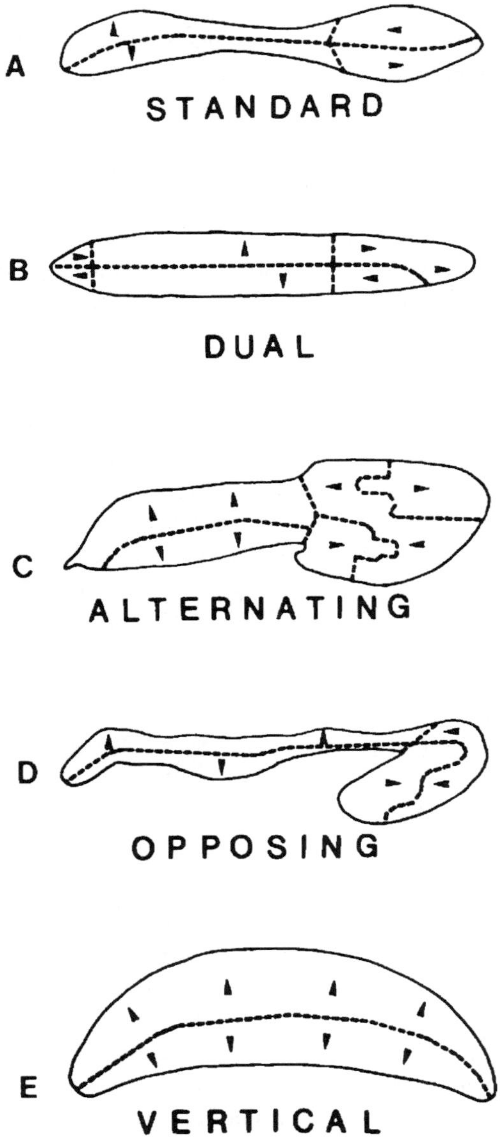

FIGURE 2.4. Saccular hair cell orientation patterns from different fishes. Anterior is to the right and dorsal to the left. The dotted lines are the areas of what is generally an abrupt transition in orientation between directions. The arrowheads indicate the direction of the kinocilia on the hair cells in each region of the epithelium (e.g., the tip of the arrow would be to the upper right in Fig. 2.3). The "Standard" pattern is typically found in fishes that are hearing generalists; the other patterns are most often found in hearing specialists. There is no taxonomic relationship for these patterns. The same basic pattern can be found in taxonomically diverse fishes. For example, the vertical pattern, which includes no rostrally and caudally oriented cells, is found in all otophysans

that each neuron from the saccule (the only end organ studied to date) innervates only hair cells oriented in a particular direction (Lu and Popper 2001) and that this information is carried to the central nervous system (CNS) where the directional response properties of neurons from sensory cells with different orientations are compared (Edds-Walton 1998) and the direction "calculated" (Popper et al. 1988; Rogers et al. 1988).

2.3.2 Comparative Ears

There is striking diversity in the inner ear structures of different fish species (Fig. 2.4) (e.g., Popper et al. 2003; Ladich and Popper 2004), yet the functional significance of the diversity is not known and there is only the most limited understanding of any correlations between ear structure and function (Schellart and Popper 1992; Ladich and Popper 2004).

There is, however, an apparent correlation between the general orientation of the sensory hair cells on the saccular epithelium and whether fishes are hearing generalists or specialists (Fig. 2.4). The hearing specialists show more "complex" saccular hair cell orientation patterns than generalists, which often have only the "standard" saccular pattern (Fig. 2.4) (Popper and Coombs 1982; Popper et al. 2003). The diversity in saccular hair cell orientation pattern in hearing specialists appears to be correlated, at least to some degree, with the acoustic coupling between the swim bladder and the saccule. And, most significantly, the diversity in saccular hair cell orientation patterns associated with hearing specializations shows functional convergence across taxonomically diverse species.

The other aspects of inner ear structure that show substantial diversity but with unknown function are the size and shape of the otoliths and, particularly, of the saccular otolith (Popper et al. 2005a). Popper et al. (2005a) pointed out that very little is known about the specific function of the otoliths other than they provide a body with a different density than the rest of the fish for stimulation of the sensory cells (see also Rogers and Zeddies, Chapter 7). However, it has been suggested that the very diverse shapes of the otoliths may be related to hearing and/or vestibular function of the ear (e.g., Popper et al. 2003). Moreover, there are differences in the percentage of area of the sensory epithelium of the saccule that is covered by the otolith. Whereas in most species studied, the otolith covers the whole epithelium and may even extend beyond it, there are species such as some myctophids and other deep-sea fishes in which the otolith may only

FIGURE 2.4. (Continued) (goldfish, catfish, and relatives) and in the elephantfishes (mormyrids). The alternating pattern is found in fishes as diverse as many eels and deep-sea gadids. (From Popper and Coombs 1982.)

cover half the epithelium. In these fishes, the only connection to the rest of the epithelium is through the otolith membrane, which lies between the otolith and epithelium and extends out to the uncovered areas (Popper 1980; Popper et al. 2005a).

2.4 The Vestibular System

The vestibular senses mediate body orientation, balance, and accelerations (e.g., Platt 1983; Popper et al. 2003) and thus play a major role in fish behavior. Although a detailed discussion of the vestibular system is beyond the scope of this chapter, it is important to note that the sensory receptors of the inner ear that mediate the vestibular senses involve the same kind of sensory hair cells found in the otolithic end organs and the lateral line. Moreover, the receptor organs involved in the vestibular senses in fishes not only include those of the semicircular canals but also of the three otolithic end organs. Indeed, nothing is known about how the nervous system of fish separates vestibular from auditory signals from the otolithic end organs. It is possible that there are different populations of sensory cells on the end organs that mediate the different senses or the difference may be in the frequency of the stimulation, with very low frequency signals being sent to the vestibular part of the brain while higher frequency signals are sent other places.

3. Applied Aspects of Fish Bioacoustics

There are a number of different issues to be considered when discussing applied fish bioacoustics. The first is the use of fish-produced sounds and hydrodynamic disturbances to assay fish distribution, abundance, and behavior. This is discussed in detail in Chapter 9 by Mann, Hawkins, and Jech and involves the use of a transducer to detect and record fish-produced sounds or hydrodynamic phenomena. This "listening-to-fish" aspect of acoustic biology is sometimes called "passive" acoustics to distinguish it from the "active" acoustics fields of fisheries acoustics (sonar, which is used to sample fish abundance and distribution; also discussed in Chapter 9) and from acoustic tagging and telemetry (which permit remote tracking of individual fish). Because these "active" acoustics categories do not involve the hearing, lateral line, or vestibular systems of the fish, they are outside the present discussion. Instead, the discussion of "active" bioacoustics in this chapter focuses on the use of anthropogenic sounds and water motions to affect fish behavior, usually to influence local distribution.

3.1 Use of Flows and Turbulence to Control Fish Distribution

A potentially important applied use of sound involves using sound or water motion to manipulate local fish distributions. This manipulation might be for a variety of uses including aggregating fish for harvest in aquaculture environments, capture of wild fish, bycatch reduction in commercial fishing operations, fish protection at industrial sites, and the exclusion of unwanted or invasive fishes from some waterways. This approach involves the development and deployment of sound, flow, or turbulence-producing systems to attract fish (perhaps toward a fishway or other fish bypass system) or repel them (e.g., away from a turbine or cooling-water intake). The production of sound stimuli can be fairly straightforward for reasonably high-frequency sound but is far more challenging for lower frequencies. The generation of sound and hydrodynamic flows is unavoidable in the operation of industrial water management facilities. Understanding the effects of natural or human-caused flow patterns and sound in attracting or repelling fish as well as the use of designed stimuli to direct fish movements is what is of interest here.

3.2 Control Local Fish Distributions

Wherever humans divert large volumes of water for industrial, municipal, or other uses, there are potential costs to fish populations. The most obvious costs are in entrainment or impingement of fish, including eggs and larvae. Entrainment refers to drawing fish into a water withdrawal route such as a cooling-water intake. In contrast, impingement refers to fish striking or being trapped by flow against screens or other engineered structures. Entrainment and impingement are important sources of environmental impact at many industrial water facilities. The need to improve facility design and operations to reduce fish losses by impingement and entrainment has long been recognized (Schuler and Larson 1975; Hocutt 1980), and methods to resolve these problems have often involved the use of sound to control the movement of fish away from areas where they could be impinged or entrained.

Any structure in an aquatic environment may attract fish by providing cover, shade, aggregated prey, artificial light, or other stimuli (Love et al. 2000; Dempster and Kingsford 2003). Engineered structures inevitably produce mechanosensory and other stimuli that potentially are detectable by fish. Some industrial sites, such as large hydropower dams, can be very noisy across wide frequency ranges, including those detectable by most species of fish, and also involve powerful and complex hydrodynamic flows that can move fish directly into areas of danger (e.g., turbines).

The effects of ambient noise, in terms of either sound pressure or hydrodynamic flow, on the hearing, vestibular, and lateral line systems are little studied. It is likely that a large industrial project, like a lock-and-dam project or a cooling-water intake, is rich in many kinds of acoustic signals that may stimulate, interfere with, overwhelm, or even damage a fish's orientation, navigation, and locomotion systems. Mitigation of some applications may simply involve fish exclusion from an avoidable hazard such as a water intake. Where fish passage is required, such as populations migrating through the world's increasingly dammed river systems, and the best available passage route is a small proportion of the total project water passage, as in the case with a hydropower dam, it is important to provide opportunities for migrants to discover, enter, and take the more benign passage routes (Rainey 1997).

Besides making systemwide changes in watersheds (Freeman et al. 2001) and violating the interconnected nature of river systems (Dynesius and Nilsson 1994; Pringle et al. 2000), engineered structures such as locks and dams can interfere with fish migrations (Dadswell et al. 1987; McAllister et al. 2000; Dixon 2003) that may be requisite for the fish's life cycles (Dadswell et al. 1987; Dixon 2003; Limburg and Waldman 2003). Sometimes even small overflow dams can impede or stop fish migration (Beasley and Hightower 2000; Zigler et al. 2004). Conversely, some engineered changes in waterways have opened new migration routes to invasive fishes (Fuller et al. 1999; Chick and Pegg 2001) that can cause unpredictable perturbations of aquatic ecosystems. In many cases, on highly regulated rivers such as the Columbia-Snake River System in the US Pacific Northwest, juvenile fish may encounter many dams in their migration to the sea and cumulative stresses may be important (Budy et al. 2002).

Construction projects in or near water bodies that involve blasting or pile or pipe driving as well as offshore seismic exploration (e.g., Engås et al. 1996; Slotte et al. 2004) may stress, injure, or kill fish. The impacts of anthropogenic sound on fish and fisheries are discussed in Section 4. Here the point is that sometimes, where local fish distribution presents particular challenges, it would be desirable to exclude fish from the vicinity of job sites, facilities, or dangerous passage routes. In these cases, bioacoustics may sometimes be useful for fish exclusion or protection. Being able to either attract or repel fish, especially if it were reasonably inexpensive and reliable, would have a number of uses to benefit both industry and fish conservation at industrial and other water management sites. However, although this use of sound still appears to hold potential, there has been almost no data in the peer-reviewed literature that point to any successes in achieving these goals other than for the use of ultrasonic sound (see Section 3.8). Indeed, data in the peer-reviewed and gray literature are often highly equivocal, and reported "successes" in using sound to control fish are very limited and in prescribed environments and may not work under other, even slightly different, conditions. Moreover, even when there may be successes, data are often limited to very few species and limited age classes within those species. As a consequence, applicability to animals of different ages, maturity, etc. is not known.

3.3 Mechanosensory Stimuli for Fish Control

It is reasonable to consider using sound or water motions to control fish distributions in engineered environments. Fish of all the species tested so far can detect both sound pressure and hydrodynamic stimuli. The interest in having stimulus systems for control of free-ranging fish goes back several decades (reviewed in Popper and Carlson 1998). Early views were rather simplistic "command and control" models that lacked appreciation for the complexity, mutability, and unpredictability of the aquatic environment and the fish response to it. The response of a fish or any animal to stimuli depends on many physiological, temporal, and environmental factors (Schilt and Norris 1997). Some of these may be evident or, at least, measurable (sound, current, light, turbidity, temperature), but others (fish motivation and condition, hunger, predation threat) may be less accessible. The response may be specific not only to fish species but also to life stage, time of day and year, presence of predators, and countless other known and unknown variables. Perhaps because stimuli are presented against different backgrounds in different places, stimulus efficacy may be site specific. Habituation to a stimulus is important, especially with resident populations but also with migrants, which may be near a given site for hours or days.

There are a number of approaches to improve fishing efficiency and reducing bycatch. These are sensory-based aspects of methods for both small scale (artisanal) and industrial fisheries that use fish response to stimuli (Parrish 1999), and mechanosensory responses may sometimes be involved in modern fish capture (Wardle 1993). However, acoustic aspects of fishing gear, such as sounds made by fish trawls, have not been extensively studied. Orbach (1977) briefly discusses the use of small explosive charges and even the practice of banging on the side of a tuna seiner to prevent the escape of fish before the purse seine can be closed. Finneran et al. (2000) have suggested that wild yellowfin tuna (*Thunnus albacares*) might be attracted for harvest at sea with the sounds produced by the dolphin schools with which they travel, although it is not clear that the sounds produced by the dolphins are in the frequency range detectable by tuna (Iversen 1967, 1969). Clearly, increased understanding of fish sensory response might be used to make fish capture methods more efficient. Still, Parrish (1999) argued that using behavioral and sensory research to increase catches requires caution and may not be sustainable. Indeed, there is evidence that the sounds of fishing boats and trawls may actually result in fish moving away, thereby decreasing catches (reviewed in Mitson 1995; Mitson and Knudsen 2003). Thus, knowledge of fish hearing could conceivably be used to increase gear specificity so as to increase catches or to reduce bycatch of nontarget fish (Broadhurst et al. 1999).

3.4 Fish Handling in Aquaculture

Relatively little has been done in the aquaculture to use sound to control fish behavior, although there was early interest in using sound to aggregate fish (Hashimoto and Maniwa 1967; Chapman 1976). Willis et al. (2002) experimented with training triploid grass carp (*Ctenopharygodon idella*) to aggregate at a sound source so that they can be retrieved from water bodies where they have been put for weed control. Parrish's (1999) warnings regarding the use of behavioral science for fisheries applications should also apply to fish farming.

At the same time, aquaculture facilities can be relatively noisy environments as a result of the use of pumps and other devices. Little is known about whether such sounds have any impact on fish, although one study (Wysocki et al. 2007) suggests that the sounds imposed by pumps and other aquaculture equipment are below the levels that have any effect on the growth and hearing of rainbow trout. Still, these sounds may occur, the effects could range from increasing stress levels to damage to hearing, and the results may range from no effect to decreased growth and/or survival (e.g., Wysocki et al. 2006).

3.5 Control of Invasive Fishes

The use of sound to control invasive fishes is largely unexplored, although there may be an increasing need as more invasive species enter new environments. As more invasive fishes are introduced across wider new ranges, they will increasingly interfere with and jeopardize native species, communities, and ecosystems. Recently, fisheries managers in North America have been faced with a growing number of invasive and sometimes very prolific fish species (Fuller et al. 1999), which can cause severe ecological problems including extinction of native species (Lassuy 1995). An especially vexing problem involves several species of very large and prolific Asian carp that were introduced into aquaculture facilities and have escaped and spread through the major river systems of the middle of North America (Chick and Pegg 2001) and that now threaten to invade the Great Lakes. Acoustic and hydrodynamic barriers offer potential tools to control these (Taylor et al. 2005) and other unwanted species, but they remain largely untested and the work that has been done is sometimes of questionable quality and often remains outside the peer-reviewed literature. Unlike fish protection wherein any reduction in stress, delay, or mortality is beneficial, the barrier that protects a waterway from a robust, prolific, and harmful invader must be very nearly perfect because even one gravid female getting through can, as a worse case, lead to a successful invasion and establishment of a population beyond the barrier.

3.6 Fish Exclusion at Polluted or Construction Sites

The use of sound to potentially provide exclusion of fish from polluted sites or construction remains largely unrealized. In the case of pollution emergencies such as chemical spills, it might be impractical to mobilize a behavioral control system, even if one were available, in time to actually protect fish. But at construction sites, where drilling, blasting, pile driving, or other activities may be predicted to be problematic, an effective acoustic deterrent might provide at least a partial solution.

3.7 Fish Protection and Passage at Hydropower Dams and Other Industrial Sites

There is a history of successful and unsuccessful attempts at improving fish protection and passage at industrial facilities throughout the world (Haymes and Patrick 1986; Fletcher 1990; Jungwirth et al. 1998; Coutant 2001; Pavlov et al.

2002). In many cases, these efforts capitalize on the natural responses of fish to signals in the environment (e.g., natural sounds).

Efforts toward developing acoustic-based tools to enhance fish passage and protection go back at least to the early 1950s in the United States (Burner and Moore 1953). Using fish mechanosensory (ear- and lateral line-mediated) behavioral responses to direct fish movement is especially appealing for several reasons (reviewed in Schilt and Nestler 1997; Popper and Carlson 1998). However, attempts to reduce fish entrainment and impingement at industrial water intakes or to otherwise redistribute fish over long time periods using sound stimuli have largely proved unsuccessful (reviewed in Popper and Carlson 1998). Positive results have been reported at one site (e.g., Hanson Environmental, Inc. 1996; tests were conducted at a slough in California), but similar treatments do not work at other times and places (Ploskey et al. 2000; tests were done at a large main-stem dam on the Lower Columbia River). Unfortunately, failures are less likely to be published than are successes. On the other hand, sometimes a study that finds no effect for a specific sound treatment may come to a very general conclusion. For example, Goetz et al. (2001) found no effect in an attempt to use a 200- to 300-Hz signal to change juvenile salmon distributions in a large navigation lock and concluded that "low-frequency sound is not an effective means of guiding salmon smolts." Of course, there might be many sound characteristics including amplitude, duration, rise time, and repetition rate, which might influence efficacy, and it may be unwise to infer that all "low-frequency sound" is ineffective from one series of experiments. Unfortunately, in many studies involving sound and fish behavior, the stimulus and noise fields are poorly described if they are described at all.

Effective reductions of fish entrainment at power-generating sites have been reported for pneumatic guns (Haymes and Patrick 1986), electronic sound sources (Hanson Environmental, Inc. 1996), and a mechanical "hammer." Even in cases in which a sound source is found to be efficacious at a given site, some sound-production systems, especially low-frequency impulse generators such as air and water "guns" and electric "sparkers" used in seismic exploration, may still have important dependability and (human) safety issues.

Beyond the use of sound, there has been considerable work with the use of hydrodynamic flows and turbulence to protect fish in hydropower applications, with the assumption that fish detect such signals with the lateral line. Industrial water impoundment and withdrawal systems often involve spectacularly large, powerful, and turbulent water flows, some of which can be directed through fishway (also called fish ladders) or fish lift (fish elevator) routes as "attraction flow" to draw upstream migrants to their downstream entrances (Barry and Kynard 1986). The positive rheotaxis (upstream swimming) of adult anadromous fishes such as salmonids and alosine herrings as well as the upstream-migrating juveniles of the catadromous eels has enabled the development of fairly successful fishway architectures for many of those fishes. The development of upstream passage routes has been relatively successful, although substantial challenges remain at some sites and with some species including upstream-migrating American shad (*Alosa sapidissima*) in the Canadian and US eastern seaboard and Pacific lamprey (*Lampetra tridentata*) in western Canada and the US Pacific Northwest (Moser et al. 2002).

Although many challenges remain in the upstream passage at specific sites and with particular species, the understanding of downstream orientation and enhancement of passage and survival at hydropower dams present more difficult and more recently addressed problems. There is a special interest in the use of mechanosensory information in the orientation and behavior of downstream-migrating juvenile anadromous salmonids (Knudsen et al. 1992) and catadromous eels (Richkus and Dixon 2003), which are affected by dam passage.

In general, the study and development of the downstream passage at hydropower dams have a more recent evolution and present more difficult challenges than does the upstream passage of adults. Typically, adult migrants are large, powerfully swimming fish with strong motivation to go upstream. Juvenile downstream migrants are young, small, and much less well understood. There is a good deal of computational fluid dynamics modeling, which describes and predicts water motions, done at many engineered sites such as the forebays of large hydropower dams. These may be helpful, at least, in knowing where fish might be unable to resist entrainment, but for understanding fish navigation through such systems, the spatial scale of such modeling studies is typically very large compared to the scale at which fish are likely to perceive the world with the lateral line systems wherein the fish's size absolutely limits the system array size (Coombs et al. 1988). Even if there were appropriately scaled models or measurements of the hydrodynamic environment through which fish pass, we do not know, to any great extent, what sensory stimuli guide downstream-migrating juvenile fish.

3.8 Use of Ultrasound to Control Fish Behavior

Although there generally has been little success in using sound to control fish behavior, one of the areas in which there has been considerable success has been the use of ultrasonic sound to keep herring in the subfamily Alosinae (Family Cluepidae) from entering cooling-water intakes and large power plants. Ultrasonic sensitivity in an alosine herring (American shad, *Alosa sapidissima*) was discovered by Boyd Kynard when, in 1982, he was using ultrasonic (about 160-kHz) sonar to sample down-running (spent) adult American shad in a canal associated with Holyoke Dam on the Connecticut River, MA. Subsequent work at the site indicated that the sound field was effective at temporarily concentrating down-running adults but that the fish would finally pass through or perhaps under the sonar beam. Up-running (prespawning) shad were more successfully concentrated by the sound (Kynard and O'Leary 1990).

3.8.1 Ultrasound Detection

Although these studies showed ultrasonic detection in Alosinae, it was not until Mann et al. (1997) did behavioral tests on hearing in the American shad that the ultrasonic hearing capability was quantified. In this and a subsequent study (Mann et al. 2001), it was demonstrated that several *Alosa* species can detect ultrasound up to almost 200 kHz, whereas members of the subfamily Clupeniae (the sea herrings and allies including sprats, sardines, pilchards, and relatives) are able to detect sounds only to about 4 or 5 kHz (also Enger 1967; Mann et al. 2005).

Not only can the alosine herrings hear ultrasound, they also show avoidance reactions to pulsed ultrasounds (Plachta and Popper 2003). The sensitivity to such high-frequency clicks may be adapted as a predation-avoidance response to the echolocation clicks of some marine mammals (Nestler et al. 1992; Mann et al. 1998; Astrup 1999; Plachta and Popper 2003).

A major question is how alosine herrings detect and directionalize ultrasonic sounds. Although evidence is still indirect, it appears that the utricle is involved (e.g., Higgs et al. 2004; Plachta et al. 2004). The utricles in all clupeids that have been examined are morphologically different from those found in any other vertebrate studied to date. In clupeids, the utricular sensory epithelium is divided into three distinct parts (Popper and Platt 1979), whereas other vertebrates have only a single epithelium. Most significantly, the center epithelial region in clupeids is suspended over fluid associated with an air bubble (Higgs et al. 2004) that may resonate at greater than 100 kHz (Hastings and Popper, unpublished data). Ultrasonic hearing is not found in young *Alosa* until the utricle is fully developed (Higgs et al. 2004).

3.8.2 Use of Ultrasound for Control of Fish Behavior

In 1989, net pen experiments were carried out on the upper Savannah River, GA (Nestler et al. 1992) in which captive adult blueback herring (*A. aestivalis*) were found to have significant avoidance responses over fairly short (to 15 min) time durations. The investigators found a reduction in fish abundance in the presence of the ultrasound compared to when it was off. Subsequent to this finding, ultrasound has been placed in operation to control the movement of several *Alosa* species (e.g., Dunning et al. 1992; Ross et al. 1993, 1996; Nestler et al. 1995; Ploskey et al. 1995). Gregory and Clabburn (2003) reported that the 200-kHz side-looking sonar with which they sample upstream-migrating Atlantic salmon (*Salmo salar*) must be turned off at intervals because it has the unforeseen consequence of stopping the concurrent upstream migration of the alosine twaite shad (*Alosa fallax*).

4. Anthropogenic Sound and Fish

An issue of growing interest deals with the effects of anthropogenic sound on fish (Popper et al. 2003; Popper et al. 2004; Hastings and Popper 2005). Those sounds might result from systems designed for sound production, such as offshore minerals exploration or sonar devices, or from systems for which sound is just a by-product, such as shipping or sea-based wind farms. Human-generated hydrodynamic flows that might cause stress to or otherwise harm fish include turbine, fish bypass, and spillway passage routes at hydropower dams. The possible harmful effects of anthropogenic sound on marine mammal populations have received considerable attention recently (e.g., National Research Council 2000; Popper et al. 2003; Hastings and Popper 2005; Popper et al. 2005b, 2007), but the effects on fish and other nonmammals are also of growing interest. However, to date, there are few peer-reviewed experimental studies to assess the effects of anthropogenic sounds on fishes.

In the following sections, anthropogenic sound is discussed in terms of the different types of potential effects on fish. It must be kept in mind, however, that the data are for very few species, and considering the diversity of fishes, one must be very cautious with any attempts to extrapolate to other species (Hastings et al. 1996; McCauley et al. 2003; Popper et al. 2005b, 2007).

4.1 Nonauditory Injury

Most of the concern about the effects of sound on fishes is associated with the sensory detectors because they are likely to be overstimulated by intense sounds. There has been some concern that these same sounds could produce nonauditory injuries that could range from cellular disruption to gross damage of the swim bladder and circulatory system. How such damage might occur has yet to be demonstrated, and in the few cases where there has been good pathology of exposed tissues, there has been no apparent damage (e.g., Hastings and Popper 2005; Popper et al. 2005b, 2007). Indeed, most of the data suggesting such injuries comes from studies that examined the effects of explosives on fish (e.g., Yelverton et al. 1975; see review in Hastings and Popper 2005).

At the same time, studies of the effects of sound on terrestrial mammals have resulted in some damage to the lungs and other organs as a result of sound exposure (e.g., Fletcher and Busnel 1978; Yang et al. 1996). Some gray literature reports suggest that high sound pressure levels may cause tearing or rupturing of the swim bladder of some (but not all) fish species (e.g., Gaspin 1975; Yelverton et al. 1975), and there is evidence that fish very close to the impulsive sounds from pile driving may suffer death or damage (e.g., Caltrans 2004).

4.2 Permanent Hearing Loss and Inner Ear Damage

A number of studies have examined the effects of high-intensity sound on the sensory hair cells of the ear. Loss of these cells results in permanent hearing loss in terrestrial animals (e.g., Fletcher and Busnel 1978; Saunders et al. 1991), and it may be hypothesized that comparable damage to sensory hair cells could also result in hearing loss. However, there has yet to be any study that has examined fish hearing before and after exposure to sounds that are also known to damage to sensory cells (but see Smith et al. 2006).

Several studies have examined the effects of high-intensity sounds on fish ears. In the first such study, Enger (1981) showed that exposing Atlantic cod (*Gadus morhua*) to high-intensity pure tones resulted in damage to sensory hair cells (as determined with scanning electron microscopy). Subsequently, Hastings et al (1996) showed that exposure of a generalist freshwater fish (the oscar, *Astronotus ocellatus*) to an hour-long continuous 300-Hz sound with a received level of 180 dB produced some damage to the sensory hair cells of the lagena and utricle. However, Hastings et al. did not find any damage resulting from a similar exposure to other frequencies or to noncontinuous sounds or shorter stimulation times. Significantly, damage to the 300-Hz signal only showed up several days after exposure, a result that was similar to that found in another species by McCauley et al. (2003).

McCauley et al. (2003) examined the effects on caged pink snapper (*Pagrus auratus*) of exposure to a seismic air gun with a source level at 1 m of 222.6 dB re 1 μPa (peak to peak) or 203.6 dB re 1 μPa (RMS). They found considerable damage to the ciliary bundles of the sensory hair cells of the saccular sensory epithelium (the other end organs were not examined). The extent of damage increased with an increase in the time the animals were kept postexposure. The animals maintained the longest, to 58 days postexposure, had the greatest damage to the ciliary bundles.

In contrast to these findings, recent investigations found no permanent damage to the ears of fish of three species that were exposed to a sound from a seismic device and then examined immediately after or 24 hours after exposure, although several species showed temporary hearing loss (Popper et al. 2005b). Moreover, exposure to a low-frequency (200- to 500-Hz) sonar at 193 dB re 1 μPa (RMS) did not result in damage to the ears of rainbow trout (*Onchorhynchus mykiss*) or channel catfish (*Ictalurus* species) even up to 96 hours postexposure (Popper et al. 2007).

The tentative conclusion one may reach from these studies is that there are differences in the effects of high-intensity sounds on fish of different species. However, further conclusions are premature at this point because there are so many variables in the different studies. Most importantly, the sounds used were all quite different from one another, and not enough is known about how sounds with different onsets or other characteristics might impact sensory receptors (see Hastings and Popper 2005). The aquatic environment in which experiments are conducted, whether in a laboratory tank or "in the field" where water depth can change sound propagation characteristics (see Rogers and Cox 1988), especially of low-frequency sound, can have important effects that can reduce a study's application to other environments. Moreover, one issue to be considered in any analysis of the effects of sound on the ears of fishes is that fish, unlike mammals, have the potential to regenerate sensory hair cells (Lombarte et al. 1993). If regeneration occurs after damage and the fish survives, regeneration may result in restored hearing and so there may be no long-term effects.

4.3 Temporary Loss of Hearing

Although not much is known about permanent hearing loss in fish, there is a growing body of literature showing that exposure to sounds that are well above normal ambient noise may result in a temporary change in hearing sensitivity from which the fish will recover over time. This loss of hearing, temporary threshold shift (TTS), is well known in mammals and often occurs in humans as a result of exposure to loud noises such as those encountered in a noisy workplace or at a loud concert.

The first study of hearing loss on fish was conducted on goldfish when Popper and Clarke (1976) showed that exposure to 8 continuous hours of sound at 149 dB re 1 μPa (RMS, received level) resulted in more than a 10-dB threshold shift.

Smith et al. (2004a,b) examined the effects of higher background noise on the hearing capabilities of the goldfish, a hearing specialist, and tilapia (*Oreochromis niloticus*), a hearing generalist, to determine how fish hearing might be affected as a result of exposure to somewhat elevated background noise as might be encountered in a hatchery, aquarium, or aquaculture facility or as might occur if the background noise levels rise as a result of human activity in an area. They found that goldfish showed a 5-dB TTS after only 10 minutes of exposure to band-limited noise (0.1 to 10 kHz, approximately 170 dB re 1 μPa [RMS] overall spectral sound pressure level). After 3 weeks of exposure to the same stimulus, goldfish had a 28-dB TTS, and the fish took more than 2 weeks to return to normal hearing. In contrast, tilapia showed no hearing loss to any of these sounds.

Similar results were obtained for goldfish exposed to white noise at 158 dB re 1 μPa for 24 hours by Wysocki and Ladich (2005), with recovery to normal hearing taking up to 2 weeks. Wysocki and Ladich (2005) also performed studies to determine whether the temporal resolving power of goldfish was affected by noise exposure. They found a decrease in temporal resolution capabilities that continued up to 3 days. This kind of hearing loss could be critical because fish of many species appear to use temporal patterns of sounds to discriminate between sounds (e.g., sounds of different species) (Myrberg and Spires 1980). Thus, the effects of noise exposure in fish may be not only on the level of the lowest sound detectable (threshold) but also on the way that fish resolve signals from one another.

Different results between hearing specialists and generalists were also found by Scholik and Yan (2001), who studied another hearing specialist, the fathead minnow (*Pimephales promelas*). They found a substantial hearing loss that continued for more than 14 days after termination of a 24-hour exposure to white noise (0.3–2.0 kHz) with an overall spectral sound pressure level of 142 dB re 1 μPa (RMS). In contrast, Scholik and Yan (2002) found no TTS in the bluegill sunfish (*Lepomis macrochirus*), a hearing generalist.

The studies discussed so far showed a TTS in response to increases in background levels of sound that are comparable to what a human might encounter in a noisy workplace, walking down a city street, or in a noisy classroom. Other studies have examined the effects of considerably higher intensity sounds on fish hearing such as those produced by high-intensity low-frequency sonars, pile driving, or seismic exploration using air guns (or nearby movement of large ships). Several such studies have also tested the effects of such high-intensity sound not only on hearing but also on other nonauditory structures (e.g., swim bladder, heart, brain, liver). In each case, the study was designed to provide an exposure that is far greater than any that a fish is likely to actually encounter and to have all appropriate controls to ensure that the results were from the noise and not from handling or other factors.

In one study, Popper et al. (2005b) examined the effects of exposure to a seismic air gun array on the hearing capabilities of three species of fish found in the Mackenzie River Delta near Inuvik, Northwest Territories, Canada. The species included one hearing specialist, the lake chub (*Couesius plumbeus*), and two species that are not known to have specializations that would enhance hearing, the northern pike (*Esox lucius*) and the broad whitefish (*Coregonus nasus*). The fish were caged and exposed to 5 or 20 shots from a 730-in.^{3} (12,000-ml) air gun array that produced received levels with an average mean peak SPL of 207 dB re 1 μPa (the mean 90 % RMS sound level was 197 dB re 1 μPa). Popper et al. (2005b) found a temporary hearing loss in both lake chub and adult northern pike to both 5 and 20 air gun shots. There was no hearing loss in the broad whitefish, a relative of salmon. Hearing loss was on the order of 20-25 dB at some frequencies for both the northern pike and lake chub, and recovery to normal hearing took place within 24 hours and fish hearing returned to normal. This study reinforces the view that there are potentially substantial differences in the effects of sound on the hearing thresholds of different species.

The second study using high-intensity sound examined the effects of exposure to high-intensity, low-frequency sonar on fish (Popper et al. 2007). In this study, rainbow trout (a hearing generalist) and channel catfish (a hearing specialist) were exposed to 324 seconds of low-frequency sonarlike sounds at 193 dB re 1 μPa (received level) as emitted by a sonar transducer. Interestingly, as in the Popper et al. (2005b) seismic study, there were no fish mortalities and no evidence of damage to any body tissues even 5 days postexposure. Fish of both species showed a small hearing loss. This loss recovered within 48 hours in the catfish, and preliminary evidence indicates recovery after 96 hours in rainbow trout. At the same time, there was no hearing loss in several other hearing generalists after the same exposure regimen (Halvorsen et al. 2006).

4.4 *Effects of Different Noise Levels on Hearing Loss*

Hastings et al. (1996), after reviewing their own studies and other work to that date, proposed the hypothesis that sounds 90–140 dB above a fish's hearing threshold may have the potential to injure the inner ear of a fish. This suggestion was supported in the findings of Enger (1981) who showed injury to Atlantic cod only when the stimulus was 100–110 dB above threshold. Hastings et al. (1996) derived the values of 90–140 dB above threshold by examining the sound levels that caused minimal injury in their test fish, the oscar, and then hypothesizing that extensive injury would require more energy.

This idea received support from the work of Smith et al. (2004a,b) and Scholik and Yan (2001, 2002). Smith et al. (2004b) further hypothesized that noise-induced threshold shifts in fish are linearly related to the difference in sound pressure difference (SPD) between that of the noise and the baseline hearing threshold of the fish, the *lin*ear *t*hreshold *s*hift (LINTS) hypothesis. The actual SPD required to cause TTS in a fish is very likely related to the frequency because the normal hearing levels in fishes vary by frequency. Other variables are likely to be the duration of sound exposure, whether the sound is continuous (as in Smith et al. 2004a,b), or whether they are impulsive.

Although preliminary, there is evidence that the LINTS hypothesis may also hold for impulsive as well as continuous signals. This was suggested based on an analysis of the Popper et al. (2005b) air gun results that showed the same relationship for these sounds as found by Smith et al. (2004b) for continuous noise. And although the Smith et al. (2004b) results supported the LINTS hypothesis only for hearing specialists, the much higher sound levels used by Popper et al. (2005b), which must involve a greater SPD, showed a similar effect in hearing generalists.

4.5 Behavioral Effects of Anthropogenic Sound

Another critical issue with regard to anthropogenic sound is whether it may have some impact(s) on fish behavior other than loss of hearing or damage to tissues other than in the auditory system. In other words, will such sounds affect communication capabilities (e.g., mask communication sounds), cause fish to leave prime feeding grounds, hiding places, or territories, or have other effects that could reduce individual fish survival and reproduction and thereby, potentially, jeopardize population or species survival?

As for hearing loss, there are only a few studies to date that address this issue. Using caged fish, Klimley and Beavers (1998) found no response to a 75-Hz phase-modulated signal (37.5-Hz bandwidth; 145–153 dB re 1 μPa received level) to three species of rockfish (*Sebastes flavidus*, *S. ariculatus,* and *S. mystinus*), which presumably are, but have not been demonstrated to be, hearing generalists.

There is some, although equivocal, evidence that the low-frequency sounds produced by fishing vessels and their associated gear result in fish avoiding the vessels (see Mitson 1995; Mitson and Knudsen 2003). There is also some evidence for a decrease in catch rate after seismic air gun activity (Pearson et al. 1992; Skalski et al. 1992; Engås et al. 1996; Engås and Løkkeborg 2002; Slotte et al. 2004). An issue of major importance is that, in most cases, the behavior of uncaged fish could not be observed, and so it is not known whether changes in catch rate result from damage to fish, their movement from a fishing area, or other factors. However, Slotte et al. used sonar to observe behavior and found that fishes in the vicinity of the air guns appeared to go to greater depths after air gun exposure compared to their vertical position before the firing of the air gun.

In the only study with extensive observation of behavior of uncaged fish during exposure to high noise levels, Wardle et al. (2001), using a video system mounted on a reef, showed no overt reactions or damage to fish resulting from emissions from a seismic air gun (peak level of 210 dB re 1 μPa at 16 m from the source and 195 dB re 1 μPa at 109 m from the source).

Although these studies examined specific effects of high-intensity sounds on fish behavior, there is also the possibility that sounds will have a more subtle effect that results in their not being able to detect biologically relevant sounds including communication sounds, sounds of prey, or sounds of predators (Myrberg 1981; Popper et al. 2004). The decrement in the ability to detect signals because of the presence of other sounds is called masking. Masking can take place whenever the received level of signal exceeds ambient noise levels or the hearing threshold of the animal (e.g., Fay and Megela Simmons 1999).

The studies on auditory masking in fish have been limited in the number of species studied, and none of these studies has directly tested whether there are behavioral changes that result from masking. The results show that species that have been studied are generally affected by masking signals in much the same way as are terrestrial animals for which such data are available (Fay 1988; Fay and Megela Simmons 1999). If the masking signal is of a significantly different frequency from the frequencies of importance to the fish, then much less (or no) masking may occur, although there is also some evidence that in at least some species, any noise signal will mask other signals and that the degree of masking may be frequency independent.

5. Opportunities and Challenges

In 1993, Popper and Fay suggested research questions that were pertinent from their perspective. Those questions (learning, response behaviors, sensitivity and bandwidth, the effects of noise on detection and response, fish capacities for frequency, sound level, temporal, and source localization perception) still bear investigation. We can also add several that would especially help develop potentially useful stimuli for fish management in an industrial context.

One of the most important and least studied issues is the nature and mechanism of habituation with respect to stimuli used in attempts to affect fish behavior. With any sensory-mediated response, habituation (Peeke and Petrinovich 1984) can be an important issue. In most fish passage and protection applications, fish may be present in a fairly small area (a hydropower dam's powerhouse or spillway forebay or a cooling-water intake) for hours or even days and a stimulus that habituates quickly will not be very effective for very long.

In what cases and to what extent are sensory capacities and responses similar across fishes and when are they not? All of the alosine herrings so far investigated have been sensitive to ultrasound but are the others around the world as well? It would be attractive to think so, but Bullock and Heiligenberg (1986) cautioned that, especially when a new sensory capacity is being explored (in that case electrosensing), it is important to remember that responses may be quite variable between closely related species. It is also instructive that a behavioral response (rheotaxis) has been found to be opposite in two conspecific but ecologically distinct populations of juvenile Atlantic salmon (*Salmo salar*; Nemeth et al. 2003).

Just as acoustic noise can obscure a stimulus signal, so can light level, time of day, current, temperature, the presence of other species including predators, crowding, and untold other factors affect behavioral responses (Schilt and Norris 1997). That is why laboratory, net pen, and field experiments are all important. Laboratory studies, sometimes starting with neurological work, can point the way toward better and more refined field studies. Net pen studies can allow for free-swimming but not free-ranging fish responses, habituation studies, and manipulation of signal and noise regimens. Field studies at an actual application site can discover unforeseen strengths or, more likely, weaknesses in an approach and test "real-world" responses of animals that cannot be foreseen in laboratory studies.

Acknowledgments. We are grateful to Dr. Jacqueline Webb for reviewing an earlier copy of this manuscript and Helen A. Popper for reviewing and editing the manuscript.

References

Astrup J (1999) Ultrasound detection in fish—a parallel to the sonar-mediated detection of bats by ultrasound-sensitive insects. Comp Biochem Physiol A Mol Integr Physiol 124:19–27.

Atema J, Fay RR, Popper AN, Tavolga WN (eds) (1988) The Sensory Biology of Aquatic Animals. New York: Springer-Verlag.

Barry T, Kynard B (1986) Attraction of adult American shad to fish lifts at Holyoke Dam, Connecticut River. N Am J Fish Manage 6:233–241.

Beasley CA, Hightower JE (2000) Effects of a low-head dam on the distribution and characteristics of spawning habitat used by striped bass and American shad. Trans Am Fish Soc 129:1316–1330.

Bregman AS (1990) Auditory Scene Analysis: The Perceptual Organization of Sound. Cambridge, MA: MIT Press.

Broadhurst MK, Kennelly SJ, Eayrs S (1999) Flow-related effects in prawn-trawl codends: potential for increasing the escape of unwanted fish through square-mesh panels. Fish Bull 97:1–8.

Budy P, Thiede GP, Bouwes N, Petrosky CE, Schaller H (2002) Evidence linking delayed mortality of Snake River salmon to their earlier hydrosystem experience. N Am J Fish Manage 22:35–51.

Bullock TH, Heiligenberg W (1986) Electroreception. New York: John Wiley & Sons.

Burner CJ, Moore HL (1953) Attempts to Guide Small Fish With Underwater Sound. US Department of the Interior Special Scientific Report, Fisheries No. 111.

Caltrans (2004) Fisheries and Hydroacoustic Monitoring Program Compliance Report for the San Francisco-Oakland Bay Bridge East Span Seismic Safety Project. Prepared by Strategic Environmental Consulting, Inc. and Illingworth & Rodkin, Inc. June.

Chapman CJ (1976) Some observations on the reactions of fish to sound. In: Schuijf A, Hawkins AD (eds) Sound Reception in Fish. Amsterdam: Elsevier, pp. 241–255.

Chick JH, Pegg MA (2001) Invasive carp in the Mississippi River Basin. Science 292:2250–2251.

Collin SP, Marshall NJ (eds) (2003) Sensory Processing in Aquatic Environments. New York: Springer-Verlag.

Coombs S, Montgomery JC (1999) The enigmatic lateral line system. In: Fay RR, Popper AN (eds) Comparative Hearing: Fish and Amphibians. New York: Springer-Verlag, pp. 319–362.

Coombs S, Popper AN (1979) Hearing differences among Hawaiian squirrelfishes (family Holocentridae) related to differences in the peripheral auditory system. J Comp Physiol 132:203–207.

Coombs S, Jansen J, Webb JF (1988) Diversity of lateral line systems: evolutionary and functional considerations. In: Atema J, Fay RR, Popper AN, Tavolga WN (eds) The Sensory Biology of Aquatic Animals. New York: Springer-Verlag, pp. 267–288.

Coutant CC (ed) (2001) Behavioral Technologies for Fish Guidance. American Fisheries Society Symposium 26. Bethesda, MD: American Fisheries Society.

Dadswell MJ, Kaluda RJ, Moffitt CM, Saunders RL, Rulifson RA, Cooper JE (eds) (1987) Common Strategies of Anadromous and Catadromous Fishes. American Fisheries Society Symposium 1. Bethesda, MD: American Fisheries Society.

Dale T (1976) The labyrinthine mechanoreceptor organs of the cod *Gadus morhua* L. (Teleostei, Gadidae). Norw J Zool 24:85–128.

Dempster T, Kingsford MJ (2003) Homing of pelagic fish to fish aggregation devices (FADs): the role of sensory cues. Mar Ecol Prog Ser 258:213–222.

Dijkgraaf S (1963) The functioning and significance of the lateral line organ. Biol Rev 38:51–105.

Dijkgraff S (1989) A short personal history of lateral line research. In: Coombs S, Görner P, Münz H (eds) The Mechanosensory Lateral Line Neurobiology and Evolution. New York: Springer-Verlag, pp. 7–14.

Dixon DA (ed) (2003) Biology, Management, and Protection of Catadromous Eels. American Fisheries Society Symposium 33. Bethesda, MD: American Fisheries Society.

Dunning DJ, Ross QE, Geoghegan P, Reichle JJ, Menezes JK, Watson JK (1992) Alewives avoid high-frequency sound. N Am J Fish Manage 12:407–416.

Dynesius M, Nilsson C (1994) Fragmentation and flow regulation of river systems in the northern third of the world. Science 266:753–762.

Edds-Walton PL (1998) Projections of primary afferent from regions of the saccule in the toadfish (*Opsanus tau*). Hear Res 115:45–60.

Engås A, Løkkeborg S (2002) Effects of seismic shooting and vessel-generated noise on fish behaviour and catch rates. Bioacoustics 12:313–315.

Engås A, Løkkeborg S, Ona E, Soldal AV (1996) Effects of seismic shooting on local abundance and catch rates of cod (*Gadus morhua*) and haddock (*Melanogrammus aeglefinus*). Can J Fish Aquat Sci 53:2238–2249.

Enger P (1967) Hearing in herring. Comp Biochem Physiol 22:527–538.

Enger PS (1981) Frequency discrimination in teleosts—central or peripheral? In: Tavolga WN, Popper AN, Fay RR (eds) Hearing and Sound Communication in Fishes. New York: Springer-Verlag, pp. 243–255.

Fay RR (1988) Hearing in Vertebrates: A Psychophysics Handbook. Winnetka, IL: Hill-Fay Associates.
Fay RR (2005) Sound localization by fishes. In: Popper AN, Fay RR (eds) Sound Source Localization. New York: Spinger Science+Business Media, pp. 36–66.
Fay RR (2008) Sound source perception and stream segregation in non-human vertebrate animals. In: Yost WA, Popper AN, Fay RR (eds) Auditory Perception of Sound Sources. New York: Springer Science+Business Media, pp. 307–323.
Fay RR, Edds-Walton PL (1997) Diversity in frequency response properties of saccular afferents of the toadfish (*Opsanus tau*). Hear Res 111:235–246.
Fay RR, Megela Simmons A (1999) The sense of hearing in fishes and amphibians. In: Fay RR, Popper AN (eds) Comparative Hearing: Fish and Amphibians. New York: Springer-Verlag, pp. 269–318.
Fay RR, Popper AN (eds) (1999) Comparative Hearing: Fish and Amphibians. New York: Springer-Verlag.
Fay RR, Popper AN (2000) Evolution of hearing in vertebrates: the inner ears and processing. Hear Res 149:1–10.
Finneran JJ, Oliver CW, Schaefer KM, Ridgeway SH (2000) Source levels and estimated yellowfin tuna (*Thunnus albacares*) detection ranges for dolphin jaw pops, breaches, and tail slaps. J Acoust Soc Am 107:649–656.
Fletcher RI (1990) Flow dynamics and fish recovery experiments: water intake systems. Trans Am Fish Soc 119:393–415.
Fletcher JL, Busnel RG (1978) Effects of Noise on Wildlife. New York: Academic Press.
Freeman MC, Bowen ZH, Bovee KD, Erwin ER (2001) Flow and habitat effects on juvenile fish abundance in natural and altered flow regimes. Ecol Appl 11: 179–190.
Fuller PL, Nico LG, Williams JD (1999) Nonindigenous Fishes Introduced into Inland Waters of the United States. Special Publication 27. Bethesda, MD: American Fisheries Society.
Gaspin JB (1975) Experimental investigations of the effects of underwater explosions on swimbladder fish I: 1973 Chesapeake Bay tests. Navel Surface Weapons Center Report NSWC/WOL/TR 75–58.
Goetz FA, Dawson JJ, Shaw T, Dillon J (2001) Evaluation of low-frequency sound transducers for guiding salmon smolts away from a navigation lock. Am Fish Soc Symp 26:91–104.
Gregory J, Clabburn P (2003) Avoidance behaviour of *Alosa fallax fallax* to pulsed ultrasound and its potential as a technique for monitoring clupeid spawning migration in a shallow river. Aquat Living Resour 16:313–316.
Halvorsen MB, Wysocki LE, Popper AN (2006) Effects of high-intensity sonar on fish. J Acoust Soc Am 119:3283.
Hanson Environmental, Inc. (1996) Giorgiana Slough Acoustic Barrier Applied Research Project: Results of 1994 Phase II Field Tests. Department for Water Resources and Bureau of Reclamation. FF/BIO. IATR/95–44.
Hashimoto T, Maniwa Y (1967) Research on the luring of fish shoals by utilizing underwater acoustical equipment. In: Tavolga WN (ed) Marine Bio-Acoustics. Oxford: Pergamon Press, pp. 93–104.
Hastings MC, Popper AN (2005) Effects of sound on fish. California Department of Transportation Contract 43A0139, Task Order 1. http://www4.trb.org/trb/crp.nsf/reference/boilerplate/Attachments/$file/EffectsOfSoundOnFish1–28–05(FINAL)s pdf.

Hastings MC, Popper AN, Finneran JJ, Lanford PJ (1996) Effect of low-frequency underwater sound on hair cells of the inner ear and lateral line of the teleost fish *Astronotus ocellatus*. J Acoust Soc Am 99:1759–1766.

Haymes GT, Patrick PH (1986) Exclusion of adult alewife, *Alosa pseudoharengus*, using low-frequency sound for application at water intakes. Can J Fish Aquat Sci 43:855–862.

Higgs DM, Plachta DTT, Rollo AK, Singheiser M, Hastings MC, Popper AN (2004) Development of ultrasound detection in American shad (*Alosa sapidissima*). J Exp Biol 207:155–163.

Hocutt CH (1980) Behavioral barriers and guidance systems. In: Hocutt CH, Stauffer JR Jr, Edinger JE, Hall LW Jr, Morgan RP II (eds) Power Plants: Effects on Fish and Shellfish Behavior. New York: Academic Press, pp. 183–205.

Hudspeth AJ (1985) The cellular basis of hearing: the biophysics of hair cells. Science 230:745–752.

Hudspeth AJ (1997) Mechanical amplification of stimuli by hair cells. Curr Opin Neurobiol 7:480–486.

Iversen RTB (1967) Response of the yellowfin tuna (*Thunnus albacares*) to underwater sound. In: Tavolga, WN (ed) Marine Bio-Acoustics II. Oxford: Pergamon Press, pp. 105–121.

Iversen RTB (1969) Auditory thresholds of the scombrid fish *Euthynnus affinis*, with comments on the use of sound in tuna fishing. FAO Fisheries Rep No. 62, 3:849–859.

Jungwirth S, Schmutz S, Weiss S (eds) (1998) Fish Migration and Fish Bypasses. Oxford: Blackwell.

Klimley AP, Beavers SC (1998) Playback of acoustic thermometry of ocean climate (ATOC)-like signal to bony fishes to evaluate phonotaxis. J Acoust Soc Am 104: 2506–2510.

Knudsen FR, Enger PS, Sand O (1992) Awareness reactions and avoidance responses to sound in juvenile Atlantic salmon, *Salmo salar* L. J Fish Biol 40:523–534.

Kynard B, O'Leary J (1990) Behavioral guidance of adult American shad using underwater AC electrical and acoustic fields. Proceedings of the International Symposium on Fishways '90 in Gifu, Japan, October 8–10, 1990, pp. 131–135.

Ladich F, Popper AN (2004) Parallel evolution in fish hearing organs. In: Manley GA, Popper AN, Fay RR (eds) Evolution of the Vertebrate Auditory System. New York: Springer-Verlag, pp. 95–127.

Lassuy DR (1995) Introduced species as a factor in extinction and endangerment of native fish species. In: Schramm HL Jr, Piper RG (eds) Uses and Effects of Cultured Fishes in Aquatic Ecosystems. American Fisheries Society Symposium 15. Bethesda, MD: American Fisheries Society, pp. 391–396.

Limburg KE, Waldman JR (eds) (2003) Biodiversity, Status, and Conservation of the World's Shads. American Fisheries Society Symposium 35. Bethesda, MD: American Fisheries Society.

Lombarte A, Popper AN (1994) Quantitative analyses of postembryonic hair cell addition in the otolithic endorgans of the inner ear of the European hake, *Merluccius merluccius* (Gadiformes, Teleostei). J Comp Neurol 345:419–428.

Lombarte A, Yan HY, Popper AN, Chang JC, Platt C (1993) Damage and regeneration of hair cell ciliary bundles in a fish ear following treatment with gentamicin. Hear Res 66:166–174.

Love MS, Caselle JE, Snook L (2000) Fish assemblages around seven oil platforms in the Santa Barbara Channel area. Fish Bull 98:96–117.

Lu Z, Popper AN (2001) Neural response directionality correlates of hair cell orientation in a teleost fish. J Comp Physiol A 187:453–465.

Lu Z, Popper AN, Fay RR (1996) Behavioral detection of acoustic particle motion by a teleost fish (*Astronotus ocellatus*): sensitivity and directionality. J Comp Physiol A 197:227–233.

Mann DA, Lu Z, Popper AN (1997) A clupeid fish can detect ultrasound. Nature 389:341.

Mann DA, Lu Z, Hastings MC, Popper AN (1998) Detection of ultrasonic tones and simulated dolphin echolocation clicks by a teleost fish, the American shad (*Alosa sapidissima*). J Acoust Soc Am 104:562–568.

Mann DA, Higgs DM, Tavolga WN, Souza MJ, Popper AN (2001) Ultrasound detection by clupeiform fishes. J Acoust Soc Am 109:3048–3054.

Mann DA, Popper AN, Wilson B (2005) Pacific herring hearing does not include ultrasound. Biol Lett 1:158–161.

McAllister DE, Craig JF, Davidson N, Delany, S, Seddon M (2000) Biodiversity Impacts of Large Dams. Background Paper No. 1. IUCN (The World Conservation Union)/UNEP (United Nations Environment Programme)/WCD (World Commission on Dams). http://intranet.iucn.org/webfiles/doc/archive/2001/IUCN850.

McCauley RD, Fewtrell J, Popper AN (2003) High-intensity anthropogenic sound damages fish ears. J Acoust Soc Am 113:638–642.

Mitson RB (ed)(1995) Underwater Noise of Research Vessels: Review and Recommendations. ICES Coop Res Rep No. 209, p. 61.

Mitson RB, Knudsen HP (2003) Causes and effects of underwater noise on fish abundance estimation. Aquat Living Resour 16:255–263.

Moser ML, Ocker PA, Stuehrenberg LC, Bjornn TC (2002) Passage efficiency of adult Pacific lampreys on the Lower Columbia River, USA. Trans Am Fish Soc 131:956–965.

Myrberg AA Jr (1981) Sound communication and interception in fishes. In: Tavolga WN, Popper AN, Fay RR (eds) Hearing and Sound Communication in Fishes. New York: Springer-Verlag, pp. 395–426.

Myrberg AA Jr, Spires JY (1980) Hearing in damselfishes: an analysis of signal detection among closely related species. J Comp Physiol 140:135–144.

National Research Council (2000) Marine Mammals and Low Frequency Sound: Progress Since 1994. Washington, DC: National Academy Press.

Nemeth MJ, Kreuger CC, Josephson DC (2003) Rheotactic response of two strains of juvenile landlocked Atlantic salmon: implications for population restoration. Trans Am Fish Soc 132:904–912.

Nestler JM, Ploskey GR, Pickens J, Menezes J, Schilt C (1992) Responses of blueback herring to high frequency sound and implications for reducing entrainment at hydropower dams. N Am J Fish Manage 12:667–683.

Nestler JM, Ploskey GR, Weeks G, Schneider T (1995) Development of an operational, full-scale fish protection system at a major pumped-storage hydropower dam. Waterpower ‘95, Proceedings of the International Conference on Hydropower, San Francisco, CA: American Society of Civil Engineers, pp. 152–161.

Orbach MK (1977) Hunters, Seamen, and Entrepreneurs: The Seinermen of San Diego. Berkeley, CA: University of California Press.

Parrish JK (1999) Using behavior and ecology to exploit schooling fishes. Environ Biol Fish 55:157–181.

Pavlov DS, Lupandin AI, Kostin VV (2002) Downstream migration of fish through dams of hydroelectric power plants (translated by Albert T, translation editor Cada GF). ORNL/TR-02/02. Oak Ridge, TN: Oak Ridge National Laboratory.

Available on-line from the US Department of Energy Hydropower Program http://hydropower.id.doe.gov/.

Pearson WH, Skalski JR, Malme CI (1992) Effects of sounds from a geophysical survey device on behavior of captive rockfish (*Sebastes* ssp). Can J Fish Aquat Sci 49: 1343–1356.

Peeke HVS, Petrinovich L (1984) Habitat, Sensitization, and Behavior. New York: Academic Press.

Plachta DTT, Popper AN (2003) Evasive responses of American shad (*Alosa sapidissima*) to ultrasonic stimuli. Acoust Res Lett Online 4:25–30.

Plachta DTT, Song J, Halvorsen MB, Popper AN (2004) Neuronal encoding of ultrasonic sound by a fish. J Neurophysiol 91:2590–2597.

Platt C (1983) The peripheral vestibular system in fishes. In: Northcutt RG, Davis RE (eds) Fish Neurobiology. Ann Arbor, MI: University of Michigan Press, pp. 89–124.

Platt C, Popper AN, Fay RR (1989) The ear as part of the octavolateralis system. In: Coombs S, Görner P, Münz H (eds) The Mechanosensory Lateral Line: Neurobiology and Evolution. New York: Springer-Verlag, pp. 663–651.

Ploskey G, Nestler J, Weeks G, Schilt C (1995) Evaluation of an integrated fish protection system. Waterpower '95, Proceedings of the International Conference on Hydropower. San Francisco, CA: American Society of Civil Engineers, pp.162–171.

Ploskey GR, Johnson PN, Carlson TJ (2000) Evaluation of a low-frequency sound-pressure system for guiding juvenile salmon away from turbines at Bonneville Dam, Columbia River. N Am J Fish Manage 20:951–967.

Popper AN (1976) Ultrastructure of the auditory regions in the inner ear of the lake whitefish. Science 192:1020–1023.

Popper AN (1980) Scanning electron microscopic studies of the sacculus and lagena in several deep-sea fishes. Am J Anat 157:115–136.

Popper AN (2003) Effects of anthropogenic sound on fishes. Fisheries 28:24–31.

Popper AN, Carlson TJ (1998) Application of sound and other stimuli to control fish behavior. Trans Am Fish Soc 127:673–707.

Popper AN, Clarke NL (1976) The auditory system of the goldfish (*Carassius auratus*): effects of intense acoustic stimulation. Comp Biochem Physiol 53A:11–18.

Popper AN, Coombs S (1982) The morphology and evolution of the ear in Actinopterygian fishes. Am Zool 22:311–328.

Popper AN, Fay RR (1993) Sound detection and processing by fish: critical review and major research questions. Brain Behav Evol 41:14–38.

Popper AN, Fay RR (1999) The auditory periphery in fishes. In: Fay RR, Popper AN (eds) Comparative Hearing: Fish and Amphibians. New York: Springer-Verlag, pp. 43–100.

Popper AN, Platt C (1979) The herring ear has a unique receptor pattern. Nature 280: 832–833.

Popper AN, Rogers PH, Saidel WM, Cox M (1988) The role of the fish ear in sound processing. In: Atema J, Fay RR, Popper AN, Tavolga WN (eds) Sensory Biology of Aquatic Animals. New York: Springer-Verlag, pp. 687–710.

Popper AN, Fay RR, Platt C, Sand O (2003) Sound detection mechanisms and capabilities of teleost fishes. In: Collin SP, Marshall NJ (eds) Sensory Processing in Aquatic Environments. New York: Springer-Verlag, pp. 3–38.

Popper AN, Fewtrell J, Smith ME, McCauley RD (2004) Anthropogenic sound: effects on the behavior and physiology of fishes. Mar Technol Soc J 37:35–40.

Popper AN, Ramcharitar J, Campana SE (2005a) Why otoliths? Insights from inner ear physiology and fisheries biology. Mar Freshw Res 56:497–504.

Popper AN, Smith ME, Cott PA, Hanna BW, MacGillivray AO, Austin ME, Mann DA (2005b) Effects of exposure to seismic airgun use on hearing of three fish species. J Acoust Soc Am 117:3958–3971.

Popper AN, Halvorsen MB, Kane E, Miller DD, Smith ME, Stein P, Wysocki LE (2007) The effects of high-intensity, low-frequency active sonar on rainbow trout. J Acoust Soc Am 122:623-635.

Pringle CM, Freeman MC, Freeman BJ (2000) Regional effects of hydrologic alterations on riverine macrobiota in the new world: tropical-temperate comparisons. BioScience 50:807–823.

Rainey S (1997) Opportunity for discovery: an important tool for assessing surface collection technology potential for improving juvenile salmon passage at Columbia River dams. In: Fish Passage Workshop Proceedings of a meeting held May 6–8, 1997, in Milwaukee, WI: Alden Research Laboratory.

Richkus WA, Dixon DA (2003) Review of research and technologies on passage and protection of downstream migrating catadromous eels at hydroelectric facilities. In: Dixon DA (ed) Biology, Management, and Protection of Catadromous Eels. American Fisheries Society Symposium 33. Bethesda, MD: American Fisheries Society, pp. 377–388.

Rogers PH, Cox M (1988) Underwater sound as a biological stimulus. In: Atema J, Fay RR, Popper AN, Tavolga WN (eds) Sensory Biology of Aquatic Animals. New York: Springer-Verlag, pp. 131–149.

Rogers PH, Popper AN, Cox M, Saidel WM (1988) Processing of acoustic signals in the auditory system of bony fish. J Acoust Soc Am 83:338–349.

Ross QE, Dunning DJ, Thorne R, Menezes JK, Tiller GW, Watson JK (1993) Response of alewives to high-frequency sound at a power plant intake on Lake Ontario. N Am J Fish Manage 13:291–303.

Ross QE, Dunning DJ, Menezes JK, Kenna MJ Jr, Tiller GW (1996) Reducing impingement of alewives with high-frequency sound at a power plant intake on Lake Ontario. N Am J Fish Manage 16:548–559.

Sand O, Karlsen HE (1986) Detection of infrasound by the Atlantic cod. J Exp Biol 125:197–204.

Sand O, Karlsen HE (2000) Detection of infrasound and linear acceleration in fish. Philos Trans R Soc Lond B 355:1295–1298.

Sand O, Enger PS, Karlsen HE, Knudsen FR, Kvernstuen T (2000) Avoidance responses to infrasound in downstream migrating European silver eels, *Anguilla anguilla*. Environ Biol Fish 47:327–336.

Sand O, Enger PS, Karlsen HE, Knudsen FR (2001) Detection of infrasound in fish and behavioral responses to intense infrasound in juvenile salmonids and European eels: a minireview. In: Coutant CC (ed) Behavioral Technologies for Fish Guidance. American Fisheries Society Symposium 26. Bethesda, MD: American Fisheries Society, pp. 183–193.

Saunders JC, Cohen YE, Szymko YM (1991) The structural and functional consequences of acoustic injury in the cochlea and peripheral auditory system: a five year update. J Acoust Soc Am 90:147–155.

Schellart NAM, Popper AN (1992) Functional aspects of the evolution of the auditory system of actinopterygian fish. In: Webster DB, Fay RR, Popper AN (eds) The Evolutionary Biology of Hearing. New York: Springer-Verlag, pp. 295–322.

Schilt CR, Nestler JM (1997) Fish Bioacoustics: Considerations for Hydropower Industry Workers. Hydropower '97, Proceedings of the International Conference held August 5–8, 1997, Atlanta, GA.

Schilt CR, Norris KS (1997) Perspectives on sensory integration systems. In: Parrish JK, Hamner WM (eds) Animal Groups in Three Dimensions. New York: Cambridge University Press, pp. 225–244.

Scholik AR, Yan HY (2001) Effects of underwater noise on auditory sensitivity of a cyprinid fish. Hear Res 152:17–24.

Scholik AR, Yan HY (2002) The effects of noise on the auditory sensitivity of the bluegill sunfish, *Lepomis macrochirus*. Comp Biochem Physiol A Mol Integr Physiol 133:43–52.

Schuler VJ, Larson LE (1975) Improved fish protection at intake systems. J Environ Eng Div, December 1975, pp. 11756–11775.

Skalski JR, Pearson WH, Malme CI (1992) Effects of sounds from a geophysical survey device on catch-per-unit-effort in a hook-and-line fishery for rockfish (*Sebastes* spp.). Can J Fish Aquat Sci 49:1357–1365.

Slotte A, Hansen K, Dalen J, Ona E (2004) Acoustic mapping of pelagic fish distribution and abundance in relation to a seismic shooting area off the Norwegian west coast. Fish Res 67:143–150.

Smith ME, Kane AS, Popper AN (2004a) Noise-induced stress response and hearing loss in goldfish (*Carassius auratus*). J Exp Biol 207:427–435.

Smith ME, Kane AS, Popper AN (2004b) Acoustical stress and hearing sensitivity in fishes: does the linear threshold shift hypothesis hold water? J Exp Biol 207:3591–3602.

Smith ME, Coffin AB, Miller DL, Popper AN (2006) Anatomical and functional recovery of the goldfish (*Carassius auratus*) ear following noise exposure. J Exp Biol 209: 4193–4202.

Taylor RM, Pegg MA, Chick JH (2005) Response of bighead carp to a bioacoustic behavioral fish guidance system. Fish Manage Ecol 12:283–286.

Wardle CS (1993) Fish behaviour and fishing gear. In: Pitcher TJ (ed) Behaviour of Teleost Fishes, 2nd ed. London: Chapman and Hall, pp. 609–643.

Wardle CS, Carter TJ, Urquhart GG, Johnstone ADF, Ziolkowski AM, Hampson G, Mackie D (2001) Effects of seismic air guns on marine fish. Cont Shelf Res 21: 1005–1027.

Willis DJ, Hover MV, Canfield DE Jr, Lindberg WJ (2002) Training grass carp to respond to sound for potential lake management uses. N Am J Fish Manage 22: 208–212.

Wysocki LE, Ladich F (2005) Effects of noise exposure on click detection and the temporal resolution ability of the goldfish auditory system. Hear Res 201:27–36.

Wysocki LE, Dittami JP, Ladich F (2006) Ship noise and cortisol secretion in European freshwater fishes. Biol Conserv 128:501–508.

Wysocki LE, Davidson JW III, Smith ME, Frankel AS, Ellison WT, Mazik PM, Popper AN, Bebak J (2007) Effects of aquaculture production noise on hearing, growth, and disease resistance of rainbow trout *Oncorhynchus mykiss*. Aquaculture 272: 687–697.

Yang Z, Wang Z, Tang C, Ying Y (1996) Biological effects of weak blast waves and safety limits for internal organ injury in the human body. J Trauma 40:S81–S84.

Yelverton JT, Richmond DR, Hicks W, Saunders K, Fletcher ER (1975) The relationship between fish size and their response to underwater blast. Report DNA 3677T, Director, Defense Nuclear Agency, Washington, DC.

Zelick R, Mann D, Popper AN (1999) Acoustic communication in fishes and frogs. In: Fay RR, Popper AN (eds) Comparative Hearing: Fish and Amphibians. New York: Springer-Verlag, pp. 363–411.

Zigler SJ, Dewey MR, Knights BC, Runstrom AL, Steingraeber MT (2004) Hydrologic and hydraulic factors affecting passage of paddlefish through dams in the upper Mississippi River. Trans Am Fish Soc 133:160–172.

3 Structures and Functions of the Auditory Nervous System of Fishes

RICHARD R. FAY AND PEGGY L. EDDS-WALTON

1. Introduction

Our understanding of the auditory system of fishes has advanced significantly over the past two decades, but more data have been obtained on the anatomy of the auditory pathway than on the physiology of auditory processing. In most cases, a few investigators have focused on a few species, resulting in a somewhat limited view of species diversity. Thus, we do not at present have high enough confidence to make many generalizations about the central auditory system for fishes as a group. However, a general understanding is beginning to emerge. For example, it is becoming clear that the organization of the auditory CNS in fishes is consistent with that in most other vertebrates, at levels from the lower hindbrain to the telencephalon. At most levels, auditory nuclei in amniotic vertebrates have functional analogies among the fishes investigated so far. At the same time, however, it has not been possible to identify homologies among nuclei across vertebrate taxa, and the highly analogous pathways and functions that we see must be attributed to parallel or convergent evolution (Grose et al. 2004). The anatomical information presented here does not emphasize evolutionary comparisons (see McCormick 1992, 1999), but rather provides a brief description of the general auditory pathway of fishes as a context for the physiological review that follows.

The comparative physiology of the auditory regions of the brain in fishes has only begun to be systematically investigated in a limited number of species. Adrian et al. (1938) were the first to record action potentials from the auditory nerve in several species of fishes. Lowenstein and Roberts (1951) analyzed the neural code from vestibular and putative auditory branches of the VIIIth cranial nerve in elasmobranchs. The pioneering work of Furukawa and Ishii (1967) began an important focus on the saccular nerve of goldfish (Carassius auratus). More recently, the physiology of auditory nerve units has been studied in *C. auratus*, and this work has been reviewed in some detail (e.g., Fay and Popper 1999). Therefore, the current review only briefly summarizes and updates these studies, including the extensive work of Furukawa and his colleagues on the hair cell–nerve fiber synapse in *C. auratus* saccule (e.g., Sugihara and Furukawa 1989). This summary is necessary for understanding the transformations of peripherally encoded neural representations that occur as a result of neural interactions in the brain. Studies on central auditory physiology of fishes have not been reviewed in detail recently, and this chapter focuses on these studies.

In recent years, there has been a new interest in electrophysiological studies of auditory evoked potentials in fishes (e.g., Kenyon et al. 1998). These have focused on the comparative mechanisms of sound detection, hearing bandwidth, and the effects of intense noise on fishes using the auditory brainstem response (ABR) technique. Because these studies do not identify the anatomical or physiological origins of the evoked responses, they are not reviewed here. Also not reviewed in detail are studies on the functioning of Mauthner cells and other reticulospinal circuits underlying acoustically triggered rapid escape responses (see Fay 1995; Popper and Edds-Walton 1995). Finally, lateral line systems are mentioned only briefly because the lateral line (Coombs et al. 1989) is no longer viewed as just an accessory hearing organ of the "acoustico–lateralis system" as it once was (e.g., van Bergeijk 1967; Popper et al. 1992), and is now considered to be a system uniquely sensitive to nearby hydrodynamic flows.

2. Anatomical Background

Many casual observers of fishes are surprised to learn that fish have ears. Although there are no external pinna, the internal structures common to all vertebrate ears are also present in the ear of fishes. Components of the ears of all jawed vertebrates (Gnathostomes, including bony fishes and elasmobranchs) include both vestibular and auditory structures: three semicircular canals and three or more sensory epithelia. Each of the semicircular canals has a relatively small epithelium (crista) with sensory hair cells that are responsible for encoding movement of the head or body. The three cristae are housed in three bulblike expansions of the fluid-filled semicircular canals and are oriented nearly orthogonally (x, y, z planes). As in other vertebrates, the fish ear also has individual sensory epithelia, each of which is coupled to a dense, calcareous structure called an otolith. Among adult fishes, the otolithic sensory epithelia can vary widely in their relative sizes and shapes even among members of the same taxonomic family (Popper and Fay 1999); however, in all bony fishes, three epithelia develop: the saccule, lagena, and utricle. Each may encode position/tilt and/or frequencies associated with environmental sounds. Our emphasis in this chapter is on auditory processing, and we do not consider the dual, auditory and vestibular, functioning of the otolithic end organs (but see McCormick 1999 for a discussion). Additional information on the vestibular sense also can be found in Platt (1983).

The apical surface of the sensory hair cells of fishes consists of a single kinocilium and an adjacent stair-step array of stereocilia that decrease in height with distance from the kinocilium. Some portion of the stereocilia is believed to be imbedded in a gelatinous matrix that mechanically couples the hair cell to the otolith. The gelatinous matrix is often called the otolithic membrane, but there is no histological basis for this designation as a membrane.

During stimulation of the ear, the relative motion of the sensory epithelium with respect to the otolith causes deformation of the surface structures on the sensory hair cells that can result in excitation of the afferent neuron. For detailed discussions of hair cell morphology and physiology see Hudspeth and Corey (1977) and Sugihara and Furukawa (1989). In some fishes, a region of the sensory epithelium may not be associated with the otolith, leading to speculations that those regions of the sensory epithelium might respond differently to stimuli (see Section 3.5).

Auditory functions have been ascribed to another sensory epithelium in the ear of some fishes: the macula neglecta. The macula neglecta lacks an otolith, but has a gelatinous cupula associated with the hair cells. While common in the cartilaginous fishes (Chondrichthyes: sharks, skates, and rays), where it may have a major role in auditory processing (Corwin 1981, 1989), the macula neglecta has not been implicated as an auditory structure in any bony fish species. Therefore, we do not consider the macula neglecta or its projections further.

We concentrate on the innervation and projections of the end organs whose auditory responsiveness has been demonstrated experimentally in actinopterygian (or ray-finned) fishes. We do not include a discussion of potential auditory structures in other members of the bony fishes, the dipnoans (lungfish) and the crossopterygians (coelacanths), as there are no physiological data for any of those species.

2.1 Auditory Afferents of the Ear

Most studies of innervation of the otolithic end organs have concentrated on saccular afferents because this end organ is believed to be the primary auditory end organ in most fishes. The lagena also may process auditory frequencies in some fishes, such as the goldfish (*Carassius auratus*, Fay 1984) and sleeper goby (*Dormitator latifrons*, Lu et al. 2003). Another exceptional case is the utricle of herrings (family Clupeidae), in which a subdivided utricle responds to sound (Blaxter et al. 1981).

The morphology of auditory afferents has been studied in detail by injecting label into individual neurons (e.g., *O. tau*, Sento and Furukawa 1987; *D. latifrons*, Edds-Walton et al. 1999; *C. auratus*, Lu, Song and Popper 1998) or by applying a label to cut or damaged nerve bundles for uptake and transport along the neuron (e.g., Presson et al. 1992; Edds-Walton and Popper 2000). As Edds-Walton and Popper (2000) noted, comparisons of data on the morphological characteristics of afferents are most appropriate when the same label and similar techniques are used for the collection of data because the various labels available (e.g., horseradish peroxidase [HRP], Lucifer yellow, DiI or DiO, cobaltous-lysine, biotin, neurobiotin, biotinylated/fluorescent dextran amines) vary in molecular weight, travel time, and efficiency of filling tiny processes and boutons, which can greatly affect the resultant appearance of the filled fibers. In addition, methods for fixing the tissue and visualizing the label (e.g., immunohistochemistry) can cause substantial shrinkage of the tissue, which ultimately affects apparent fiber and arbor diameter. Lastly, the location of diameter measurements is important because fiber diameter varies near the soma, at the first branch point, and along the entire length. Therefore, we compare the findings from two studies that used similar methods on two species of fish. One study (Presson et al. 1992) examined the innervation of the saccule of the oscar (*Astronotus ocellatus*), a hearing "generalist" with regard to the organization of the inner ear. The second study (Edds-Walton and Popper 2000) examined the innervation of the saccule and lagena of *C. auratus* (*Carassius auratus*), a hearing "specialist" with enhanced auditory sensitivity. As noted in the preceding text, *C. auratus* may use both the saccule and lagena for auditory processing and directional hearing (Fay 1984). Both of these studies used cobaltous-lysine that was dried on to the tip of a minutin or insect pin and poked into bundles of cranial nerve VIII, thereby damaging neurons in the bundles and permitting uptake of the label. The only neurons considered in either study were those that had a filled dendritic arbor and a filled soma. Because efferent cells have cell bodies in the medulla, this simple distinction eliminated efferent fibers and arbors from consideration.

Afferent diameters were measured for dendrites on the otolithic end organs just prior to the first branch point (Presson et al. 1992, Edds-Walton and Popper 2000). Saccular afferent fiber diameters ranged from 1 to 9 μm (median 2 μm) in *Astronotus ocellatus* and from 1 to 10 μm (median 3 μm) in *C. auratus*. Lagenar measurements were reported only in *C. auratus* study. The range of lagenar fiber diameters was 1–12 μm, with a median fiber diameter of 4 μm. It is important to note, however, that the majority of afferents were between 2 and 4 μm on both the saccule and the lagena of *C. auratus*. No attempt was made to adjust the measurements for the effects of histological processing, so these numbers do not reflect the actual size in vivo. It is clear, however, that the median afferent diameter differs little between the saccules of *A. ocellatus* versus *C. auratus* and the saccule versus lagena of *C. auratus*.

The dendritic arbors on the saccule (and lagena) vary considerably in size and shape (Presson et al. 1992; Edds-Walton and Popper 2000). Both parameters are potentially important with regard to function, as larger arbors likely innervate more hair cells, which may influence afferent sensitivity. In the studies under discussion, arbor size was measured as the maximum arbor width (MAW) across the epithelium. The MAW of saccular afferent arbors in *A. ocellatus* had a range of 17–305 μm, while arbors on *C. auratus* saccule had a range of 16–155 μm and arbors on *C. auratus* lagena had a range of 40–165 μm. Terminal boutons, or endings of fibers that appeared to be modified for synaptic contact (i.e., were not endings of partially filled fibers) were counted in both studies. Larger arbors tended to have more boutons, but there was much scatter in the data. Some smaller arbors may have more boutons than larger arbors. The actual number of terminals is of some interest, as they may represent the number of hair cells innervated. Saccular afferents in *A. ocellatus* and *C. auratus* had similar ranges of terminal boutons (3–45; median 10), but *C. auratus* saccule had primarily small bouton numbers (75% had 15 or fewer) and lagenar afferents did not have fewer than 15 (maximum of 47). However, these data may be misleading, as we do not know whether a single afferent in either of these species has multiple synaptic contacts with the same hair cell. Lu and Popper (2001) presented an illustration of *D. latifrons* afferent with what appears to be two boutons on a single hair cell. There has been no systematic study of what proportion of afferent boutons may be associated with the same hair cell in a fish, but, in general, the shapes and sizes of dendritic arbors in all species examined indicate that many hair cells are innervated by a single afferent.

Saccular arbors were examined in the toadfish (*Opsanus tau*) after physiological characterization and injection with neurobiotin (Edds-Walton et al. 1999). While fiber diameters and maximum arbor widths are not directly comparable to the studies described above based on the differences in methodology, the numbers of boutons, or potential synaptic sites, is of interest. In *O. tau*, a single saccular afferent had up to 111 potential synaptic sites (median 39). Given that saccular afferents have directional sensitivity, it is important to note that even an afferent with 111 potential synaptic sites had good directional selectivity. Given the known saccular hair cell orientation pattern (Edds-Walton and Popper 1995 of *O. tau*), the arbor size and location data indicated that the hair cells innervated had very similar, or the same, orientation. Although there was a continuum of arbor sizes and physiological characteristics, the extremes provided an interesting comparison. The smallest arbor had 22 terminal boutons and showed no spontaneous activity. The largest arbor had the highest spontaneous activity (143 spikes/s), 85 terminal points, and the lowest threshold. Overall, however, there was no significant positive correlation between arbor size and threshold.

Lu and Popper (2001) also used neurobiotin injections to assess arbor morphologies on the saccule of *D. latifrons*. They measured arbor area rather than maximum arbor width, and they found that the median area of the epithelium covered by the filled arbors was quite large (2894 μm^2). Of particular interest is their evidence that the larger arbors included hair cells that had opposing best directions (orientations differing by approximately 180°). The maximum number of "dendritic endings" reported in that study was 54, and the distribution indicated a similarity to the data for *C. auratus* and *A. ocellatus* described above. Their data also indicated that the number of dendritic endings of a fiber was not correlated with its sensitivity; in other words, afferents that respond well to the lowest stimulus levels do not necessarily have more boutons than afferents that require higher stimulus levels. It is clear from the arbor drawings presented by Lu and Popper (2001), Edds-Walton et al. (1999), and Edds-Walton and Popper (2000) that similar arbor morphologies and branching patterns are present among saccular afferents of teleost fishes in different taxonomic orders. The role of arbor morphology and relative sensitivity has not yet been considered systematically among fishes.

It is important to note that in the study conducted by Lu and Popper (2001), the afferents injected with neurobiotin had significantly more terminals than the controls produced by cutting the nerve and applying neurobiotin for uptake. Their results suggest that transport of neurobiotin applied to cut fibers may require more than the 15 hours permitted in that study. Clearly, injected neurobiotin can fill saccular afferents more efficiently, with as little as 3–6 hours at room temperature for transport (Lu and Popper 2001; personal observations).

All afferents from the otolithic end organs travel via the VIIIth cranial nerve to the medulla. The axons from each end organ join those from other end organs in various ways that influence the ease with which afferents from any one end organ can be accessed. In *D. latifrons*, for example, the afferents from the saccule are quite distinct from all other octaval afferents as they enter the medulla (Figure 1 in Lu, Song and Popper 1998). In *O. tau*, saccular afferents are joined by lagenar afferents as they approach the medulla (Edds-Walton et al. 1999), requiring either extracellular recording close to the saccular epithelium to be certain of the source of physiological activity or intracellular recording followed by the injection of label to confirm the origin of the afferent. *C. auratus* has one of the most challenging arrangements of saccular and lagenar afferents, since the large lagenar epithelium lies directly against the saccular epithelium, and the fibers merge immediately above the epithelia. However, in most fishes for which individual end organs of the ear have been labeled, lagenar and saccular projections have some degree of overlap in what are presumed to be primary auditory processing areas in the medulla (see Section 2.2) (Highstein et al. 1992; Braford and McCormick 1994; O'Marra and McCormick 1999). To date, the exceptions are the clupeids, in which the utricle is subdivided, and in which at least one of the divisions is an auditory end organ. In clupeids, utricular auditory projections appear where saccular projections are found in other fishes, with some overlap with more lateral saccular projections (McCormick 1982, 1997).

Also within the VIIIth nerve are the axons of efferent cells whose somata are located in the efferent nuclei in the medulla (e.g., Highstein and Baker 1986). The role of these efferent cells is likely to be modulation of response characteristics. Only one study (on *O. tau*) has labeled fibers believed to be individual efferents on an auditory end organ based on both physiological responses and morphology (Edds-Walton et al. 1999). The filled efferent axons had widely separated, large arborizations and many branches, potentially contacting many hair cells across a far greater area of the saccule than even the largest afferents. Those findings indicate that in *O. tau*, the activity of different regions of the end organ may be modulated by a single efferent cell. In addition, Highstein and Baker (1985, 1986) suggested that a single efferent can project to multiple end organs.

2.2 Primary Auditory Nuclei of the Medulla

Afferent fibers from the otolithic end organs project to four or five nuclei that extend along the length of the lateral medulla, rostral and caudal to the entrance of the VIIIth cranial nerve. A detailed review of the neuroanatomy and circuitry of the medulla in fishes has been presented by McCormick (1999).

Only a summary of points pertinent to auditory processing is presented here. In more primitive fishes (e.g., the sturgeon, *Scaphirhynchus* and the bowfin (*Amia calva*)) four nuclei receive input from VIII: the anterior (AON), magnocellular (MON), descending (DON), and posterior (PON) octaval nuclei, in rostrocaudal sequence. In all other fishes investigated to date, there are five octaval nuclei, the four found in primitive fishes, plus the tangential octaval nucleus (TON), which lies ventral to the DON along the lateral edge of the medulla, adjacent to and ventral of the descending trigeminal tract (Fig. 3.1). Taken together, these nuclei have been called the "octaval column" (Northcutt 1980).

In general, the nuclei of the octaval column lack distinctive boundaries when viewed in horizontal sections with standard counterstains such as cresyl violet. In general, the nuclei are loose groupings of similarly shaped cells. The most easily distinguished are the MON and the TON, as both have relatively large cells, and both are located near the entrance of VIII along the lateral side of the medulla. In some fishes (e.g., toadfish [*Opsanus tau*] and midshipman [*Porichthys notatus*]), small cells are mixed among the large cells of the MON (Highstein et al. 1992; Bass et al. 2001). The magnocellular nucleus is primarily dorsal to the descending trigeminal and secondary gustatory tracts, which also run along the lateral edge of the medulla (Fig. 3.1). The tangential nucleus is sometimes lateral to those two tracts, but always ventral of the other nuclei in the octaval column. Replacing the MON and TON caudally are the dorsal and ventrolateral regions, respectively, of the DON, and continuing caudally, the small, dorsally located PON. Rostrally, the MON transitions into the AON. The moderately sized AON also does not have components that extend around the descending tract of V, but AON may extend more medially than the MON.

Along the octaval column, there is a general organization that is fairly consistent among teleosts. Afferents from the otolithic end organs project more dorsally than the afferents from the semicircular canal cristae. The degree of interdigitation of otolithic afferents and cristae afferents varies among the octaval nuclei and among fish species, but in theory, four of the octaval nuclei have the potential to process aspects of both auditory and vestibular senses: AON, DON, MON, and PON (Fig. 3.1). Of those four octaval nuclei, only the AON and DON are known to send projections to the auditory midbrain. Although the MON is intriguing due to extensive merging of inputs from VIII and the lateral line system, it is not considered to be an auditory nucleus. The MON may be involved in reflex responses to auditory stimuli in some fishes based on its descending projections (Prasada Rao et al. 1987; Highstein et al. 1992, and see McCormick 1999). To date, no studies have characterized the frequency response characteristics of cells in the MON nor assessed whether directional auditory responses occur there. The few anatomical studies that have been able to assess inputs to the PON have indicated massive overlap among afferent inputs there as well (McCormick 1999), indicating that the PON may play some role in sensory integration. Projections from PON are not known.

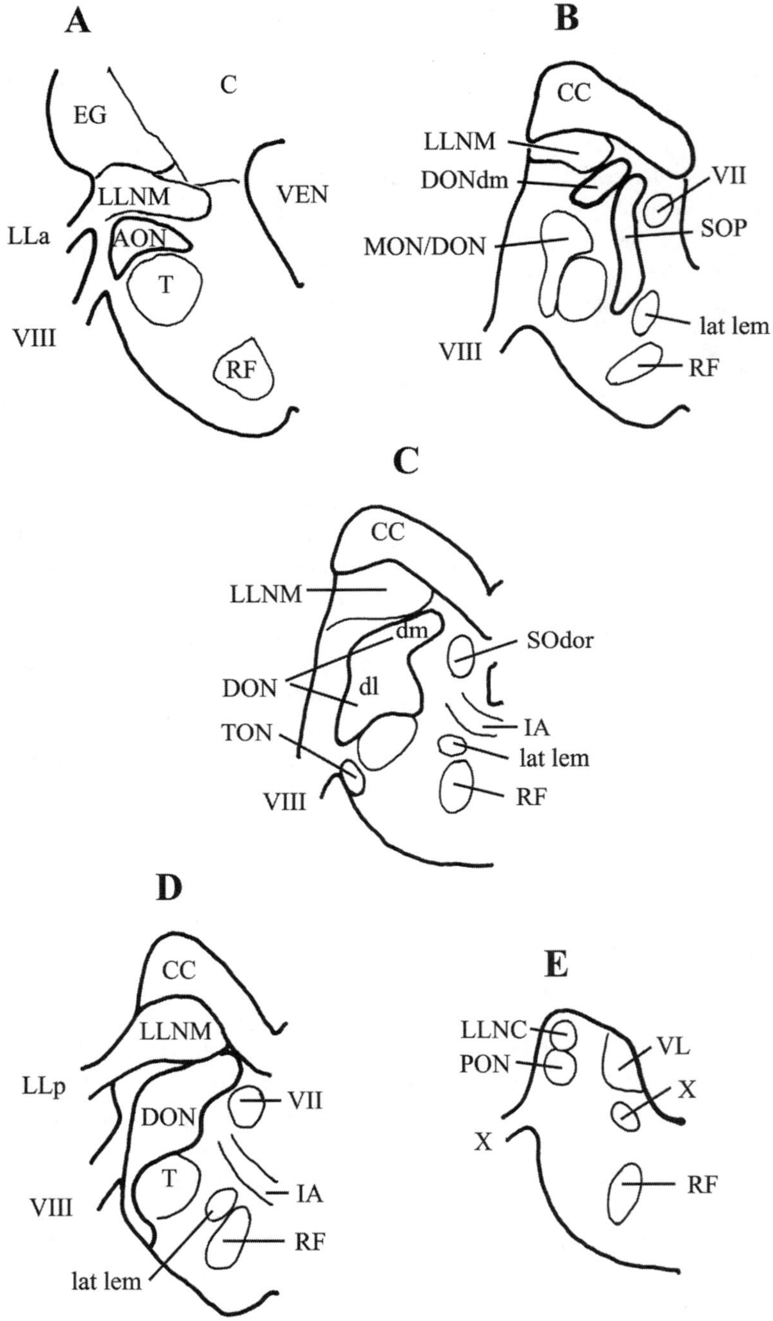

FIGURE 3.1. Cross-sections of a generalized octaval column illustrating the relationships among the octaval and lateral line nuclei in the medulla. Section **A** is most rostral. Note that incoming neurons form tracts along the lateral edge of each section that are not shown. Sites confirmed physiologically as auditory include the dorsolateral and dorsomedial areas

2.2.1 Organization of the Descending Octaval Nucleus

The DON is by far the largest of the octaval nuclei, in both the lateromedial and rostrocaudal axes. In addition, the DON may be the most compartmentalized, having distinct subdivisions of various relative sizes. Those divisions include the dorsomedial zone, with cells that have dendrites in the overlying cerebellar crest; the dorsolateral or intermediate zone that overlies the descending trigeminal tract; and a lateroventral zone that lies lateral to the descending trigeminal and/or secondary gustatory tracts (Fig. 3.1D). Of these subdivisions, the dorsomedial (DONdm) may be the most extensive, continuing along the medulla and overlapping with the presence of the MON in some fishes, as illustrated in Fig. 3.1B.

The shape of the DON varies from a relatively distinct concentration of cells lying dorsal and lateral to the descending trigeminal tract (e.g., *Crenicichla*, McCormick 1983; *Anguilla*, Meredith et al. 1987) to what appears to be an hypertrophied form with components reaching dorsally up to the molecular layer of medialis (lateral to the medial nucleus) and/or medial components stretching across the medulla to lie adjacent to the ventricle (McCormick and Braford 1993; best illustrated in Figure 5.7 of McCormick 1999). The widespread occurrence of these dorsal and medial DON subdivisions among fish species was not appreciated until recently (McCormick 1999; O'Marra and McCormick 1999; Bass et al. 2000). Contributing to the difficulty of identifying the nuclear affinity of those DON cells is the fact that their somata may not be contiguous with the more commonly recognized DON components around the descending trigeminal tract. These cells often have a dorsally oriented, fusiform soma. Their ventral

FIGURE 3.1. (Continued) of the descending octaval nucleus (DON dl and dm) and the AON. The secondary octaval population (SOP) consists of two or three nuclei that may be found medial and ventral to the anterior octaval nucleus (AON in **A**) or near the magnocellular octaval nucleus and descending octaval nucleus transition (MON/DON in **B**). The illustration indicates the general area where SOP nuclei may be found with respect to other landmarks. In some species, VII may lie lateral to components of SOP. The most consistent subpopulation of SOP, the vertically oriented cells of the SOdor (in **C**), can replace the most dorsomedial components of the descending octaval nucleus (DONdm in **B** and **C**) or can lie medial to DONdm. The function of the posterior octaval nucleus (PON in **E**) is not known. AON, Anterior octaval nucleus; C, cerebellum; CC, cerebellar crest; DON, descending octaval nucleus; DONdl, dorsolateral descending octaval nucleus; DONdm, dorsomedial descending octaval nucleus; lat lem, lateral lemniscus; LLa, LLp, lateral line nerve branches, anterior and posterior; LLNC, caudal nucleus of lateral line system; LLNM, medial nucleus of the lateral line system; MON, magnocellular octaval nucleus; PON, posterior octaval nucleus; RF, reticular formation; SOP, secondary octaval population; T, descending V and gustatory tracts; TON, tangential octaval nucleus; VL, vagal lobe; VII, facial cranial nerve/tract; VIII, acoustic cranial nerve; X, vagal cranial nerve and motor nucleus. (Derived from McCormick and Braford 1987; McCormick 1982, McCormick 1999; and Popper and Edds-Walton 1995.)

dendrites extend into neuropil receiving afferent input from VIII, and the dorsal dendrites extend into the molecular layer of nucleus medialis (a lateral line nucleus), a region also called the cerebellar crest. The cerebellar crest overlies the rostral medulla and consists of axons from cells in the vestibulolateral lobe of the cerebellum. The function of the input to the medulla via the cerebellar crest is believed to be modulation of neural activity (Montgomery et al. 1995), although such modulation has not been demonstrated in the DON cells. Because the dorsomedial DON cells reach into the cerebellar crest, this distinct subdivision of the DON is only present rostrally, coincident with the cerebellar crest. The range of morphologies for the DON among fishes is illustrated by the following anatomical studies: Northcutt (1979; longjaw mudsucker *Gillichthyes mirabilis*), McCormick (1981; bowfin *Amia calva*; 1983, cichlid *Crenicichla lepidota*), McCormick and Braford (1993; catfish *Ictalurus punctatus*; 1994; goldfish *Carassius auratus*), McCormick (1997; gizzard shad *Dorosoma cepedianum*), Highstein et al. (1992; oyster toadfish, *Opsanus tau*) and Edds-Walton (1998; *Opsanus tau*), Koslowski and Crawford (1998; the mormyrid *Pollimyrus* sp.), and Tomchick and Lu (2005, sleeper goby *Dormitator latifrons*).

Given the distinct location and orientation of the cells of the dorsomedial DON, the function of the dorsomedial DON cell group may be different from that of the other auditory cell group that is dorsal to the descending trigeminal tract (called the dorsolateral DON by Bass et al. 2000, 2001). No physiological recordings have been made in the most dorsal and medial regions of the DON; however, the axons of a subset of those cells project to midbrain auditory sites in *Crenicichla lepidota* (McCormick 1983), *Cyprinus carpio* (Echteler 1984), *Porichthys notatus* (Bass et al. 2000), and the *O. tau* (Edds-Walton and Fay 2005).

McCormick (1999) suggested that the dorsomedial extension of DON may be due to hypertrophy of a smaller dorsal, auditory processing region, as occurs in extant, primitive fishes such as *Amia* (see Figure 5.7 in McCormick 1999). McCormick also has suggested that this hypertrophy reflects specialized auditory processing sites, based on the preponderance of saccular input in hearing specialists, like the *C. auratus* (McCormick and Braford 1994). Further support for the regional specialization hypothesis is the preponderance of utricular afferent input to the more dorsal and medial regions of the DON in those herring species for which at least a portion of the utricle is auditory (McCormick 1997).

One final variation in projections from the otolithic end organs occurs in the DON. In hearing generalists, primary auditory afferents project ipsilaterally only, but in some hearing specialists (e.g., mormyrids, herrings) afferents from the saccule, lagena, and/or utricle project bilaterally to the dorsomedial zone. This projection may be relatively small, as in *D. cepedianum*, in which the bilateral projections from the auditory portion of the utricle are only present in the caudal region of the dorsomedial DON. The physiological importance of the bilateral projection is not known, but the location of those inputs is interesting in view of the observations of Edds-Walton (1998b) on *O. tau*. Primary auditory afferents from the saccule of *O. tau* do not project bilaterally (Highstein et al. 1992; personal observations), but Edds-Walton (1998b) observed that some dorsally located DON cells project to the contralateral dorsal DON. Therefore, binaural auditory processing may be occurring in dorsolateral and/or dorsomedial locations in the DON of some teleosts, which certainly warrants attention from physiologists.

2.2.2 Lateral Line System and Hearing

Directly dorsal and sometimes medial to nuclei of the octaval column are the nuclei that receive afferents from the lateral line sensory system: n. medialis and n. caudalis. Lateral line afferents do not travel with VIII, but may be associated with the Vth or VIIth cranial nerves (anterior lateral line nerve) and IX or X (posterior lateral line nerves). Although some authors use the phrase "octavolateralis system" because of the similarities in the sensory hair cells of both systems, we prefer to distinguish the octaval system from the lateralis system given that the origins and projections of the nerve bundles serving these two senses are separate, and their target nuclei in the medulla largely differ (McCormick 1999). However, there is evidence that some auditory afferents projecting to the DON may send collaterals dorsally to the lateral line nucleus medialis, and a small lateral line projection has been traced to the DON (e.g., McCormick and Braford 1993; McCormick 1997; Bass et al. 2000; Weeg and Bass 2000). Based on both anatomical and physiological evidence, the auditory and lateral line systems may provide complementary information at frequencies below 100 Hz (McCormick and Braford 1993; Edds-Walton and Fay 1999, 2003, 2005; Weeg and Bass 2000, and see discussion by Braun et al. 2002).

2.3 Secondary Auditory Nuclei of the Medulla

Although additional functions are possible, the superior olivary complex (SOC) of mammals is a critical component of the directional auditory pathway (discussed in detail by Yin 2002). The complex is a group of nuclei that receive secondary input, meaning that the input to the SOC originates from medullary nuclei rather than from primary auditory afferents of the ear. The cells of the SOC project to the auditory midbrain via the lateral lemniscus. Similar circuitry exists in birds and reptiles with similar functions (Carr and Code 2000). The existence of secondary auditory nuclei in the medulla of teleost fishes was indicated in earlier studies, but it was not until the work of McCormick and Hernandez (1996) that the secondary octaval nuclei were described in detail.

McCormick and Hernandez (1996) found three secondary nuclei in *Carassius auratus* and *I. puntatus*: a dorsal population (SOdor) of relatively large, vertically oriented fusiform cells; an intermediate population (SOint) of spherical cells between the descending trigeminal tract and the internal arcuate tract, and a ventral population (SOven) of fusiform cells located ventral to the internal arcuate tract, adjacent to reticular cells in some fishes. These nuclei have been called the secondary octaval complex by some authors. That designation can be misleading because "SOC" implies that the fish and mammalian complex may be homologous. At present there are no data to support evolutionary or functional similarities between the SOC and the secondary octaval nuclei of fishes. Therefore, McCormick (1999) suggested that the secondary nuclei in fishes be called the secondary octaval population (SO or SOP), and we will use this terminology. The SOP of fishes may have one to three subdivisions (see McCormick 1999). The most consistently present subdivision is the SOdor population. To be considered a secondary octaval site like the vertebrate SOC, tract-tracing studies must confirm that cells at the site receive input from the auditory regions of the primary octaval nuclei (e.g, dorsal DON) and those cells must project to the auditory midbrain. Some portion of the SOP of fishes may also be involved in lateral line processing, with cells that receive input from the LLNM and send projections to the lateral line or bimodal areas of the midbrain (McCormick 1999; Edds-Walton and Fay 2005).

2.4 Auditory Midbrain

Axons from auditory sites in the medulla travel via the lateral lemniscus to the midbrain in fishes as they do in other vertebrates. The majority of the projection axons in the lateral lemniscus of teleosts are from the dorsal and medial DON and divisions of the secondary octaval population, but there may also be projections from AON and a paralemniscal (or isthmoreticular) nucleus. At present, too few tract tracing studies have been conducted at this level of the auditory circuit to make many generalizations about the nature of these projections. Thus far, all studies have shown that projections from the DON have contralateral predominance, as is the case with the auditory nuclei of the medulla that project to the midbrain of other vertebrates, and projections from the SOP can be quite variable as noted in the preceding section.

The auditory region of the fish midbrain lies within the torus semicircularis (TS). The actual size and shape of the auditory region varies among fishes (Fig. 3.2), but in all cases investigated thus far, auditory sites tend to be located dorsally in the TS and to a large extent, overlie the area believed to be the lateral line center. Electrosensory areas are also present in the midbrain of electroreceptive species, but they are not discussed here since there is no evidence to date that the electrosensory system is involved in auditory processing. Integration of electrosensory information may play a role in source localization in sound producing electric fish, and integration among the sensory systems (auditory, lateral line, electrosensory) in the midbrain is very likely (Braun et al. 2002; Coombs and New 2002; Nelson et al. 2002).

Specific divisions of the TS are of interest with regard to auditory processing. Most dorsally there is a shallow layer of darkly staining cells, the periventricular (PV) cell layer. This layer is believed to consist of efferent cells that project to the forebrain from the TS, but the nature of the projection, whether it is auditory, lateral line, or an integrated sensory projection is not known. Both Bass et al. (2000) and Edds-Walton and Fay (2003) have found that the processes of those

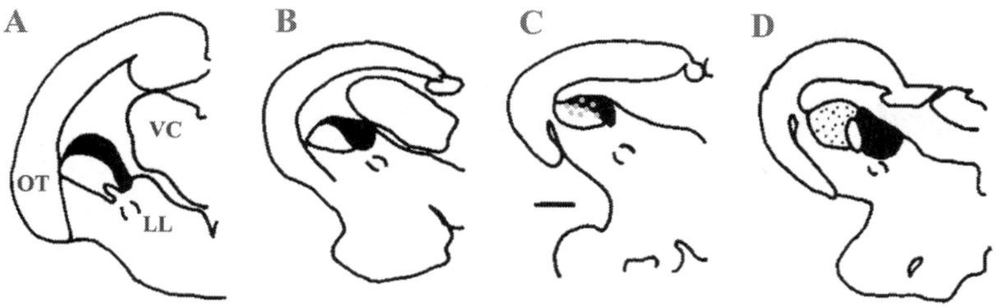

FIGURE 3.2. Organization of the midbrain torus semicircularis in various bony fishes. The auditory regions (black-fill area, nucleus centralis NC) usually lie dorsal and/or medial to the lateral line processing regions (open area, nucleus ventralis NV) as shown in (**A**) and (**B**). Sensory integration sites may exist between or within the subdivisions of the midbrain, e.g., bimodal cells that respond to both auditory and lateral line stimuli (indicated by gray circles in **C**). Even in fishes with electrosensory systems (processed in stippled area of midbrain in **D**), auditory processing areas are concentrated in the more dorsal or medial sites of the torus semicircularis. Species represented: (**A**) hearing nonspecialist, *Notopterus notopterus*; (**B**) a hearing specialist, *Carassius auratus*; (**C**) hearing nonspecialist with bimodal cells, *Opsanus tau*; (**D**) a weakly electric hearing specialist, *Ictalurus punctatus*. LL, Lateral lemniscus; OT, optic tectum; VC, valvula of the cerebellum. (**A, B, D** modified from McCormick and Braford 1988 and McCormick 1999; **C** derived from Edds-Walton and Fay 2005). Scale bar in **C** = 500 μm, for **C** only.

cells extend across the lateromedial axis or the dorsoventral axis of the TS. Therefore, a sensory integration role is possible for the PV cells.

The TS is divided into two nuclei in nonelectroreceptive fishes. The auditory nucleus centralis (NC) and the lateral line nucleus ventrolateralis or ventralis (NV) do not have a distinct anatomical border in all fishes (e.g., *P. notatus*, Bass et al. 2000; Bass et al. 2001; and toadfish, Edds-Walton and Fay 2003, 2005). Despite the absence of distinct anatomical landmarks, injections into either of these subdivisions of the midbrain in nonelectric fish reveal restricted label spread that indicates two subdivisions. In general, anatomical studies have indicated that auditory and lateral line sensory processing remain distinct through the midbrain, with multimodal sensory inputs converging in the thalamus as in other vertebrates (see discussion in Striedter 1991 and McCormick 1999). The assumption has been that there are parallel pathways for the auditory and lateral line sensory systems up to the level of the midbrain. More recently, however, sensory integration has been indicated physiologically in the torus semicircularis of batrachoidids (*O. tau*), and neurobiotin injections into NC or NV have indicated that the processes of some NC cells and NV cells can extend into the other nucleus (NV and NC, respectively), as illustrated in Edds-Walton and Fay (2003). Physiological evidence for the convergence of auditory and lateral line sensory processing in the midbrain is described in Section 4.2.1.

Integration of auditory and visual information may also occur in the midbrain, as bimodal cells responding to both auditory stimuli and visual stimuli have been described (see Schellart and Kroese 1989) and implicated in orientation

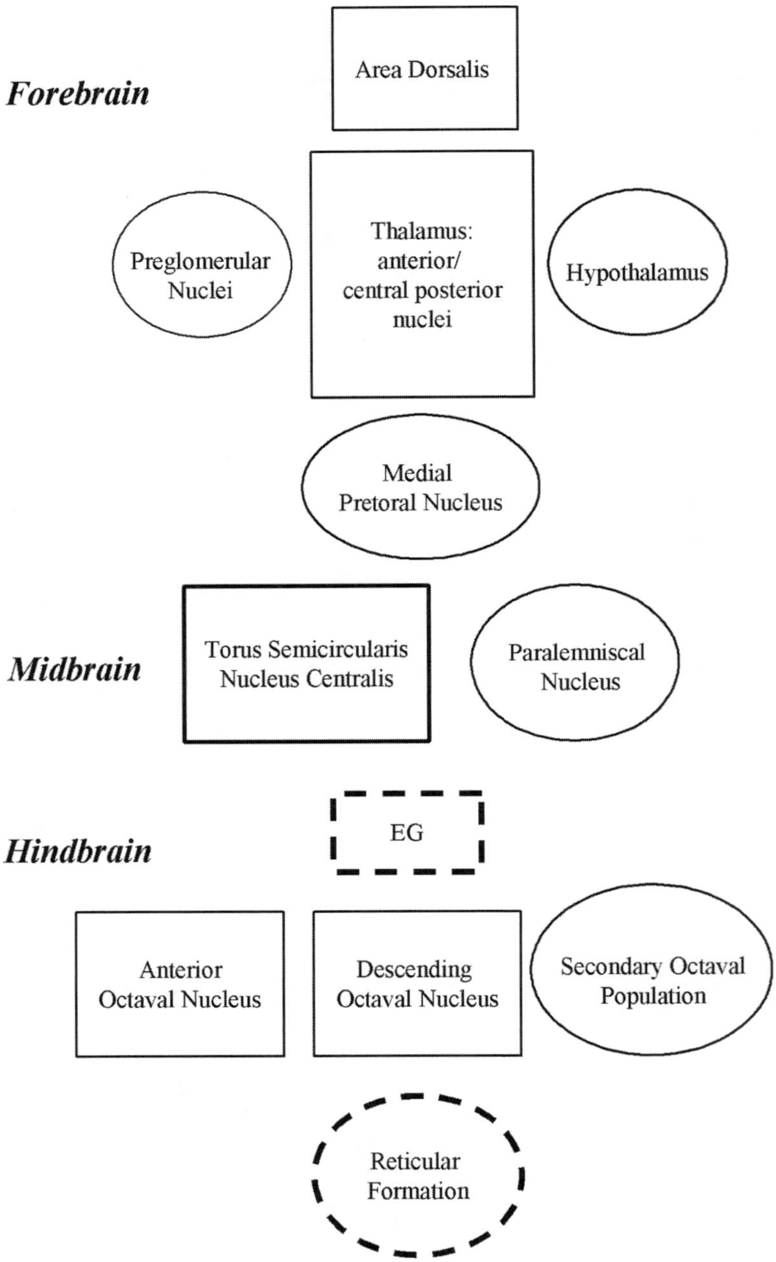

FIGURE 3.3. Auditory components in the brains of teleost fishes. This diagram is generalized to indicate areas that are implicated in auditory processing, and it does not apply to a particular species or to all species. The interconnections of these areas are not shown, nor are ipsilateral versus contralateral contributions to the circuit, as much work remains to clarify the interconnections. The areas enclosed by solid lines are widely accepted as sites for auditory processing. The dashed lines indicate areas that may be components of

responses (Echteler 1984). Interconnections between the TS and the optic tectum have been demonstrated anatomically in *Porichthys notatus* by Bass et al. (2000). In all cases, the locations of the sites of auditory and visual integration were the midbrain tegmentum or the optic tectum.

2.5 Forebrain

The forebrain consists of both the diencephalon and the telencephalon. Very few acoustical studies have been done in the diencephalon, and none have been conducted in identified areas of the telencephalon. Therefore, we limit our discussion here to the auditory sites identified in the diencephalon. The diencephalon of the teleost fishes examined to date (e.g., *Danio rerio* and *Ictalurus punctatus*) has subdivisions that are recognized in all vertebrates. The only sites that have identified auditory responses are within the dorsal thalamus and the preglomerular complex. For details about the organization of the forebrain, see McCormick (1999).

Reciprocal connections exist between the auditory nucleus centralis of the midbrain torus semicircularis and the central posterior nucleus of the dorsal thalamus in *Amia calva* (Braford and McCormick 1979), *Cyprinus carpio* (Echteler 1984), *Ictalurus punctatus* (Striedter 1991), and *Porichthys notatus* (Bass et al. 2000). Other potential auditory sites based on projections from NC include the ventromedial nucleus of the ventral thalamus, the preglomerular complex, and the anterior tuberal nucleus of the hypothalamus in both otophysines ("auditory specialists") and auditory generalists (e.g., Echteler 1984; McCormick 1999; Bass et al. 2000).

A summary diagram of the components of the ascending auditory system is presented in Fig. 3.3. The components are listed by their level in the central nervous system, but there are no interconnections shown to make the diagram more general. For example, the SOP is shown in the medulla, but the known inputs from the contralateral and ipsilateral DONs are not shown. In addition, no nuclei associated with the lateral lemniscus in the medulla or midbrain are shown as their roles in the auditory circuit are unclear. The most important purpose for this figure is to indicate which components of the auditory pathway are

FIGURE 3.3. (Continued) reflex response circuits to acoustic stimulation rather than the ascending auditory pathway. Areas enclosed by boxes have been confirmed as auditory sites based on anatomical and physiological data. The areas enclosed by ovals are believed to be acoustic based on anatomical data only. Auditory responses have been reported from the SOP of an electric fish (Koslowski and Crawford 2000), but confirmation is required in nonelectric fishes. Cells in the eminentia granularis (EG) respond to acoustic signals in a carp (Echteler 1985) and in at least one species of herring (Plachta et al. 2004). Note that the reticular formation does not include the Mauthner cell, which has been shown to receive auditory inputs and is part of the startle/escape circuitry in many, but not all fishes (Zottoli et al. 1995).

known from anatomical versus physiological studies and to provide inspiration for further research on the areas for which little or no physiological data are available.

3. Neurophysiology of the Auditory Periphery

We define the auditory periphery as the most peripheral neural structures of the auditory system, including the synapse between hair cells and primary afferents, and the primary afferents themselves. The data on the periphery are extensive only for one species (goldfish *Carassius auratus*).

3.1 Hair Cell–Nerve Fiber Synapse in the Goldfish Saccule

Excitatory postsynaptic potentials (EPSP) intracellularly recorded in saccular afferents of *C. auratus* have been studied quantitatively by Furukawa and his colleagues (e.g., Ishii et al. 1971; Furukawa 1981; Furukawa et al. 1982; Suzue et al. 1987). Furukawa (1986) has summarized a multiple-release-site model of the hair cell synapse. EPSPs are graded in amplitude and adapt during a stimulus (Furukawa and Matsuura 1978). A statistical analysis showed that the rundown of EPSP amplitude was due to a reduction in the number of transmitter quanta available and not to the probability of a given quantum being released. These and other studies have shown that: (1) there are numerous presynaptic release sites; (2) each release site has a different threshold and is activated only if the transmembrane voltage level reaches that threshold; (3) a single synaptic vesicle is allocated to each release site; and (4) once a vesicle is released, the site remains empty until replenished from a larger store.

This model also explains why an increment in sound level results in a robust spike response from highly adapted afferents (the existence of release sites having higher thresholds), and why a small sound level decrement may result in a transient loss of all spikes (only empty, low-threshold sites are activated). Furukawa et al. (1982) also obtained evidence that vacant release sites are replenished in an order from high-threshold to low, and that sites with thresholds below the stimulus level are not replenished as long as the stimulus remains on and above their threshold.

3.2 Physiology of Auditory Nerve Afferents

The responses of otolithic organs to sound are encoded in the response patterns of the eighth nerve neurons that innervate them. Studies of their activity patterns help reveal the acoustic response properties of the otolithic organs and more peripheral sound conducting structures, the functional characteristics of hair cells and their synapses on primary afferents, and the dimensions of neural activity that represent acoustic features of sound sources such as level, frequency, and source location.

The responses of primary otolithic organ afferents to sound and head motion have been systematically studied in only a few fish species. These include *C. auratus* (reviewed in detail below), catfish (*Ictalurus punctatus*: Moeng and Popper 1984), bullhead (a sculpin, *Cottus scorpius*: Enger 1963), Atlantic cod (*Gadus morhua*: Horner et al. 1981), tench (*Tinca tinca*: Grözinger 1967), a mormyrid (Suzuki et al. 2002), sleeper goby (*Dormitator latifrons*: Lu et al. 1998), and the oyster toadfish (*Opsanus tau*, e.g., Fay and Edds-Walton 1997a,b). The following discussion focuses on *C. auratus* with treatment of other species investigated where there are sufficient data for comparison.

Single-cell recordings from *C. auratus* VIIIth cranial nerve can be made from visually identified branches that innervate the utricle (anterior branch) and the saccule and lagena (posterior branches). Saccular afferents have cell bodies scattered throughout the visible portion of the nerve (large-diameter afferents), and grouped near the sensory epithelium (small-diameter afferents). Cross sections of the lagenar and utricular nerve branches have not been quantitatively studied. Much of what we know about the physiology of *C. auratus* saccular afferents has come from the numerous studies of Furukawa and his colleagues (e.g., Furukawa and Ishii 1967; Sento and Furukawa 1987), and from the work of Fay and colleagues (e.g., Fay, 1978a; Fay and Ream 1986; Fay 1997).

3.3 Spontaneous Activity

As in all vertebrate auditory systems investigated, primary afferents of the several fish species investigated show varying degrees and patterns of spontaneous activity. In *C. auratus* (Fay 1978a,b), catfish (Moeng and Popper 1984), *G. morhua* (Horner et al. 1981), *D. latifrons* (Lu et al. 1998), and *O. tau* (Fay and Edds-Walton 1997a), saccular afferents generally fall into four spontaneous pattern groups: those that demonstrate (1) zero spontaneous firing, (2) approximately random interspike-interval distributions, (3) random bursts of spikes giving bimodal distributions of interspike intervals, and (4) regular spontaneous patterns. Spontaneous rates are found up to 250 spikes/s. Regular spontaneous afferents are very insensitive to sound and may serve a vestibular function or may be efferents. Afferents showing no spontaneous activity tend to be less sensitive than those with low, irregular spontaneous activity, as is also the case for mammals (Ruggero 1992).

3.4 Frequency Selectivity of Auditory Afferents

In all fish species investigated so far, primary afferents from the saccule and other otolithic organs are necessarily band-limited in their frequency response, and are thus frequency selective to some degree. The degree of frequency selectivity is important for understanding the mechanics of hair cells and for understanding the central processes that help in the detection and determination of sound sources based on their frequency composition. Fishes encode sound signals through phase-locking in the time domain (e.g., Fay 1978a) and through selectivity in the frequency domain (e.g., Furukawa and Ishii 1967; Fay and Edds-Walton 1997b; Fay 1997). The importance of these two representations has been a matter of debate for mammals and other amniotes (Wever 1949), and the issue has now extended to all vertebrate groups, including fishes.

A macromechanical, von Békésy-type (1960) frequency-to-place transformation does not occur in otolithic organs. It has been argued (von Frisch 1938) that any frequency discrimination that occurs among fishes (e.g., Fay 1970) probably depends on time-domain computations on phase-locked inputs (Fay 1978). However, it is now clear that other factors such as hair cell tuning (Crawford and Fettiplace 1981) and micromechanical factors (reviewed by Patuzzi 1996; Fay 1997) at the levels of hair cell stereovilli and their attachments to restraint structures could produce frequency selectivity and a spatial frequency map without a macromechanical traveling wave (Holton and Weiss 1983).

Quantitative data on frequency selectivity of primary auditory afferents are available for a few fishes including *C. auratus* (e.g., Furukawa and Ishii 1967; Fay and Ream 1986; Fay 1997), *I. punctatus* (Moeng and Popper 1984), *G. morhua* (Horner et al. 1981), a *Pollimyrus sp.* (Suzuki et al. 2002), *P. notatus* (Weeg et al. 2002), and oyster *O. tau* (Fay and Edds-Walton 1997b). For these species, saccular afferents are diverse with respect to frequency response characteristics.

Furukawa and Ishii (1967) described two categories of frequency selectivity of saccular afferents of *C. auratus* (S1 and S2). According to their study, S1 afferents respond best at high frequencies (>500 Hz), have large-diameter axons with low spontaneous activity, and primarily innervate the rostral portion of the saccule. The hair cells they innervate have short cell bodies, short stereovilli (Platt and Popper 1984), and tend to exhibit a damped oscillation, or low-quality resonance, of the membrane potential in response to depolarizing current steps (Sugihara and Furukawa 1989). The S2 afferents respond best at low frequencies (<300 Hz), have smaller axon diameters, and innervate the caudal region of the saccule where hair cells are tall with stereovilli longer than those of the rostral hair cells. Tall hair cells produce a "spike-plateau" response when depolarized. More recent anatomical studies have indicated that there is not a simple dichotomy of saccular afferent morphologies on *C. auratus* saccule (Edds-Walton and Popper 2000) and a range of hair cell morphologies (Lanford and Popper 1996, Lanford et al. 2000) but the S1, S2 designations persist in the literature.

The frequency tuning properties of *C. auratus* saccular afferents have been quantitatively described using several methods including frequency threshold (tuning) curves based on phase-locking (Fay 1978b) and spike rate (Fay and Ream 1986), frequency-by-level response areas (RA) based on spike rate criteria (Fay 1990, 1991; Lu and Fay 1996), and by the reverse correlation (revcor) method (Fay 1997) in response to spectrally flat noise. Revcor tuning is measured by averaging hydrophone recordings of the acoustic noise, triggered by spike times. This average produces an impulse response that reflects the linear filtering preceding spike generation. The Fast-Fourier Transform (FFT) of the revcor

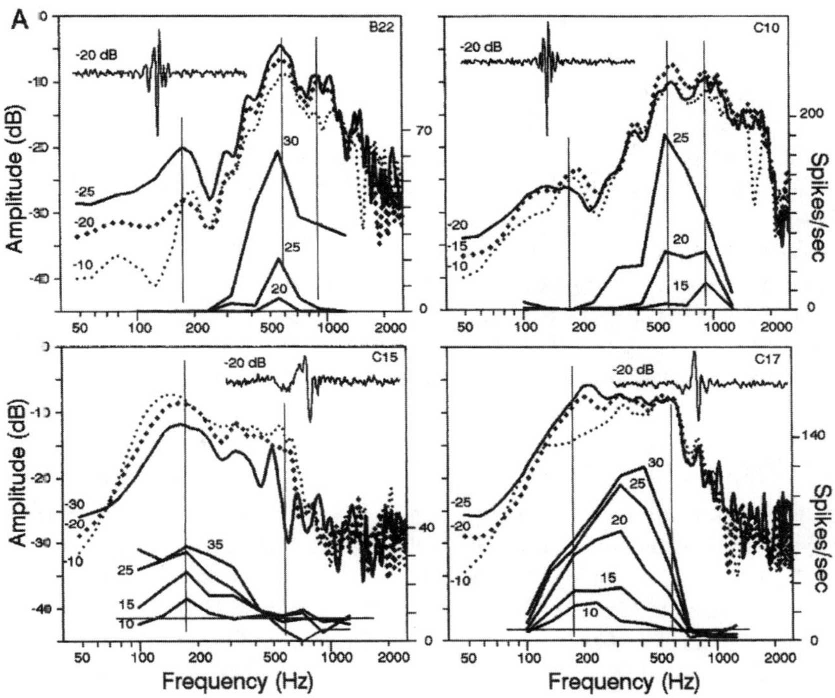

FIGURE 3.4. Frequency response of four representative saccular afferents of *Carassius auratus*. The impulse response (40 ms in duration) is shown as an inset. The filter shapes (smooth solid and dotted curves) are FFTs of the impulse responses. The left ordinate applies to the revcor filters and is in dB with an arbitrary reference. Noise spectrum levels in dB re: 1 dyne cm^{-2} are given for each filter function. Iso-level spike rate functions of tone burst frequency (RAs) are shown as straight lines connecting data points referred to the right ordinate. Numbers are sound pressure levels in dB re: 1 dyne cm^{-2}. (From Fay 1997.)

impulse response estimates the afferent's filter shape.These two measures for four representative saccular afferents of *C. auratus* are compared in Fig. 3.4 (Fay 1997). The continuous lines are the revcor FFTs at three noise levels, and the functions plot spike rate as a function of frequency (RA) at several overall levels.

Revcor FFTs for C15 and C17 show relatively broad tuning with corners at about 150–200 Hz and 600 Hz. Afferent B22 (upper left) has a characteristic frequency (CF) near 600 Hz while C10 has a CF between 600 and 900 Hz.

These tuning data are rather complex, illustrating that different experimental paradigms (e.g., RA and revcor) may produce somewhat different views of tuning. The revcor filter functions present a simple view indicating two major categories: those with a major peak at about 200 Hz and with a plateau extending to 600 Hz, and those with prominent peaks at about 600–900 Hz. In any case, there is no doubt that *C. auratus* saccule divides the sound spectrum into at least two frequency regions. Thus, frequency selectivity exists for *C. auratus* saccule as it does for the auditory receptor organs investigated in all other vertebrate classes (e.g., Sachs and Kiang 1968; Manley and Gleich 1984; Köppl and Manley 1992; Lewis 1992).

There are limited data on frequency selectivity of saccular afferents in other fish species, including other hearing specialists (e.g., a mormyrid – Suzuki et al. 2002), and two hearing generalists: *Gadus morhua* (Horner et al. 1981) and *O. tau* (e.g., Fay and Edds-Walton 1997b). In all species investigated, there appear to be a small number of differently tuned primary afferents. *O. tau* and *G. morhua* hear only up to several hundred Hz and their tuned channels are restricted to relatively low frequencies. Data for *O. tau* using the revcor method (Fay and Edds-Walton 1997b) reveal two filters with peaks at 74 and 140 Hz, and a third population that seems to be the sum of these two filters.

3.5 Origins of Frequency Selectivity

Frequency selectivity in primary otolithic afferents may arise from hair cell resonance (Sugihara and Furukawa 1989 for *C. auratus*, Steinacker and Romero 1992 for *O. tau*) and micromechanical mechanisms that are local to hair cells and their ciliary attachments to the otoliths (Fay 1997). In *C. auratus* and *O. tau*, two general classes of hair cells have been found: those that produce a damped oscillation (low-quality resonance) to a current step; others that produce a spike. In *C. auratus*, resonance frequencies measured in isolated hair cells do not correspond to the characteristic frequencies of primary afferents measured in vivo. How, then, can differently tuned primary afferents be explained?

A possible explanation is illustrated in Fig. 3.5, which shows averaged revcor filter functions for the two classes of saccular afferents in *C. auratus* (Fay 1997) and *O. tau* (Fay and Edds-Walton 1997b). The dotted line in each panel is the high-frequency filter function that has been spectrally "tilted" about the frequency point at which the two filter functions intersect. The low-frequency filter function in each species may be the result of a simple transformation of the high-frequency filter (or vice versa). One hypothesis is that low-CF afferents may innervate hair cells responding to otolith displacement while high-CF afferents contact hair cells responding to otolith acceleration. These differing response properties could arise as a result of differences among hair cells in bundle stiffness and the mode of coupling to the otolith (cf. Rogers and Cox 1988) and would result in differently shaped (tilted) frequency response functions. A detector that is sensitive to acceleration would be equivalent to an up tilt toward higher frequencies because acceleration increases with high frequencies (compared to displacement). In any case, micromechanical processes may be combined with hair cell resonance to contribute to peripheral frequency selectivity in fishes. Similar hypotheses have been suggested to account for tuning in amphibians (Lewis 1992), reptiles (Köppl and Manley 1992), and birds (Manley and Gleich 1984).

The *C. auratus* saccule is crudely tonotopically organized; high-CF afferents originate primarily from the rostral region (Furukawa and Ishii 1967). However, there is no evidence that the saccule of the *O. tau* (Fay and Edds-Walton

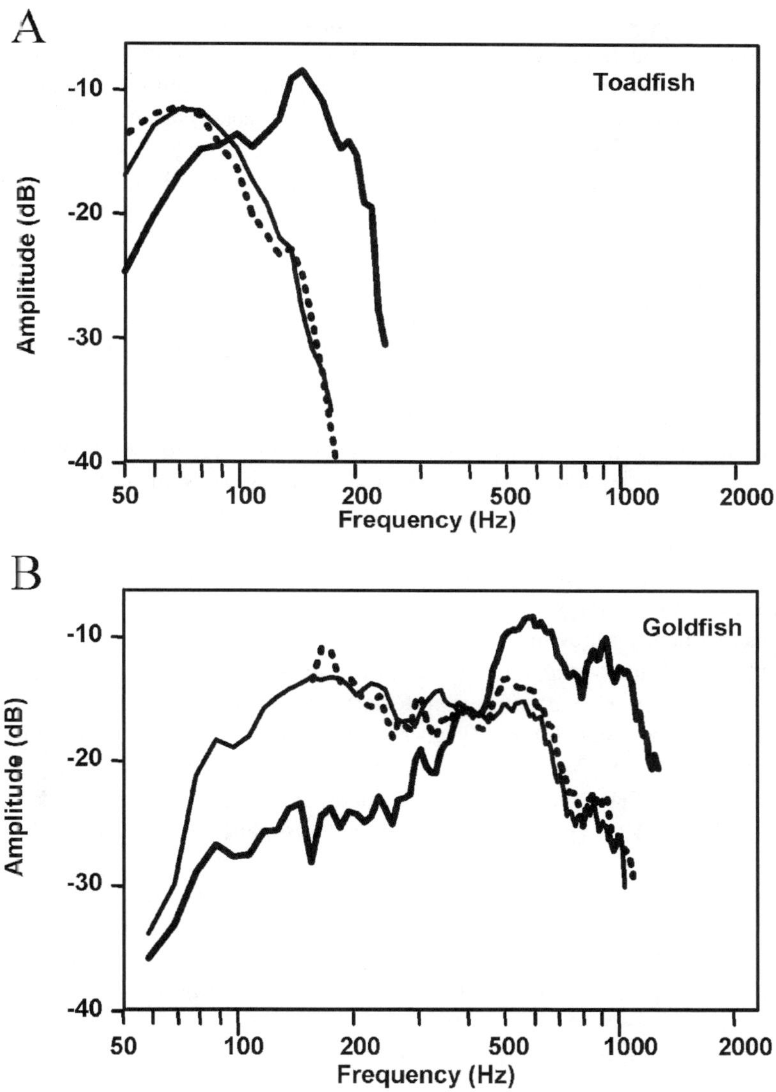

FIGURE 3.5. Averaged primary afferent filter functions for *Carassius auratus* (Fay 1997) and *Opsanus tau* (Fay and Edds-Walton 1997b) derived from revcor experiments. The two solid lines in each panel are averaged filter functions for a low- and high-frequency element measured for fibers of the saccular nerve. In both panels, the dotted line function is the high-frequency filter function after it had been "tilted" by −15 dB per octave (*C. auratus*) and −24 dB per octave (*O. tau*). The pivot points around which the functions were tilted were 400 Hz for *C. auratus* and 100 Hz for *O. tau*.

1997a; Steinacker and Romero 1992) or Atlantic cod (Horner et al. 1981) are tonotopically organized. The sort of frequency analysis observed in saccular afferents of *C. auratus* and *O. tau* (Fig. 3.5) may be the simplist mechanism for parsing the acoustic spectrum yet observed in vertebrates (Fay 1997). *C. auratus* and *O. tau* have a small number of differently tuned channels (two or three), while the auditory nerves of anuran amphibians, reptiles, birds, and mammals demonstrate continuously variable tuning.

3.6 Phase-Locking

In all fishes investigated to date, all sound-responsive otolithic afferents phase-lock, or synchronize, to all acoustic waveforms within the frequency range of hearing (Fay et al. 1983; Fay 1997). Since all saccular afferents in fishes phase-lock within their tuning curve, the details of the stimulating waveform are preserved in the times between spikes. In addition, because phase-locking is ubiquitous among low-frequency afferents (for effective frequencies below about 4 kHz) in all vertebrate auditory systems investigated, it is apparently a primitive characteristic of the neural code for hearing. Note that phase-locking occurs even in afferents for which spike rate can be very low, and in cases without frequency following.

In *C. auratus* and *O. tau*, the stimulus phase angles to which afferents lock vary widely from afferent to afferent, even for afferents with the same best frequency stimulated at the same frequency at a comparable level above threshold (Fay 1981; Fay and Edds-Walton 1997a). The origin of this variability is not clear, but it is possible that the angles are determined by a high dynamic order filtering process (Lewis 1992) produced by hair cell resonance and micromechanical processes.

Phase-locking has been hypothesized to play a role in pitch perception and other aspects of frequency analyses in tetrapods (Wever 1949) and fishes (Fay et al. 1983). In *C. auratus*, the temporal error with which saccular afferents synchronize to tones predicts behavioral frequency discrimination acuity (Fay 1978b).

Saccular afferents also synchronize to the envelopes of amplitude-modulated tones and noise (Fay 1980; McKibben and Bass 1999). For modulated tones, each *C. auratus* afferent has a modulation rate to which it is most sensitive, ranging between 20 Hz and greater than 200 Hz. Sensitivity to amplitude modulated tones can be quite high (Furukawa et al. 1982; Fay 1985). Responses of *C. auratus* saccular afferents to temporally asymmetrical envelopes, qualitatively predicted by Furukawa's (1986) model of the hair cell synapse, can account for the perceptual distinctiveness of envelope shapes (Fay et al. 1996).

Sisneros et al. (2004) have observed that the accuracy or magnitude (vector strength) of phase-locking is dependent on steroid hormone levels (or season) in female *P. notatus*. In these fish, primary afferents phase-locked robustly at higher frequencies (up to 300 Hz) during the spring courtship season than at other times of the year (up to 200 Hz). This enhancement in peripheral encoding could be induced in nonreproductive females by injections of both testosterone and estradiol. These effects were interpreted as adaptations for more robust encoding of the multiharmonic male advertisement call during the breeding season, when the shallow water depth in which mating occurs makes reception of higher frequency components of the multiharmonic hum more important biologically.

3.7 Masking

Background noise raises rate-increment and synchronization thresholds to tone bursts with a slope of 1 dB per dB noise level in accordance with Weber's Law. Signal-to-noise spectrum level ratios at rate increment threshold are about 17 dB for low-frequency afferents, and increase at about 3 dB/octave for afferents with higher best frequencies. These rate increment thresholds correspond closely to behavioral masked thresholds and are significantly higher than neural thresholds based on a synchronization criterion (Fay et al. 1983). Thus, behavioral detection is most likely determined by spike rate increments, and not by synchronization. Thresholds for some nonspontaneous afferents with best frequencies above 600 Hz are paradoxically lowered in the presence of low-level background noise (e.g., Fay 1990).

3.8 Hypoxia and Temperature Effects

Temperature change and hypoxia produce consistent, reversible effects on the response of primary saccular afferents in *C. auratus* (Fay and Ream 1992). Cooling and hypoxia reduce a cell's spontaneous activity, sensitivity, best frequency (BF, or the most excitatory frequency at a given signal level), and overall responsiveness to acoustic stimulation. Warming above ambient temperatures increases a cell's spontaneous activity, sensitivity, BF, and overall responsiveness. Adaptation increases during hypoxia and cooling, and decreases during warming. In *C. auratus*, single-tone suppression (see later) remains robust during transient hypoxia and cooling of the animal (Fay and Ream 1992). The effects of temperature change and hypoxia on the neural response are probably due in part to changes in the release and replenishment of neurotransmitter at the synapses between hair cells and auditory nerve afferents (Suzue et al. 1987).

3.9 Adaptation

Adaptation patterns of saccular afferents have been characterized using the shapes of peri-stimulus-time histograms (PSTH) (Fay 1978a; Coombs and Fay 1985, 1987; McKibben and Bass 1999). In general, adaptation patterns vary among and within afferents. Some afferents show an essentially tonic response for several hundred milliseconds (ms), while others show essentially phasic responses lasting less than 25 ms. In response to tones at the best frequency, low-frequency afferents tend to produce more tonic responses, and high-frequency afferents tend to produce more phasic responses. At the same time, an afferent's adaptation pattern may vary with stimulus frequency, often becoming more phasic as stimulus frequency rises. Part of the phasic nature of some saccular responses can be due to suppression (Fay 1991; see later) as well as to adaptation (Furukawa and Matsuura 1978).

3.10 Suppression in Saccular Afferents

Saccular afferents in the *C. auratus* reveal a set of nonlinear phenomena that are common to auditory afferents in most vertebrate classes, and thus may be primitive vertebrate functional characteristics. The most striking are the phenomena of two-tone rate suppression (TTRS) and single-tone suppression (STS).

Two-tone rate suppression (TTRS) is defined as the reduction in evoked spike rate to one stimulus as a result of the addition of a second stimulus. TTRS has been observed in some afferents of the anuran amphibian papilla (Capranica and Moffat 1980), afferents of the basilar membrane of reptiles (Manley 1990) and birds (Hill et al. 1989), the cochlea of mammals (Sachs and Kiang 1968), and most recently in saccular afferents of *C. auratus* (Lu and Fay 1996).

In *C. auratus*, spontaneous activity can also be suppressed in some low-CF saccular afferents by single tones presented at frequencies well above CF (Fay 1990, 1991). This is a controversial type of suppression that has been termed "single-tone suppression" (STS) to distinguish it from TTRS. STS has also been reported for mammals (Henry and Lewis 1992), some reptiles (Manley 1990), birds (e.g., Hill et al. 1989), and anuran amphibians (Lewis 1986). STS is difficult to demonstrate because the background activity that is suppressed may be acoustically driven and thus not truly "spontaneous." In this case, the suppression could be defined as TTRS. Hill et al. (1989) proposed that STS could result from hyperpolarization at the spike initiation zone due to positive, extracellular fields produced by receptor currents through nearby hair cells. One of the consequences of single-tone suppression is that the frequency response areas (RA) for some saccular afferents are truncated or "sharpened" at some frequencies above CF. So far, suppression has been demonstrated only in primary saccular afferents of *C. auratus*.

3.11 Directional Responses to Whole-Body Acceleration

The primitive and shared mode of sound detection in fishes results from the ability of otolithic organs to respond to acoustic particle motion as inertial accelerometers (de Vries 1950; Dijkgraaf 1960; Sand 1974; Fay and Olsho 1979). Whole-body acceleration can activate primary afferents innervating all otolithic organs (saccule, lagena, and utricle) in *C. auratus* with the most sensitive afferents having thresholds as low as 0.1 nanometers (nm) at 140 Hz (Fay 1984). Saccular afferents in *O. tau*, *D. latifrons*, and *P. notatus* have similar thresholds (Fay and Edds-Walton 1997a,b; Lu et al. 1998; Weeg et al. 2002). In general, the sensitivity and frequency response functions for lagenar afferents of *C. auratus* resemble those for saccular afferents of *O.tau* and *G. morhua* (Horner et al. 1981) having best frequencies between 100 and 200 Hz. Lagenar afferents in the *D. latifrons* appear to be less sensitive than their saccular afferents (Lu et al. 2003). In *C. auratus* (Fay 1984), *O. tau* and *O. beta* (Fay and Edds-Walton 1997a; Weeg et al. 2002), *D. latifrons* (Lu et al. 1998); and *G. morhua* (Hawkins and Horner 1981) the responses of most otolithic afferents vary as a function of the axis of translatory motion according to a cosine function. The directional response of typical units of *O. tau* saccular nerve (SN), descending octaval nucleus (DON), and torus semicircularis (TS) are illustrated in Fig. 3.6. The saccular nerve unit is typically cosine-shaped. Since a cosinusoidal directional response function has been measured for individual hair cells (Hudspeth and Corey 1977), it appears that most otolithic afferents receive effective input only from similarly oriented hair cells, as was demonstrated for a saccular afferent in a *D. latifrons* by Lu, Song and Popper (1998).

3.12 Neural Directionality and Sound Source Localization

It is assumed that directional hearing by fishes depends upon the convergence of peripheral inputs and central computations (e.g., Schuijf 1975; Fay 2005). The most detailed studies of peripheral inputs have been conducted in *C. auratus* and *O. tau*. Otolithic afferents in *C. auratus* (Fay 1984) and *O. tau* (Fay and Edds-Walton 1997a) are widely distributed with respect to orientation of their characteristic axis (CA) in spherical coordinates. The best axes of otolithic afferents vary predictably given the orientation of hair cells in the otolithic end organ and the overall orientation of the end organ with respect to the vertical and horizontal axes of the fish (Lu, Song and Popper 1998).

We can compare the directional responsiveness among the end organs that are oriented differently. In *C. auratus*, the saccule is largely oriented vertically, with a slight twist (Platt 1977). The hair cells are oriented in a bidirectional array down the length of the saccule, and the best axes for saccular afferents cluster near 90° in elevation. The lagena, also oriented vertically, has a very different hair cell orientation pattern that is essentially able to encode all angles in the vertical plane. The best axes of *C. auratus* lagenar afferents are more widely scattered in elevation and in azimuth than those of the saccule. Lastly, the utricle of *C. auratus*, like most fishes, has a continuous sweep of hair cell orientations; however, the utricular epithelium is oriented substantially horizontally, and the best axes of utricular afferents tend to have elevations close to 0°, but azimuths that vary between 0° and 360° (Fay 1984).

Generalizations about directional responses among the different otolithic end organs are difficult because hair cell orientations vary among the otolithic end organs within and among species, and the spatial orientations of the sensory epithelia can vary significantly as well. In general, the diverse patterns of hair cell orientation determine the range of elevations for saccular and lagenar fibers and the range of azimuths for utricular fibers. For the saccule of the *O. tau* and *D. latifrons*, the best azimuths tend to correspond with the overall azimuthal

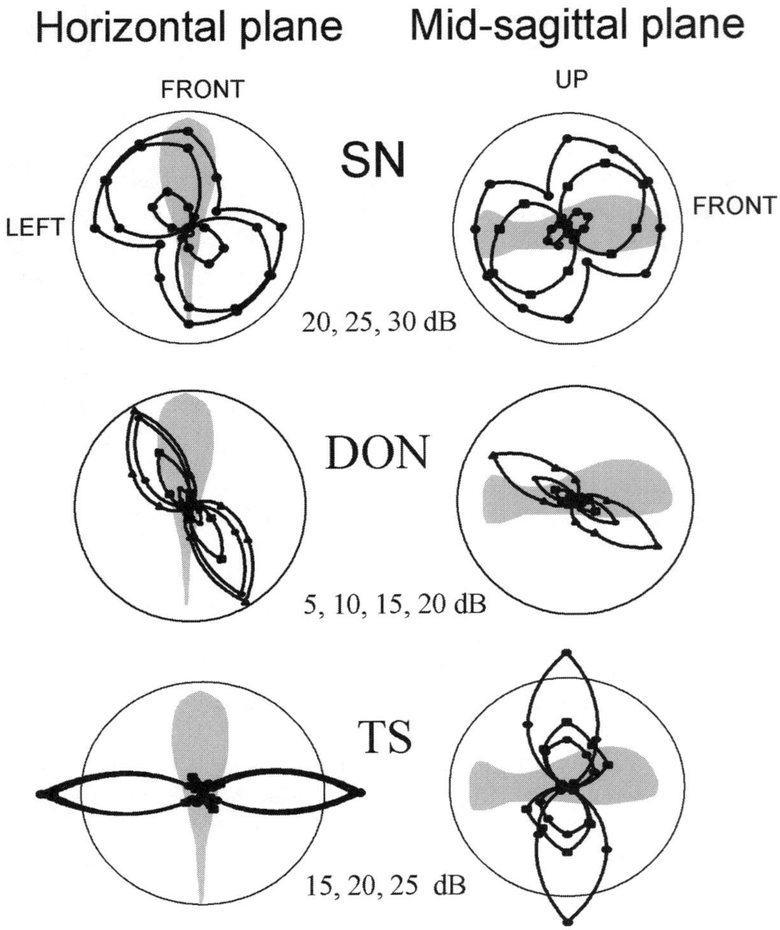

FIGURE 3.6. Directional response patterns for three typical cells of *Opsanus tau*, one each from the saccular nerve, the descending octaval nucleus, and torus semicircularis, in two planes. Most units at all levels of the brainstem have directional preferences. The dB values of each panel are stimulus levels in displacement units (dB re: 1 nm). Saccular nerve data are phase-locked spikes per sec. DON and TS data are in spikes/s. The saccular nerve unit shows the tendency toward a circular (dipole) shape in both planes, and the beginning of saturation (flattening) at the highest level. The DON and TS cell are directionally sharpened and do not show the tendency for a saturated response. (and see Edds-Walton and Fay 2005b)

orientations of the otoliths and sensory epithelia of the respective organs. For instance, in *O. tau*'s left saccule, best axes tend to cluster on the horizontal plane axis near –45°, which is qualitatively consistent with the oblique angle of the left saccular epithelium in the horizontal plane (Edds-Walton et al. 1999). This correspondence between neural directionality and otolithic organ orientation results from the nearly vertical orientation of the saccular and lagenar sensory epithelia, and the constraints this has on an afferent's best azimuth.

Utricular afferents tend to have best axes clustering at 0° elevation, corresponding to the essentially horizontal orientation of the utricular epithelium in most vertebrate species (Hawkins and Horner 1981; Fay 1984). However, while anatomical analyses can measure the directional orientations of hair cells on the sensory epithelium, physiological studies are required to reveal the actual functional responsiveness in an intact fish.

In the saccular nerves of *O. tau* (Fay and Edds-Walton 1997a) and *G. morhua* (Horner and Hawkins 1981), there is sufficient diversity in best axis elevation to account for directional hearing in the vertical planes. However, this is not true for the saccule of *C. auratus*, in which afferents are tightly clustered near vertical best elevations (Fay 1984). Thus, *C. auratus* (and probably other Otophysi) would appear to require inputs from both the saccule and lagena, or from the lagena alone, for encoding the elevation of sound sources. Encoding sound source azimuth in fishes probably depends on binaural processing, as it does in tetrapods (Fay 2005). Directional hearing in fishes depends on two, intact labyrinths (Moulton and Dixon 1967; Schuijf 1975; Schuijf and Siemelink 1974), and it has been demonstrated that excitatory–inhibitory binaural interactions take place in the brainstem of the *G. morhua* (Horner et al. 1980).

4. Neurophysiology of the Central Nervous System

The auditory central nervous system (CNS) in fishes is organized similarly to that of other vertebrates, including mammals. Neurophysiological studies of the auditory CNS in fishes have focused on torus semicircularis in the midbrain, with less attention paid to analysis at the levels of the hindbrain and forebrain. Some studies have measured multiunit and gross evoked responses, while others have focused on single-unit responses. In the discussion to follow, we focus on the single unit studies.

4.1 Auditory Nuclei of the Medulla

Several of the studies discussed here could best be understood by comparing the response of central cells in the CNS with those of primary afferents. However, peripheral recordings are missing for several of the studies discussed in this section. Perhaps the earliest study of presumably first-order medullar nuclei in fishes was Enger's (1967) study on hearing in a herring (*Clupea harengus*). Multi- and single-unit recordings were made using metal and glass microelectrodes from what was presumed to be the "central acoustic lobe" of the medulla, but without histological verification. The many single units recorded were presumed to be second or third order because of the complexity of the response in many cases, although the possibility remains that some units were primary afferents, presumed to be from the saccule or utricle. The utricle is thought to be an auditory organ in this family (e.g., Blaxter et al. 1981). Spontaneous spike rates ranged between 0 and 200 spikes/s with most between 10 and

80 spikes/s. Two remarkable observations were made (Fig. 3.7). First, frequency tuning curves for single units showed continuously distributed characteristic frequencies (CF) between about 100 and 1200 Hz. One unit shown was very sharply tuned with a CF at 1000 Hz. Second, many units that responded to frequencies above 500 Hz showed a frequency- and level-dependent tone-evoked suppression or inhibition of the response to below spontaneous spike rates. Enger interpreted these as inhibitory responses and did not evaluate the possibility that the responses were due to suppression (demonstrated and defined later in *C. auratus* saccular afferents by Fay [1990]). However, Enger's Figures 2, 4, and 5 show that at some frequencies, tone presentation caused a reduction in what was presumably spontaneous activity. This could be an example of single-tone suppression, or alternatively, neural inhibition. Assuming with Enger that these are examples of neural inhibition, these data were the first to demonstrate central frequency sharpening effects in a fish auditory system (demonstrated later in *C. auratus* auditory midbrain [Lu and Fay 1993]). Phase-locking and adaptation were not described, but one example was presented in which inhibition of spontaneous activity was replaced by excitation about 500 ms after tone onset. Enger combined several unit tuning curves and estimated the audiogram for this species to be quite sensitive (75 dB re: 1 μPa) and relatively flat between 30 and 1200 Hz (see Fig. 3.7).

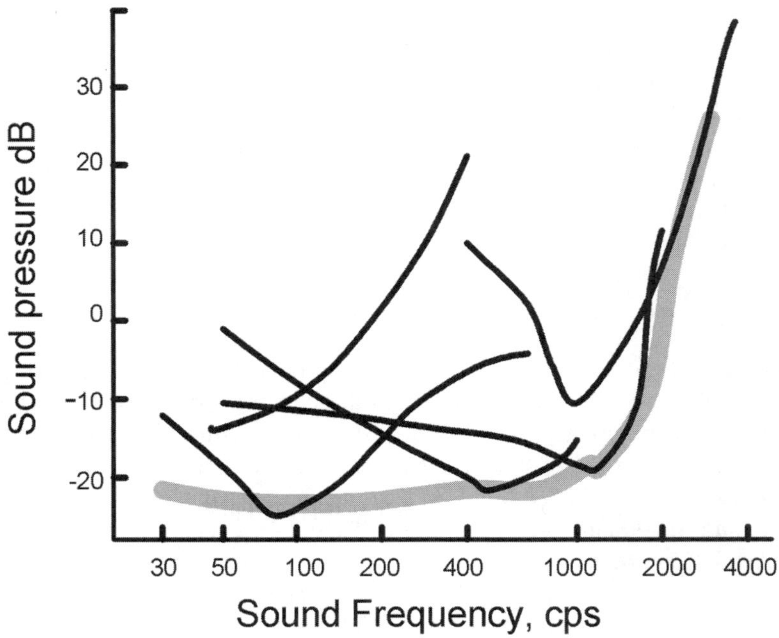

FIGURE 3.7. Single-unit tuning curves and a frequency response envelope (gray line) for units of the *Clupea harengus* medulla, redrawn from Enger (1967). Note the diversity of best frequencies.

Wubbels et al. (1993) recorded auditory responses from medullar units in the rainbow trout (*Oncorhynchus mykiss*) in a region that they identified as the descending octaval nucleus. Units were classified as "low-frequency" (<50–150 Hz response) or as "high-frequency" (responding up to 150–400 Hz). Most units showed significant phase-locking at all frequencies to which they responded. Based on several criteria, the low-frequency units were classified as higher-order lateral line cells, and the high-frequency units were classified as auditory.

Page (1970) recorded 35 single auditory units in the medulla of *C. auratus* from a region surrounding the nucleus of sensory nerve VII, just dorsal and medial to the descending root of V. Spontaneous activity was low (<5 spikes/s), and most units had latencies of less than 10 ms. Both phasic and tonic responses were observed and adaptation was described as slight or absent in tonic units. Some units apparently showed little or no phase-locking while about half of those recorded showed a phase-locked response with frequency following at approximately one spike per cycle, at least at the lower frequencies. All units were frequency selective with many having rather wide frequency-response bandwidths (>1000 Hz). Frequency threshold curves exhibited diverse characteristic frequencies (CF) from below 100 Hz to about 1200 Hz. The tuning curves with the most narrow bandwidths had CFs at or below 200 Hz. Unit thresholds at CF were quite high, ranging between 110 and 130 dB re: 1 μPa.

Sawa (1976) recorded from sound-responsive regions of *C. auratus* medulla using metal microelectrodes, again without presenting histological verification of electrode tip locations. He described spontaneous spike patterns as silent (zero rate), irregular, regular, and bursting, with most sound-sensitive units having irregular (presumably, approximately random) patterns at rates between 5 and 200 spikes/s. Units with regular spontaneous patterns did not respond to sound. Units were found to be tuned in two frequency ranges with CFs between 200 and 300 Hz, and between 600 and 700 Hz. Diverse adaptation patterns were described as slow, intermediate, and fast. Within one unit, adaptation patterns of all rates could be observed at different frequencies with slow (or zero) adaptation occurring to the lowest frequencies, and very fast adaptation occurring at the highest frequencies. Sawa described "follow up" responses that appeared to be phase-locked to the stimulus tone, but quantitative descriptions were lacking. In general, Sawa's description of medullar units in *C. auratus* correspond closely with descriptions of primary saccular afferents (e.g., Fay 1978) with respect to spontaneous patterns, tuning curve CFs, and adaptation. It is possible that Sawa was recording from primary afferents within the medulla or in the anterior octaval nucleus and/or divisions of the secondary octaval population (see Section 2.3).

Plassman (1985) recorded single units of what he believed to be the anterior octaval nucleus (AON) in a catfish (*I. nebulosus*). In these experiments, however, no clear distinction was made between units recorded in the medulla or in the midbrain. All together, these units were categorized as Type I (onset or phasic units) or Type II (more tonic units). Type II units showed responses to amplitude-modulated tones that were nearly independent of modulation rate. In contrast, Type I units displayed bandpass modulation frequency response functions having diverse best modulation rates between 3 and 60 Hz. Similar diversity in responses to amplitude-modulated tones was described in primary saccular afferents of *C. auratus* (Fay 1980), so it is not clear whether these response differences arise peripherally or are due to neural interactions in the brain.

Koslowski (1998) surveyed the auditory pathway and physiological functions extensively for *Pollimyrus isidori*, a sound-producing, weakly electric fish. Recordings have been made at the level of the descending octaval nucleus (DON – Koslowski and Crawford 1998) and presumably from subdivisions of the secondary octaval population (Koslowski and Crawford 2000) of the medulla. Units recorded in DON originated in each of two subdivisions recognized for this species; the dorsal zone (dzD) and the intermediate zone (izD). Multiunit recordings in dzD indicated moderate spontaneous activity, robust phase-locking, and rather sharp frequency tuning centering on about 600 Hz. The degree of phase-locking in some DON units exceeded that typical of primary saccular afferents.

Units recorded in the SOP of *Pollimyrus* were thought to originate in three subdivisions in this species: the dorsal (dSO), intermediate (iSO) and ventral (vSO) secondary octaval populations. Unit responses were analyzed in detail with an emphasis on chopper units ($n = 18$) compared with "primary-like" units ($n = 232$). Primary-like units showed diverse spontaneous activity rates between 0 and 45 spikes/s, robust phase-locking, best frequencies between 50 and 1500 Hz, and diverse sharpness of tuning based on iso-level frequency response areas. These recordings are the only data published for fishes that are likely to be from the SOP.

Chopper units in *Pollimyrus* (presumably from the SOP) were defined as those exhibiting a sustained, quasiperiodic spike response to tones with a single and narrow inter-spike interval mode that was independent of the stimulus tone period (i.e., showing little or no phase-locking). Two types of choppers were identified: stationary and nonstationary. Both chopper types showed low spontaneous rates (0–15 spikes/s), rather low slopes for spike rate versus intensity functions (2–4 spikes/s per dB), relatively wide dynamic ranges (up to 38 dB), best frequencies between 100 and 1300 Hz, and relatively broad frequency tuning compared with primary-like units. Stationary choppers responded to tones with a periodic burst of spikes having an interspike interval mode that remained constant over many tone burst repetitions. Nonstationary choppers showed an unusual, long-term change in the mode for interspike interval, primarily over the first 100 s of tone burst stimulation. These different temporal response patterns are illustrated in Fig. 3.8. Choppers responded to trains of clicks with a tendency to produce one or fewer spikes per click depending on the interclick interval. Long interclick intervals (e.g., 90 ms) tended to produce one spike per click while the shortest intervals (10 ms) tended to evoke 0.4 spikes per click. Koslowski and Crawford (2000) speculated that these chopper units might process the sounds that this species produces, and further hypothesized that they may be part of a bifurcating input stream that is especially well suited for intensity processing. The second stream was identified with primary-like DON units that are better suited for waveform analysis in the time domain. Chopper units are commonly recorded in brainstem nuclei of other vertebrates (e.g., Rhode and Greenberg 1992) and were previously described in the midbrain torus semicircularis of *C. auratus*, a nonvocal species (Lu

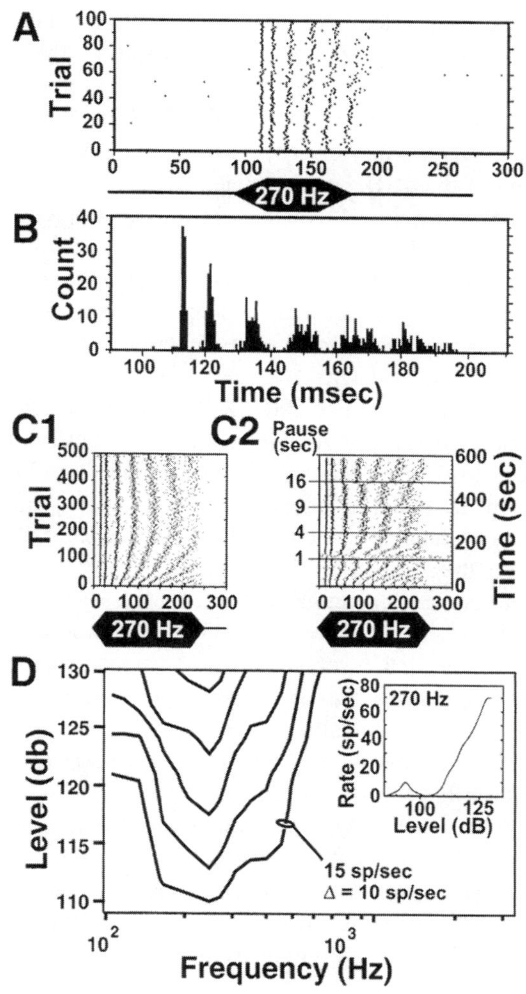

FIGURE 3.8. Illustration of the response of nonstationary chopper cells recorded from the region of the secondary octaval complex of *Pollimyrus adspersus* by Koslowski and Crawford 2000. (**A**) Dot raster pattern in response to a brief tone. (**B**) Peristimulus–time histogram (PSTH). (**C1**) Dot raster pattern for 500 stimulus presentations. (**C2**) Dot raster patterns as in **C1** with pauses introduced. (**D**) Frequency response area. The periodic nature of the PSTH response is typical of choppers. The nonstationarity is illustrated by the changes in the PSTH response that occurs over trials.

and Fay 1993; see later). Therefore, the role of chopper units in auditory processing may not be restricted to the encoding of specific communication sounds.

Most single units recorded in the dorsal subdivision of the DON (dDON) in *O. tau* show simple directional preferences for the axis of whole-body translational acceleration. The maintenance of directionality in the dDON (and in other auditory nuclei investigated) indicates that excitatory convergence from auditory neurons having different directional preferences tends not to occur in the brain; i.e., directional selectivity originating at the periphery is maintained throughout the auditory brainstem. Edds-Walton et al. (1999) searched for, but did not find, indications of a topographical organization of the dDON with respect to stimulus directionality. The axons of directional primary saccular afferents enter the dDON anteriorly and project caudally throughout the entire rostrocaudal extent of the dDON, with collaterals directed medially. Thus, there is no physiological or anatomical evidence for a topographic directional map in the rostrocaudal axis of this large nucleus in *O. tau*.

In *O. tau*, the sensitivity, frequency response, and phase-locking of dDON units are similar to those of saccular afferents, but the directional response patterns (DRP) of most units tend to be more directionally selective than saccular afferents. This increased selectivity has been termed "sharpening" (Fay and Popper 1999; Edds-Walton and Fay 2003, 2005). Sharpening has been hypothesized to arise when a central cell receives excitatory input from one directional cell, and inhibitory input from another directional cell, both having cosine-like DRPs with different best axes in azimuth or elevation (Fay and Popper 1999). This sort of excitatory–inhibitory convergence appears to be the most common sort of interaction in the auditory brainstem and inevitably results in some degree of directional sharpening, depending on the best axes and weights associated with each input. Interestingly, directional sharpening was not coincident with frequency tuning in the dDON of *O. tau* (e.g., highly sharpened cells could be broadly tuned), suggesting that directional sharpening is not associated with sharpening in the frequency domain.

Plachta et al. (2004) investigated the occurrence of ultrasonic responses throughout the brainstem of American shad (*Alosa sapidissima*). This species had been shown to be sensitive to ultrasound (70–110 kHz) in behavioral studies (Mann et al. 1998, 2001). The authors state that lesions marking recording sites were found in nuclei of the afferent acoustic pathway, the secondary octaval population, and the DON. One recording location with ultrasonic responses was confirmed by lesion in the eminentia granularis of the cerebellum. However, electrophysiological response properties were not correlated with recording site. Ultrasonic responses were a mixture of excitation and inhibition, with 80 % of units showing phasic responses at either onset or offset. The response characteristics of some units (12 %) changed in a frequency-dependent manner. No ultrasound-sensitive units responded to the whole range of test frequencies (20–90 kHz). Ultrasonic best frequencies were found between 50 and 90 kHz, and very few best frequencies were in the range between 20 and 40 kHz where behavioral sensitivity was still good. There are many unanswered questions about ultrasound detection in herrings. Are sonic and ultrasonic responses from the same receptor organ? Are these responses found together or are they segregated in the auditory pathway?

4.2 Midbrain (Torus Semicircularis)

The midbrain torus semicircularis (TS) has been known as an important auditory nucleus in fishes for a considerable time and it attracted the attention of early

workers. Page (1970) recorded from the midbrain of *C. auratus*, and Page and Sutterlin (1970) recorded from the midbrain's tegmentum (deep to the torus semicircularis). They found that toral units showed little spontaneous activity, and no phase-locking. Focusing on the tuning of units, Page (1970) described frequency response functions that were wideband (400–2200 Hz), narrowband (800–1200 Hz), low frequency (BF below 200 Hz), and high frequency (BF at 1200 Hz). Page (1970) damaged the lateral line system, but found no effect on low-frequency responses, which indicated that the lateral line was not responsible for the low-frequency responsiveness. He then deflated the gas bladder and found a reduction in sensitivity across the entire frequency range of hearing. This indicated that the saccule, via the ossicular connection with the gas bladder, was responsible for the entire frequency range encoded in the midbrain. Page and Sutterlin (1970) found bimodality (auditory and visual sensitivity) in 20 % of the units recorded in the tegmentum. These bimodal units had very narrow frequency response functions and low BFs and were described as responding to "a very specific sound stimulus." Interactions were described as linear and additive for some units, and suppressive for others. This interesting work has not been followed up by other investigators (but see work by Fay and Edds-Walton 2003 on *O. tau*).

Knudsen (1977) recorded multiunit responses in the midbrain of a catfish (*Ictalurus nebulosus*) and concluded that there are distinct and separate nuclei serving the auditory and lateral line systems, the nucleus centralis and nucleus lateralis respectively. More recently, Fay and Edds-Walton (2001), Edds-Walton and Fay (2004, 2005) have shown in *O. tau* (*Opsanus tau*) that many of the units recorded in toral nuclei are bimodal, responding to both hydrodynamic and acoustic stimuli. The bimodality appears to be derived from the convergence of projections from the auditory descending octaval nucleus (dDON) and from the lateral line nucleus medialis.

Only one study has examined relative responsiveness to sound pressure and particle motion in the torus semicircularis (Fay et al. 1982). Simultaneous recording was done in the medial and lateral torus in the midbrain of *C. auratus* during stimulation with sound pressure or particle motion. Swim bladder deflation was used to reduce the sound pressure component conveyed to the ear and sound-evoked and then head vibration-evoked activity were compared. Swim bladder deflation caused a flat loss in sensitivity of about 20–50 dB as recorded at the medial electrode, but resulted in a more complex effect at the lateral electrode, which sometimes exhibited increased sensitivity with swim bladder deflation. The authors concluded that both sound pressure and particle motion are encoded in *C. auratus* midbrain, but that the TS of *C. auratus* is not homogeneous with respect to relative pressure and motion sensitivity.

Echteler (1985a,b) recorded multiunit activity in the TS of carp (*Cyprinus carpio*) and found a crude tonotopy. High-frequency neurons were located medially and rostrally in the torus. Mid-frequency neurons were found more laterally and caudally. Low-frequency neurons were found most caudally and laterally in the TS. The result was a high-to-low distribution located in the horizontal plane from medial and rostral (high) to lateral and caudal (low). Whether this is a purely computational map or a reflection of anatomical mapping at the level of the saccule or the medulla of the carp is not known.

Wubbels and Schellart and colleagues have presented a series of studies on directional sound encoding in the midbrain of the rainbow trout (*Oncorhynchus mykiss*). This species is a hearing generalist and was assumed to receive both direct motion and reradiated, pressure-dependent motion inputs from the swim bladder to the ears under normal conditions (Schellart et al. 1987). Fish were stimulated by whole-body acceleration at various angles in the horizontal plane using a vibrating platform that could be rotated to any angle (Schellart et al. 1995). Some units were classified as directional (about 44 %), and some nondirectional (Wubbels and Schellart 1998). Directional units were described as roughly mapped in the TS with the medial TS containing rostrocaudal orientations and the lateral TS containing all possible orientations (Wubbels et al. 1995).Based on individual electrode tracks, the TS was described as having a columnar organization with similar best axes of horizontal motion tending to be constant within vertical columns (Wubbels et al. 1995; Wubbels and Schellart 1998). Some phase-locked units had phase angles of synchronization that did not vary with the stimulus axis angle (except for the expected 180° shift at one angle around the circle), while others showed a phase shift that varied continuously with stimulus angle (Wubbels and Schellart 1998).

Wubbels and Schellart (1998) concluded that these and other results strongly supported the phase model (Schuijf 1976), which states that the 180° ambiguity is resolved by comparing the phase angles of sound pressure and particle motion. Further, they speculated that the rostrocaudally oriented units of the medial TS were channels activated by swim bladder-dependent motion input, while the diversely oriented units of the lateral TS represented direct motion input to the otolithic organs. The utricle was hypothesized to be the most important otolithic organ supplying the direct motion-dependent input because of its horizontal orientation. The authors further speculated that the units with synchronization angles independent of stimulus direction represented pressure-dependent swim bladder inputs while the units with variable synchronization phase angles represented direct motion inputs. Wubbels and Schellart (1998, p. 3073) then concluded that "...the phase difference between the(se) two unequivocally encodes the stimulus direction (0–360°)..." (i.e., solves the 180° ambiguity problem). This conception relies on the faithful encoding of phases to be compared, but work on *G. morhua* (Hawkins and Horner 1981), *O. tau* (Fay and Edds-Walton 1997), and *C. auratus* (Fay 1984) shows that response phase varies widely among neurons and therefore does not represent one or a small range of phase angles. Furthermore, Wubbels and Schellart's (1998) explanation for directional hearing lacks a mechanistic basis for the direction-dependent variation in synchronization angle shown by some units and does not suggest a testable model for the final step that solves the 180° ambiguity.

The auditory midbrain of *C. auratus* has been investigated by Lu and Fay (1993, 1996). Single units were characterized in terms of frequency response functions and best frequencies (BF), peristimulus-time histogram (PSTH) patterns, and phase locking to tones. Spontaneous rates were generally distributed below 20 spikes/s, and phase-locking varied from nonexistent, to very highly phase-locked. Frequency response areas were band-pass and varied from those with BF at just below 200 Hz, to those with BFs just below 1000 Hz (Fig. 3.9). Some response areas were continuous and monotonic, while some others were discontinuous, having multiple excitatory areas separated by bands of inhibition (Fig. 3.10). Some nonmonotonic response areas were closed figures, showing suppression or inhibition at higher levels. For saccular afferents, BF shifts upward at higher levels. But for toral units, this tendency disappeared; BF was independent of sound level. This effect appeared to be due to the sculpting of the frequency response area by inhibition, usually at frequencies above BF. Most midbrain units exhibited inhibition at some frequencies, and all frequency

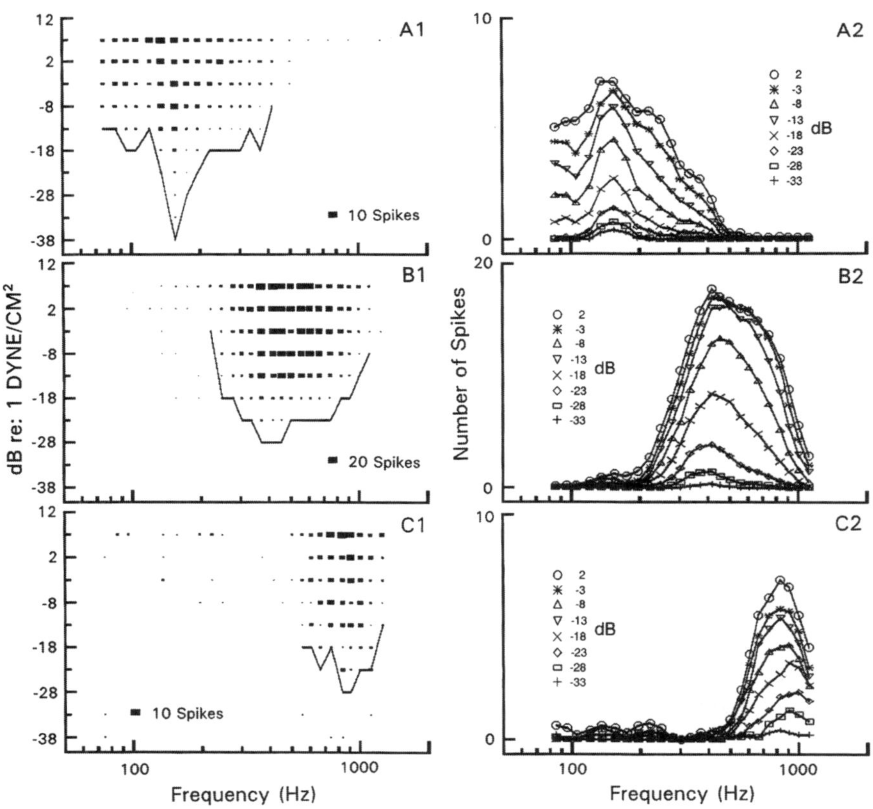

FIGURE 3.9. Typical frequency response areas of low-CF (**A1–2**), mid-CF (**B1–2**), and high CF (**C1–2**) units recorded at the torus semicircularis of *Carassius auratus*, from Lu and Fay (1993). Left panels: Sound level versus frequency functions with tuning curves (solid lines). The size of the filled boxes represents the number of spikes. Right panels: The smoothed number of spikes versus frequency functions for eight levels.

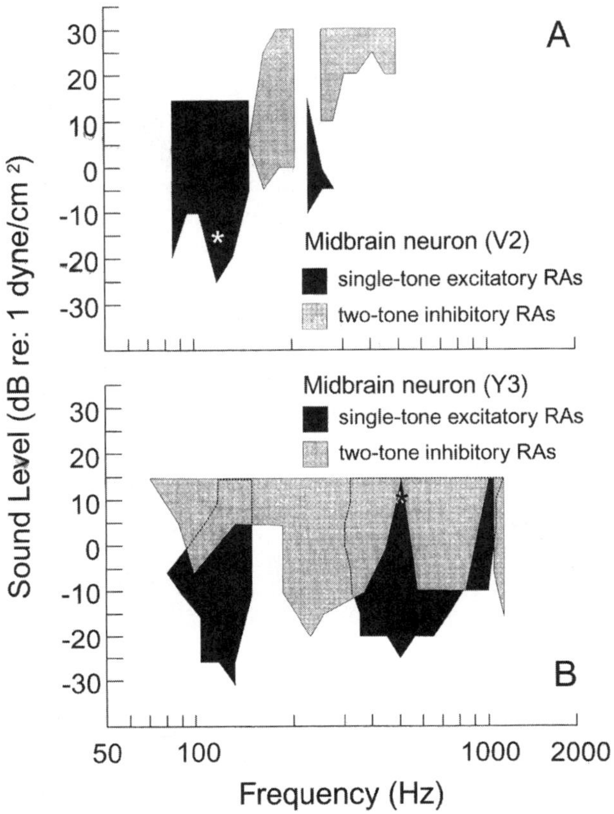

FIGURE 3.10. Island frequency response areas for two neurons of *Carassius auratus* torus semicircularis. Black areas represent excitatory response to single tones. Gray areas represent inhibitory responses to two-tone stimuli with fixed probe tones marked by asterisks. Sound levels are the levels of single tones (black areas) and the levels of suppressor tones (gray areas).

response areas resulted from a balance between excitation and inhibition. Thus, all frequency selectivity at the level of the midbrain was synthesized or computed and did not simply reflect saccular frequency selectivity.

In a quantitative investigation of these effects, Lu and Fay (1996) used two-tone interaction to investigate inhibition. Results for three toral units, compared with three saccular afferents, are shown in Fig. 3.11. In these experiments, a BF tone was presented at a level above threshold to establish a baseline response (horizontal shaded bar in the figure). Then a second tone was presented simultaneously at various frequencies and levels to trace out the frequency contours of its excitatory or inhibitory response. Saccular afferents showed excess excitation except at very low frequencies where entrainment of a phase locked response caused suppression (Fig. 3.11A) and at very high frequencies where two-tone rate suppression reduced the response (Fig. 3.11B). In contrast, midbrain units

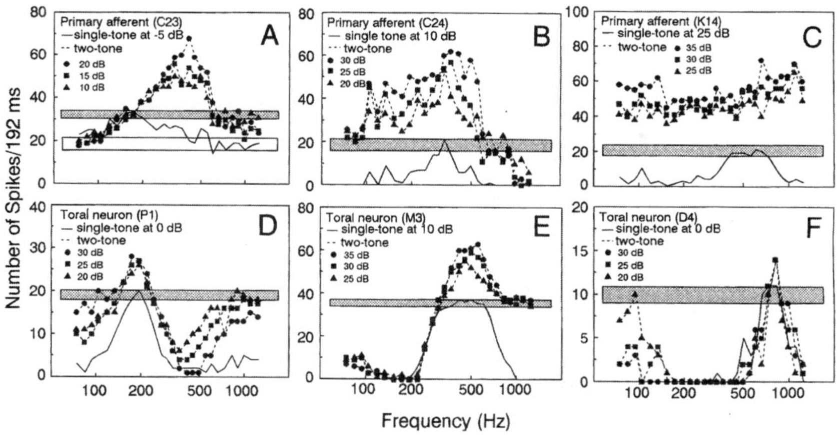

FIGURE 3.11. Typical iso-level response rate functions for *Carassius auratus* primary afferents (**top**) and midbrain neurons (**bottom**) in response to single tones (solid lines) and two-tone (dashed lines with symbols) stimuli. The shaded horizontal bars represent mean response rates to the probe tone, plus and minus one standard deviation. The two-tone stimuli consisted of (1) a tone fixed in frequency near CF presented at the same levels as the single tone response function and (2) a second tone presented at various frequencies as indicated on the abscissa and at the three levels indicated on each panel. Suppressive two-tone interaction is defined as a response to two-tone stimulation that falls below the shaded bar (response to the probe tone alone). Note the large frequency response areas within which the second tone has a suppressive effect for the midbrain neurons. (From Lu and Fay 1996.)

often showed inhibition throughout the frequency response area (Fig. 3.11D, E, and F). Thus, the diversity of tuning and frequency response functions in *C. auratus* midbrain is a result of neural computation, and the outcomes resemble those of many other vertebrate auditory systems, including those of mammals.

The same conclusion can be reached with respect to the PSTH patterns observed. Units of *C. auratus* midbrain are remarkably diverse, showing patterns that can be described as onset, chopper, onset with notch, primary-like, and buildup (Fig. 3.12). These descriptions have also been applied to units of the mammalian brainstem (Rhode and Greenberg 1992). Apparently, these response types are as useful for auditory processing by fishes as they are for mammals and other vertebrates. Exactly what their usefulness is remains a mystery in many cases, but they may help solve the same problems in fishes that they do in tetrapods.

The auditory midbrain of a mormyrid fish (*Pollimyrus adspersus*) contains neurons that are strongly selective to particular temporal intervals. Approximately 30 % of the cells respond preferentially or selectively to trains of impulses at particular interclick intervals (5–80 ms intervals) (Crawford 1993) with a high spike rate. The vocalizations of *Pollimyrus* are courtship calls consisting of grunts, moans, and growls. These sound components are differentiated from one

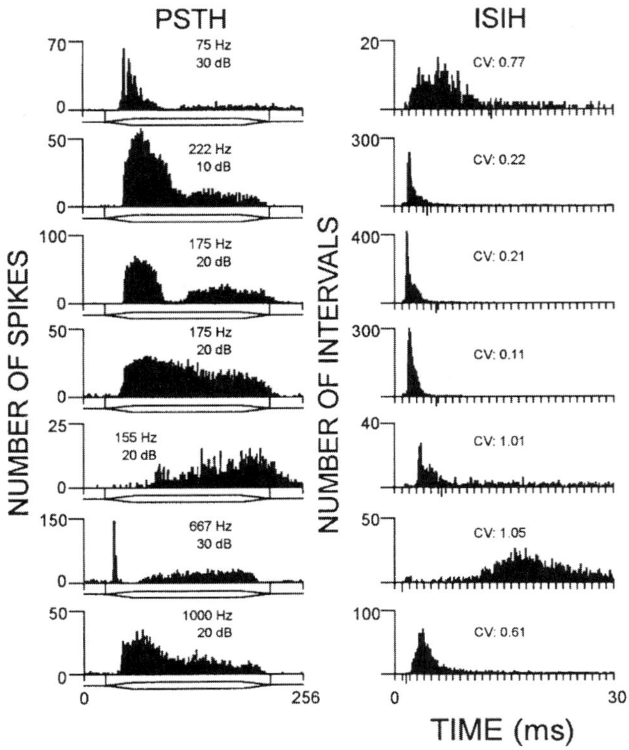

FIGURE 3.12. Dot raster patterns and peristimulus–time histograms (PSTH) for six distinctive temporal response patterns for units of the torus semicircularis without phase-locking. (**B1, B2**) Interspike interval histogram and a "regularity" analysis for the PSTH shown in **B**. The units are classified as: (**A**) onset, (**B**) chopper, (**C**) pauser, (**D**) sustained chopper, (**E**) primary-like, (**F**) buildup. (From Lu and Fay 1993.)

another by their periodicities; grunts have a 21-ms period, moans have 4.5-ms period, and growls have a 39-ms period. Crawford (1993) has hypothesized that these interval-selective neurons are a mechanism by which these periodicities are analyzed and recognized. Large and Crawford (2002) hypothesized a simple network that could underlie the interval selectivity. Essentially, a model is constructed such that each individual impulse evokes a subthreshold EPSP followed by an IPSP, and a brief period of postinhibitory rebound excitation. The correspondence of successive click excitation with the postinhibitory rebound creates the interval preferences (higher spike probability), and variation in the time course of these events creates variability in interval selectivity.

Recordings from the torus semicircularis (TS) of the midbrain of *O. tau* (Fay and Edds-Walton 2001; Edds-Walton and Fay 2003) show unit sensitivity and frequency response similar to units of the dDON, but with dramatically reduced phase-locking and augmented directional sharpening. Directional auditory responses were found both in the nucleus centralis (nominally, the "auditory" nucleus), and the nucleus ventrolateralis (nominally, the "lateral line" nucleus) in the TS of *O. tau*. In addition, many units recorded in both nuclei showed interactions of auditory and lateral line inputs (excitatory and inhibitory) (Fay and Edds-Walton 2001; Edds-Walton and Fay 2003). It is not known whether such bimodal interactions play a role in sound source localization; there are no major theories of source localization that require auditory–lateral line interactions. At the same time, however, source localization is likely a multimodal function (Braun et al. 2002), and the lateral line system could play an important role at short ranges (cf. Weeg and Bass 2002).

In general, the distributions of best axes for midbrain auditory units are more widely distributed in azimuth and elevation than the same distributions for saccular afferents in *O. tau* (Edds-Walton and Fay 2003). Thus, the across-neuron or population representations of the axis of acoustic particle motion appear to be enhanced by excitatory–inhibitory interactions in the medulla or midbrain, particularly in azimuth. It is not known whether this processing is based on binaural, monaural, or both types of neural interactions, but it is known that excitatory–inhibitory binaural interactions take place in the medulla of G. morhua (Horner et al. 1980).

The directional characteristics of TS units also have been studied in *C. auratus*, a hearing specialist (Ma and Fay 2002). In general, most units responded best to nearly vertical whole-body motion, in accord with the nearly uniform vertical orientation of saccular hair cells in *C. auratus* and other Otophysi. Thus, excitatory inputs to the TS appear to be primarily, if not exclusively, from the saccule in *C. auratus*. Nevertheless, deviations from cosine directionality among unit DRPs (i.e., sharpening) was also observed in *C. auratus* TS, and could be accounted for by simple excitatory–inhibitory interactions as in *O. tau*. This suggests that sound source localization in Otophysi, if it occurs at all (see Schuijf et al. 1977), may be based on computations taking place elsewhere in the ascending auditory system where lagenar or utricular inputs could be used to help resolve the axis of acoustic particle motion in a population code comprised of a wide distribution of best axes among neurons. In any case, the representation of acoustic particle motion appears at present to be organized quite differently in the midbrains of *O. tau* and *C. auratus*.

A unique auditory response was investigated in *P. notatus*. The auditory midbrain of the midshipman was analyzed with respect to detection of "beats" produced by concurrent advertisement calls ("hums") from multiple individuals in a breeding area (Bodnar and Bass 1997, 1999). The "hum" is a long duration, multiharmonic call with a fundamental frequency near 100 Hz. Often these hums occur in choruses, with many individual males humming simultaneously. Because all of the hums are similar in frequency content, "beats" may be produced. Bodnar and Bass (1997) found that there was very little synchronization to single frequency components (little phase-locking) among midbrain cells, but that there was synchronization to the beat frequency at the midbrain level. Further, cells of the midbrain were selective to particular beat frequencies. The functional significance of beat responsiveness is not clear, as beats do not represent an individual fish, but the combined output of multiple fish. The presence of beats may signal a concentration of males, which may be of interest to either sex, or cells that are sensitive to beats may be part of a circuit to eliminate "noise" and permit the receiver to sort out individual hums from a chorus.

4.3 Forebrain (Thalamus)

Only one study has been published on the response properties of cells of the forebrain (central posterior nucleus of the thalamus) (Lu and Fay 1996). In general, thalamic units of *C. auratus* resemble toral units in response properties with the exception that thalamic units had broader tuning and less phase locking than toral units. The same diversity of peristimulus–time histogram profiles was seen (Fig. 3.13), including phasic, onset chopper-like, pauser chopper-like, sustained chopper-like, buildup, pauser, and primary-like. Thus, chopper-like cells have been observed in the supplemental octaval population (SOP) of mormyrids (Koslowski and Crawford 1998), and in the TS and thalamus of *C. auratus* (Lu and Fay 1993, 1996). Acoustic analysis appears to take place up to the midbrain level (e.g., sharper frequency selectivity), but at the thalamic level, further analysis (e.g., frequency analysis) seems not to take place and

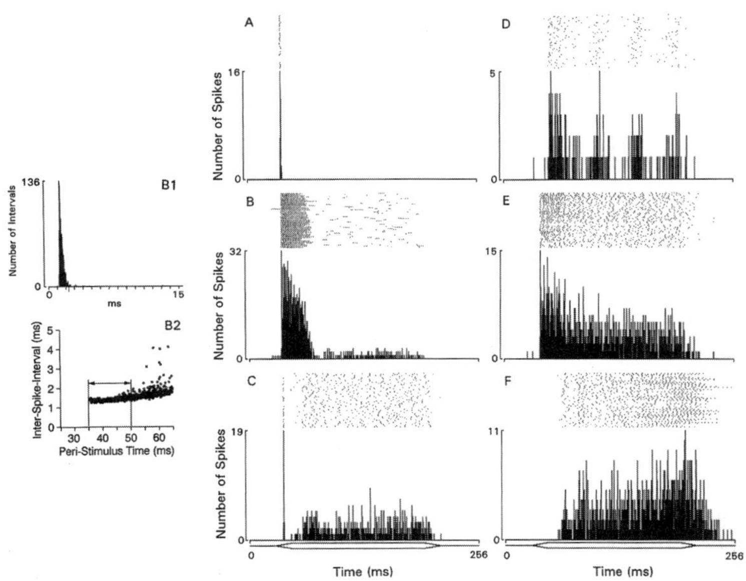

FIGURE 3.13. Peristimulus–time histograms and interspike interval histograms in response to tone bursts for seven units recorded in the central posterior nucleus of the thalamus of *Carassius auratus*. From top to bottom, unit classified as phasic, onset chopper-like, pauser chopper-like, sustained chopper-like, buildup, pauser, primary-like. (From Lu and Fay 1996.)

steps appear to be taken toward synthesis. Integration of acoustic input and other sensory inputs is a distinct possibility, based on sensory processing in other vertebrates.

4.4 Summary and Conclusions

Structurally, the inner ear of teleost fishes is similar in many ways to the ears of other vertebrates. The distinctive difference is the presence of three maculae associated with calcareous otoliths. Individually, the function of a particular end organ may vary among species; thus, one cannot assume an auditory or nonauditory function for any one of the three. In general, the saccule encodes auditory information, but the utricle and/or lagena may also serve an auditory function in some species. No one experimenter has examined a single macula for dual functionality. Therefore, the idea that a single macula may encode both auditory and positional information remains only a hypothesis. Recent anatomical and physiological studies have revealed similarities in the general flow of auditory information from the periphery to the midbrain of teleost fishes (Fig. 3.11). Physiological and psychophysical studies have shown that the auditory system of teleosts has many of the same response properties found in other vertebrates. Anatomical studies also have indicated sites where additional physiological research should be directed: (1) the response properties of (and inputs to) the large, dorsomedial vertical cells of the rostral medulla that may have cerebellar connections or lateral line input unique to fishes; (2) the response properties of cells in the various divisions of the secondary octaval population that may or may not be analogous to the superior olive of mammals; (3) the inputs to and response properties of periventricular cells in the torus semicircularis that may integrate auditory and lateral line sensory systems; (4) response properties of auditory cells in the diencephalon; (5) the location and response properties of multimodal sensory cells in the telencephalon; and (6) documentation of sensory weighting and confirmation of the presence or absence of combinatorial sensory maps.

The acoustic features revealed by single unit and multiunit studies indicate that basic acoustic features are encoded by the auditory afferents of teleost fishes with much in common with terrestrial vertebrates, including highly selective and discontinuous tuning curves not seen in the periphery, a gradual loss of phase-locking as the auditory system is ascended, and many of the PSTH response properties known for tetrapods. Units of the central nervous system encode temporal patterns and frequency via phase-locking, with frequency selectivity and source direction at all levels, as in all other vertebrate brains so far investigated. Limited evidence indicates that frequency tuning and directional sharpening occur in the midbrain, probably through excitatory–inhibitory interactions within auditory nuclei, as occurs in other vertebrates.

The data gathered to date are consistent with the hypothesis that many of the basic functions of auditory processing found in terrestrial vertebrates also are found in fishes. The most important of these functions must be the common fundamental auditory capacity of all vertebrates: i.e., the capacity to determine and perceptually segregate sources of sound so that appropriate behavior may occur with respect to them (Lewis and Fay, 2004).

References

Adrian ED, Craik KW, Sturdy RS (1938) The electrical responses of the auditory mechanism of cold blooded vertebrates. Proc Roy Soc B 125:435–455.

Bass AH, Bodnar DA, Marchaterre MA (2000) Midbrain acoustic circuitry in a vocalizing fish. J Comp Neurol 419:505–531.

Bass AH, Bodnar DA, Marchaterre MA (2001) Acoustic nuclei in the medulla and midbrain of the vocalizing gulf toadfish (*Opsanus beta*). Brain Behav Evol 57:63–79.

Blaxter, JHS, Denton EJ Gray JAB (1981) Acousticolateralis system in clupeid fishes. In: Tavolga WN, Popper AN, Fay RR (eds) Hearing and Sound Communication in Fishes. New York: Springer-Verlag, pp. 39–56.

Bodnar DA, Bass AH (1997) Temporal coding of concurrent acoustic signals in auditory midbrain. J Neurosci 17:7553–7564.

Bodnar DA, Bass AH (1999) A midbrain combinatorial code for temporal and spectral information in concurrent acoustic signals. J Neurophysiol 81:552–563.

Braford MR, McCormick CA (1979) Some connections of the torus semicircularis in the bowfin, *Amia calva*: a horseradish peroxidase study. Soc Neurosci Abstr 5:193.

Braford MR, McCormick CA (1994) Organization of inner ear endorgan projections in the goldfish, *Carassius auratus*. Brain Behav Evol 43:189–205.

Braun C, Coombs S, Fay R (2002) What is the nature of multisensory interaction between octavolateralis sub-systems? Brain Behav Evol 59:162–176.

Capranica RR, Moffat AJM (1980) Nonlinear properties of the peripheral auditory system of anirans. In: Popper AN, Fay RR (eds) Comparative Studies of Hearing in Vertebrates. New York: Springer-Verlag, pp. 139–165.

Carr CE, Code RA (2000) The central auditory system of reptiles and birds. In: Dooling RJ, Fay RR, Popper AN (eds) Comparative Hearing: Birds and Reptiles. New York: Springer-Verlag, pp. 197–248.

Coombs S, Fay RR (1985) Adaptation effects on amplitude modulation detection: behavioral and neurophysiological assessment in the goldfish auditory system. Hear Res 19:57–71.

Coombs SL, Fay RR (1987) Response dynamics of goldfish saccular fibers: effects of stimulus frequency and intensity on fibers with different tuning, sensitivity, and spontaneous activity. J Acoust Soc Am 81:1025–1035.

Coombs S, New JG (2002) Preface to Special Issue of Brain Behav Evol Sens Integr 59:159–161.

Coombs S, Gorner P, Munz H (1989) A brief overview of the mechanosensory lateral line system and the contributions to this volume. In: Coombs S, Goner P, Munz H (eds) The Mechanosensory Lateral Line: Neurobiology and Evolution. New York, Springer-Verlag, pp. 3–5.

Corwin JT (1981) Peripheral auditory physiology in the lemon shark: evidence of parallel otolithic and non-otolithic sound detection. J Comp Physiol 142:370–390.

Corwin JT (1989) Functional anatomy of the auditory system in sharks and rays. J Exp Zool Suppl 2:62–74.

Crawford AC, Fettiplace R (1981) An electrical tuning mechanism in turtle cochlear hair cells. J Physiol 312:377–412.

Crawford JD (1993) Central auditory neurophysiology of a sound producing fish: the mesencephalon of *Pollimyrus isidori* (Mormyridae). J Comp Physiol A 172:139–152.

de Vries HL (1950) The mechanics of labyrinth otoliths. Acta Oto-Laryngol 38:262–273.

Dijkgraaf S (1960) Hearing in bony fishes. Proc Roy Soc B 152:51–54.

Echteler SE (1984) Connections of the auditory midbrain in a teleost fish, *Cyprinus carpio*. J Comp Neurol 230:536–551.

Echteler SE (1985a) Organization of central auditory pathways in a teleost fish, *Cyprinus carpio*. J Comp Physiol A 156:267–280.

Echteler SE (1985b) Tonotopic organization in the midbrain of a teleost fish. Barin Res 338:387–391.

Edds-Walton PL (1998a) Projections of primary afferents from regions of the saccule in toadfish (*Opsanus tau*). Hear Res 115:45–60.

Edds-Walton PL (1998b) Anatomical evidence for binaural processing in the descending octaval nucleus of the toadfish (*Opsanus tau*). Hear Res 123:41–54.

Edds-Walton P, Fay RR (2003) Directional selectivity and frequency tuning of midbrain cells in the oyster toadfish, *Opsanus tau*. J Comp Physiol 189:527–543.

Edds-Walton PL, Fay RR (2005a) Projections to bimodal sites in the torus semicircularis of the toadfish, *Opsanus tau*. Brain Behav Evol 66:73–87.

Edds-Walton PL, Fay RR (2005b) Sharpening of directional responses along the auditory pathway of the oyster toadfish, *Opsanus tau*. J Comp Physiol 191: 1079–1086.

Edds-Walton PL, Popper AN (2000) Dendritic arbors on the saccule and lagena in the ear of the goldfish, *Carassius auratus*. Hear Res 141:229–242.

Edds-Walton PL, Fay RR, Highstein SM (1999) Dendritic arbors and central projections of auditory fibers from the saccule of the toadfish (*Opsanus tau*). J Comp Neurol 411:212–238.

Enger PS (1963) Masking of auditory responses in the medulla oblongata of goldfish. J Exp Biol 59:415–424.

Enger PS (1967) Hearing in herring. Comp Biochem Physiol 22:527–538.

Fay RR (1970) Auditory frequency discrimination in the goldfish (*Carassius auratus*). J Comp Physiol Psychol 73:175–180.

Fay RR (1978a) Coding of information in single auditory-nerve fibers of the goldfish. J Acoust Soc Am 63:136–146.

Fay RR (1978b) Phaselocking in goldfish saccular nerve fibres accounts for frequency discrimination capacities. Nature 275:320–322.

Fay RR (1980) Psychophysics and neurophysiology of temporal factors in hearing by the goldfish: amplitude modulation detection. J Neurophysiol 44:312–332.

Fay RR (1981) Coding of acoustic information in the eighth nerve. In: Tavolga WN, Popper AN, Fay RR (eds) Hearing and Sound Communication in Fishes. New York: Springer-Verlag, pp. 189–219.

Fay RR (1984) The goldfish ear codes the axis of acoustic particle motion in three dimensions. Science 225:951–954.

Fay RR (1985) Sound intensity processing by the goldfish. J Acoust Soc Am 78: 1296–1309.

Fay RR (1990) Suppression and excitation in auditory nerve fibers of the goldfish, *Carassius auratus*. Hear Res 48:93–110.

Fay RR (1991) Masking and suppression in auditory nerve fibers of the goldfish, (*Carassius auratus*). Hear Res 55:177–187.

Fay RR (1995) Physiology of primary saccular afferents of goldfish: implications for Mauthner cell response. Brain Behav Evol 46:141–150.

Fay RR (1997) Frequency selectivity of saccular afferents of the goldfish revealed by revcor analysis. In: Lewis ER, Long GR, Lyon RF, Narins PM, Steele CR, E. Hecht-Poinar E (eds) Diversity in Auditory Mechanics. Singapore: World Scientific Publishers, , pp. 69–75.

Fay RR, Coombs S (1983) Neural mechanisms in sound detection and temporal summation. Hear Res 10:69–92.

Fay RR, Edds-Walton PL (1997a) Directional response properties of saccular afferents of the toadfish, *Opsanus tau*. Hear Res 111:1–21.

Fay RR, Edds-Walton PL (1997b) Diversity in frequency response properties of saccular afferents of the toadfish (*Opsanus tau*). Hear Res 113:235–246.

Fay RR, Edds-Walton PL (1999) Sharpening of directional auditory responses in the descending octaval nucleus of the toadfish (*Opsanus tau*). Biol Bull 197:240–241.

Fay RR, Edds-Walton PL (2001) Bimodal units in the torus semicircularis of the toadfish (*Opsanus tau*). Biol Bull 201:280–281.

Fay RR, Edds-Walton P (2003) Directional selectivity and frequency tuning of midbrain cells in the oyster toadfish, *Opsanus tau*. J Comp Physiol A 189:527–543.

Fay RR, Edds-Walton PL (2005) Sharpening of directional responses along the auditory pathway of the oyster toadfish, *Opsanus tau*. J Comp Physiol 191:1079–1086.

Fay RR, Olsho I (1979) Discharge patterns of lagenar and saccular neurons of the goldfish eighth nerve: displacement sensitivity and directional characteristics. Comp Biochem Physiol 62:377–386.

Fay RR, Popper AN (1999) The auditory periphery in fishes. In: Fay RR, Popper AN (eds) Comparative Hearing: Fish and Amphibians. New York: Springer-Verlag, pp. 43–100.

Fay RR, Ream TJ (1986) Acoustic response and tuning in saccular nerve fibers of the goldfish (*Carassius auratus*). J Acoust Soc Am 79, 1883–1895.

Fay RR, Ream TJ (1992) The effects of temperature change and transient hypoxia on auditory nerve response in the goldfish (*Carassius auratus*). Hear Res 58:9–18.

Fay RR, Hillery CM, Bolan K (1982) Representation of sound pressure and particle motion information in the midbrain of the goldfish. Comp Biochem Physiol 71:181–191.

Fay RR, Yost WA, Coombs S (1983) Psychophysics and neurophysiology of repetition noise processing in a vertebrate auditory system. Hear Res 12:31–55

Fay RR, Chronopoulos M, Patterson RD (1996) The sound of a sinusoid: perception and neural representations in the goldfish (*Carassius auratus*). Audit Neurosci 2:377–392.

Furukawa T (1981) Effects of efferent stimulation on the saccule of goldfish. J Physiol 315:203–215.

Furukawa T (1986) Sound reception and synaptic transmission in goldfish hair cells. Jpn J Physiol 36:1059–1077.

Furukawa T, Ishii T (1967) Neurophysiological studies on hearing in the goldfish. J Neurophysiol 30:1337–1403.

Furukuwa T, Matsuura S (1978) Adaptive rundown of excitatory post-synaptic potentials at synapses between hair cells and eight nerve fibres in the goldfish. J Physiol 276: 193–209.

Furukawa T, Kuno M, Matsuura S (1982) Quantal analysis of a decremental response at hair cell-afferent fibre synapses in the goldfish sacculus. J Physiol 322:181–195.

Grose B, Carr CE, Casseday JH, Fritzsch B, Koppl C (2004) The evolution of central pathways and their neural processing patterns. In: Manley GA, Popper AN, Fay

RR (eds) Evolution of the Vertebrate Auditory System. New York: Springer-Verlag, pp. 289–359.

Grözinger B (1967) Elektro-physiologische Untersuchugen an der Hirbahn der Schleie (*Tinca tinca* [L.]). Z Vergl Physiol 57:44–76.

Hawkins AD, Horner K (1981) Directional characteristics of primary auditory neurons from the cod ear. In: Tavolga WN, Popper AN, Fay RR (eds) Hearing and Sound Communication in Fishes. New York: Sprenger-Verlag, pp. 311–328.

Hawkins AD, Sand O (1977) Directional hearing in the median vertical plane by the cod. J Comp Physiol 122:1–8.

Henry KR, Lewis ER (1992) One-tone suppression in the cochlear nerve of the gerbil. Hear Res 63:1–6.

Highstein SM, Baker R (1985) Action of the efferent vestibular system on primary afferents in the toadfish, *Opsanus tau*. J Neurophysiol 54:370–384.

Highstein SM, Baker R (1986) Organization of the efferent vestibular nuclei and nerves of the toadfish, *Opsanus tau*. J Comp Neurol 243:309–325.

Highstein SM, Kitch R, Carey J, Baker R (1992) Anatomical organization of the brain stem octavolateralis area of the oyster toadfish, *Opsanus tau*. J Comp Neurol 319: 501–518.

Hill KG, Stange G, Gummer AW, Mo J (1989) A model proposing synaptic and extrasynaptic influences on the responses of cochlear nerve fibers. Hear Res 39:75–90.

Holton T, Weiss T (1983) Frequency selectivity of hair cells and nerve fibers in the alligator lizard cochlea. J Physiol 345:241–260.

Horner K, Sand O, Enger PS (1980) Binaural interaction in the cod. J Exp Biol 85: 323–331.

Horner K, Hawkins AD, Fraser PJ (1981) Frequency characteristics of primary auditory neurons from the ear of the cod, *Gadus morhua*. In Tavolga WN, Popper AN, Fay RR (eds) Hearing and Sound Communication in Fishes. New York: Springer-Verlag, pp. 223–241.

Hudspeth AJ, Corey DP (1977) Sensitivity, polarity, and conductance change in the response of vertebrate hair cells to controlled mechanical stimuli. Proc Natl Acad Sci USA 74:2407–2411.

Ishii Y, Matsuura S, Furukawa T (1971) Quantal nature of transmission at the synapse between hair cells and eighth nerve fibers. Jpn J Physiol 21:79–89.

Kenyon TN, Ladich F, Yan HY (1998) A comparative study of hearing ability in fishes: the auditory brain stem response approach. J Comp Physiol A 182:307–318.

Knudsen EI (1977) Distinct auditory and lateral line nuclei in the midbrain of catfishes. J Comp Neurol 173:417–431.

Köppl C, Manley G (1992) Functional consequences of morphological trends in the evolution of lizard hearing organs. In: Webster DB, Fay RR, Popper AN (eds) The Evolutionary Biology of Hearing. New York: Springer-Verlag, pp. 489–510.

Koslowski J, Crawford JD (1998) Functional neuroanatomy of auditory pathways in the sound-producing fish *Pollimyrus*. J Comp Neurol 401:227–252.

Koslowski J, Crawford JD (2000) Transformations of an auditory temporal code in the medulla of a sound-producing fish. J Neurosci 20:2400–2408.

Lanford PJ, Popper AN (1996) Novel afferent terminal structure in the crista ampullaris of the goldfish, *Carassius auratus*. J Comp Neurol 366:572–579.

Lanford PJ, Popper AN, Platt C (2000) Structure and function in the saccule of the goldfish (*Carassius auratus*): a model of diversity in the non-amniote ear. Hear Res 143:1–13.

Large EW, Crawford JD (2002) Auditory temporal computation: interval selectivity based on post-inhibitory rebound. J Comput Neurosci 13:125–42.

Lewis ER (1986) Adaptation, suppression, and tuning in amphibian acoustical fibers. In: Moore BCJ, Patterson R (eds) Auditory Frequency Selectivity. New York: Plenum, pp. 129–136.

Lewis ER (1992) Convergence and design in vertebrate acoustic sensors. In: Webster DB, Fay RR, Popper AN (eds) The Evolutionary Biology of Hearing. New York: Springer-Verlag, pp. 163–184.

Lewis ER, Fay RR (2004) Environmental variables and the fundamental nature of hearing. In: Manley G, Popper AN, Fay RR (eds) Evolution of the Vertebrate Auditory System. New York: Springer-Verlag, pp. 27–54.

Lowenstein O, Roberts TDM (1951) The localization and analysis of the responses to vibration from the isolated elasmobranch labyrinth: a contribution to the problem of the evolution of hearing in vertebrates. J Physiol (Lond) 114:471–489.

Lu Z and Fay RR (1993) Acoustic response properties of single units in the torus semicircularis of the goldfish, *Carassius auratus*. J Comp Physiol 173:33–48.

Lu Z, Fay RR (1996) Two-tone interaction in auditory nerve fibers and midbrain neurons of the goldfish, *Carassius auratus*. Audit Neurosci 2:257–273.

Lu Z, Popper AN (1998). Neural response directionality correlates of hair cell orientation in a teleost fish. J. Comp. Physiol. A 187:453–465.

Lu Z, Song J, Popper AN (1998) Encoding of acoustic directional information by saccular afferents of the sleeper goby *Dormitator latifrons*. J Comp Physiol A 182:805–815.

Lu Z, Xu Z, Buchser WH (2003) Acoustic response properties of lagenar nerve fibers in the sleeper goby, *Dormitator latifrons*. J Comp Physiol A 189:889–905.

Ma W-L, Fay RR (2002) Neural representations of the axis of acoustic particle motion in the auditory midbrain of the goldfish, *Carassius auratus*. J Comp Physiol [A] 188:301–313.

Manley G (1990) Peripheral Hearing Mechanisms in Reptiles and Birds. Berlin: Springer-Verlag.

Manley G, Gleich O(1984) Avian primary auditory neurons. The relationship between characteristic frequency and preferred intervals. Naturwissenschaften 71:592–594.

Mann DA, Lu Z, Hasting MC, Popper AN (1998) Detection of ultrasonic tones and simulated dolphin echolocation clicks by a teleost fish, the American shad (*Alosa sapidissima*). J Acoust Soc Am 104:562–568.

Mann DA, Higgs DM, Tavolga WN, Souza MJ, Popper AN (2001) Ultrasound detection by clupeiform fishes. J Acoust Soc Am 109:3048–3054.

McCormick CA (1981) Central projections of the lateral line and eighth nerves in the bowfin, *Amia calva*. J Comp Neurol 197:1–15.

McCormick CA (1982) The organization of the octavolateralis area in actinopterygian fishes: a new interpretation. J Morphol 171:159–181.

McCormick CA (1983) Central connections of the octavolateralis nerves in the pike cichlid, *Crenicichla lepidota*. Brain Res 265:177–185.

McCormick CA (1992) Evolution of central auditory pathways in anamniotes. In: Webster DB, Fay RR, Popper AN (eds) The Evolutionary Biology of Hearing. New York: Springer-Verlag, pp. 323–350.

McCormick CA (1997) Organization and connections of octaval and lateral line centers in the medulla of a clupeid, *Dorosoma cepedianum*. Hear Res 110:39–60.

McCormick CA (1999) Anatomy of the central auditory pathways of fish and amphibians. In: Fay RR, Popper AN (eds) Comparative Hearing: Fish and Amphibians. New York: Springer-Verlag, pp. 155–217.

McCormick CA, Braford MR Jr (1988) Central connections of the octavolateralis system. In: Atema J, Fay RR, Popper AN, Tavolga WN (eds) Sensory Biology of Aquatic Animals. New York: Springer-Verlag, pp. 733–756.
McCormick CA, Braford MR Jr (1993) The primary octaval nuclei and inner ear afferent projections in the otophysan *Ictalurus punctatus*. Brain Behav Evol 42:48–68.
McCormick CA, Braford MR Jr (1994) Organization of inner ear endorgan projections in the goldfish, *Carassius auratus*. Brain Behav Evol 43:189–205.
McCormick CA, Hernandez DV (1996) Connections of octaval and lateral line nuclei in the medulla of the goldfish, including the cytoarchitecture of the secondary octaval population in goldfish and catfish. Brain Behav Ecol 47:113–137.
McKibben JR, Bass AH (1999) Peripheral encoding of behaviorally relevant acoustic signals in a vocal fish: single tones. J Comp Physiol A 184:563–576.
Meredith GE, Roberts BL, Maslam S (1987) Distribution of afferent fibers from end organs in the ear and lateral line in the European eel. J Comp Neurol 265:507–520.
Moeng R, Popper AN (1984) Auditory responses of saccular neurons of the catfish, *Ictalurus punctatus*. J Comp Physiol 155:615–624.
Montgomery JC, Coombs S, Conley RA, Bodznick D (1995) Hindbrain sensory processing in lateral line, electrosensory, and auditory systems: a comparative overview of anatomical and functional similarities. Audit Neurosci 1:207–231.
Moulton JM, Dixon RH (1967) Directional hearing in fishes. In: Tavolga WN (ed) Marine Bio-acoustics, Vol. II. New York: Pergamon Press, pp. 187–228.
Nelson ME, MacIver MA, Coombs S (2002) Modeling electrosensory and mechanosensory images during the predatory behavior of weakly electric fish. Brain Behav Evol 59:199–210.
Northcutt RG (1979) Primary projections of VIII nerve afferents in a teleost, *Gillichthys mirabilis*. Anat Rec 193:638–645.
Northcutt RG (1980) Central auditory pathways in anamniotic vertebrates. In: Popper AN, Fay RR (eds) Comparative Studies of Hearing in Vertebrates. New York: Springer-Verlag, pp. 79–118.
O'Marra SK, McCormick CA (1999) Organization and connections of the dorsal descending nucleus and other presumed acoustic areas in the brainstem of the teleost fish, *Astronotus ocellatus*. Hear Res 129:7–19.
Page CH (1970) Electrophysiological study of auditory responses in the goldfish brain. J Neurophysiol 33:116–128.
Page CH, Sutterlin AM (1970) Visual and auditory responses in the goldfish tegmentum. J Neurophysiol 33:129–136.
Patuzzi R (1996) Cochlear micromechanics and macromechanics. In: Dallos P, Popper AN, Fay RR (eds) The Cochlea. New York: Springer-Verlag, pp. 186–257.
Plachta DT, Song J, Halvorsen MB, Popper AN (2004) Neuronal encoding of ultrasonic sound by a fish. J Neurophysiol 91:2590–25977.
Plassman W (1985) Coding of amplitude-modulated tones in the central auditory system of catfish. Hear Res 17:209–217.
Platt C (1977) Hair cell distribution and orientation I goldfish otolithic organs. J Comp Neurol 172:283–298.
Platt C (1983) The peripheral vestibular system of fishes. In: Northcutt RG, Davis RE (eds) Fish Neurobiology, Vol. 1 Ann Arbor: University of Michigan Press, pp. 89–123.
Platt C, Popper AN (1984) Variation in lengths of ciliary bundles on hair cells along the macula of the sacculus in two species of teleost fishes. Scan Electron Microsc (Pt 4):1915–1924.

Popper AN, Eaton R (1995) The octavolateralis system and Mauthner cell: interactions and questions. Brain Behav Evol 46:124–130.
Popper AN, Edds-Walton PL (1995) Structural diversity in the inner ear of teleost fishes: implications for connectins to the Mauthner cell. Brain Behav Evol 46:131–140.
Popper AN, Fay RR (1999). The auditory periphery in fishes. In: Fay RR, Popper AN (eds) Comparative Hearing: Fish and Amphibians. New York: Springer-Verlag, pp. 43–100.
Popper AN, Platt C, Edds PL (1992) Evolution of the vertebrate inner ear: an overview of ideas. In: Webster DB, Fay RR, Popper AN (eds) The Evolutionary Biology of Hearing. New York: Springer-Verlag, pp. 49–57.
Prasada Rao PD, Jadhao AG, Sharma SC (1987). Descending projection neurons to the spinal cord of the goldfish. J Comp Neurol 265:96–108.
Presson JC, Edds PL, Popper AN (1992) Central-peripheral and rostral-caudal organization of the innervation of the saccule in a cichlid fish. Brain Behav Evol 39:187–207.
Rhode WS, Greenberg S (1992) Physiology of the cochlear nuclei. In: Popper AN, Fay RR (eds) The Mammalian Auditory Pathway: Neurophysiology. New York: Springer-Verlag, pp. 94–152.
Rogers PH, Cox H (1988) Underwater sound as a biological stimulus. In: Atema J, Fay RR, Popper AN, Tavolga WN (eds) Sensory Biology of Aquatic Animals. (. New York: Springer-Verlag, pp. 131–149.
Ruggero M (1992) The auditory nerve. In: Fay RR, Popper AN (eds) The Mammalian Auditory Pathway: Neurophysiology. New York: Springer-Verlag, pp. 34–93.
Sachs M, Kiang NY-S (1968) Two-tone inhibition in auditory nerve fibers. J Acoust Soc Am 43:1120–1128.
Sand O (1974) Directional sensitivity of microphonic potentials from the perch ear. J Exp Biol 60:881–899.
Sawa M (1976) Auditory responses from single neurons of the medulla oblongata in the goldfish. Bull Jpn Soc Sci Fish 42:141–152.
Schellart NAM, Kroese ABA (1989) Interrelationship of acousticolateral and visual systems in the teleost midbrain. In: Coombs S, Gorner P, Munz H (eds) The Mechanosensory Lateral Line. New York: Springer-Verlag, pp. 421–443.
Schellart NAM, Kamermans M, Nederstigt LJA (1987) An electrophysiological study of the topographical organization of the multisensory torus semicircularis of the rainbow trout. Comp Biochem Physiol 88A:461–469.
Schellart NAM, Wubbels RJ, Schreurs W, Faber A, Goossens JHLM (1995) Two-dimensional vibrating platform in nm range. Med Biol Eng Comput 33:217–220.
Schuijf A (1975) Directional hearing of cod (*Gadus morhua*) under approximate free field conditions. J Comp Physiol A 98:307–332.
Schuijf A (1976) The phase model of directional hearing in fish. In: Schuijf A, Hawkins AD (eds) Sound Reception in Fish. Amsterdam: Elsevier, pp. 63–86.
Schuijf A, Siemelink M (1974) The ability of cod (*Gadus morhua*) to orient towards a sound source. Experientia 30:773–774.
Schuijf A, Visser C, Willers A, Buwalda RJ (1977) Acoustic localization in an ostariophysine fish. Experientia 33:1062–1063.
Sento S, Furukawa T (1987) Intra-axonal labeling of saccular afferents in the goldfish *Carassius auratus*: correlations between morphological and physiological characteristics. J Comp Neurol 258:352–367.
Sisneros JA, Forlano PM, Deitcher DL, Bass AH (2004) Steroid-dependent auditory plasticity leads to adaptive coupling of sender and receiver. Science 305:404–407.

Steinacker A, Romero A (1992) Voltage-gated potassium current resonance in the toadfish saccular hair cell. Brain Res 574:229–236.

Streidter GF (1991) Auditory, electrosensory, and mechanosensory pathways through the diencephalon and telencephalon of channel catfishes. J Comp Neurol 312:311–331.

Sugihara I, Furukawa T (1989) Morphological and functional aspects of two different types of hair cells in the goldfish sacculus. J Neurophysiol 62:1330–1343.

Suzue T, Wu GB, Furukawa T (1987) High susceptibility to hypoxia of afferent synaptic transmission in the goldfish sacculus. J Neurophysiol 58:1066–1079.

Suzuki A, Koslowski J, Crawford JD (2002) Temporal encoding for auditory computation: physiology of primary afferent neurons in sound-producing fish. J Neurosci 22: 6290–301.

Tomchik SM, Lu Z (2005) Octavolateral projections and organization in the medulla of a teleost fish, the sleeper goby (*Dormitator latifrons*). J Comp Neurol 481:96–117.

van Bergeijk WA (1967) The evolution of vertebrate hearing. In Neff WD (ed) Contributions to Sensory Physiology, Vol. 2. New York: Academic Press, pp. 1–49.

von Békésy G (1960) Experiments in Hearing. New York: McGraw-Hill.

von Frisch K (1938) The sense of hearing in fish. Nature 141:8–11.

de Vries HL (1950) The mechanics of labyrinth otoliths. Acta Oto-Laryngol 38:262–273.

Weeg MS, Bass AH (2000) Central lateral line pathways in a vocalizing fish. J Comp Neurol 418:41–64.

Weeg M, Fay RR, Bass A (2002) Directional response and frequency tuning in saccular nerve fibers of a vocal fish, *Porichthys notatus*. J Comp Physiol 188:631–641.

Wever EG (1949) Theory of Hearing. New York: John Wiley & Sons.

Wubbels RJ, Schellart NAM (1998) Neural coding of sound direction in the auditory midbrain of the rainbow trout. J Neurophysiol 77:3060–3074.

Wubbels RJ, Kroese ABA, Schellart NAM (1993) Response properties of lateral line and auditory units in the medullar oblongata of the rainbow trout (*Oncorhynchus mykiss*). J Exp Biol 179:77–92.

Wubbels RJ, Schellart NAM, Goossens JHLM (1995) Mapping of sound direction in the trout lower midbrain. Neurosci Lett 199:179–182.

Yin TCT (2002) Neural mechanisms of encoding binaural localization cues in the auditory brainstem. In: Oertel D, Fay RR, Popper AN (eds) Integrative Functions in the Mammalian Auditory Pathway. New York: Springer-Verlag, pp. 99–159.

Zottoli SJ, Bently AP, Prendergast BJ, Reiff HI (1995) Comparative studies on the Mauthner cell of teleost fish in relation to sensory input. Brain Behav Evol 46:151–164.

4
Evolution of Peripheral Mechanisms for the Enhancement of Sound Reception

CHRISTOPHER B. BRAUN AND TERRY GRANDE

> The external linkages of vertebrate ears have been the kind of material that brings joy to the comparative morphologist.
>
> —Carl Gans, 1992 (p. 3)

1. Introduction

The evolutionary history of hearing is a rich and fascinating pageant. The inner ear and the closely related mechanosensory lateral line show a tremendous diversity among living and fossil vertebrates. This chapter documents how these systems have evolved new functions by juxtaposing novel linkages (i.e., transduction mechanisms) between fundamentally conservative hair cell sensors and the outside world. These linkages dictate the ear's function, and are so diverse that the functions of the ear (and lateral line) have changed repeatedly in vertebrate history. The linkages of the vertebrate ear do indeed bring joy to the comparative biologist, and the evolution of these linkages is the evolution of new sensory functions, many of which may have led to the rapid diversification of individual taxa (e.g., Otophysi) and the expansion of behavioral repertoires.

To discuss the evolution of enhanced hearing capabilities, one must understand the primitive functions of the octavolateralis systems, and ask what new functions has evolution wrought, and what are the new stimuli to which the ear responds in derived taxa? To constrain the discussion of the diversity of inner ear linkages, this chapter reviews the evolution of specializations that alter the function of the inner ear in teleost fishes and grant the ability to detect fluctuations in the ambient pressure (i.e., sound). Several instances of lateral line specialization that may provide this system with pressure sensitivity are also described. When the distribution of these specializations is compared to our best estimates of teleost relationships (Fig. 4.1), it appears that the detection of pressure fluctuations (what terrestrially chauvinistic vertebrates call hearing) has evolved dozens of times! This chapter describes some of these novel morphologies in detail and attempts

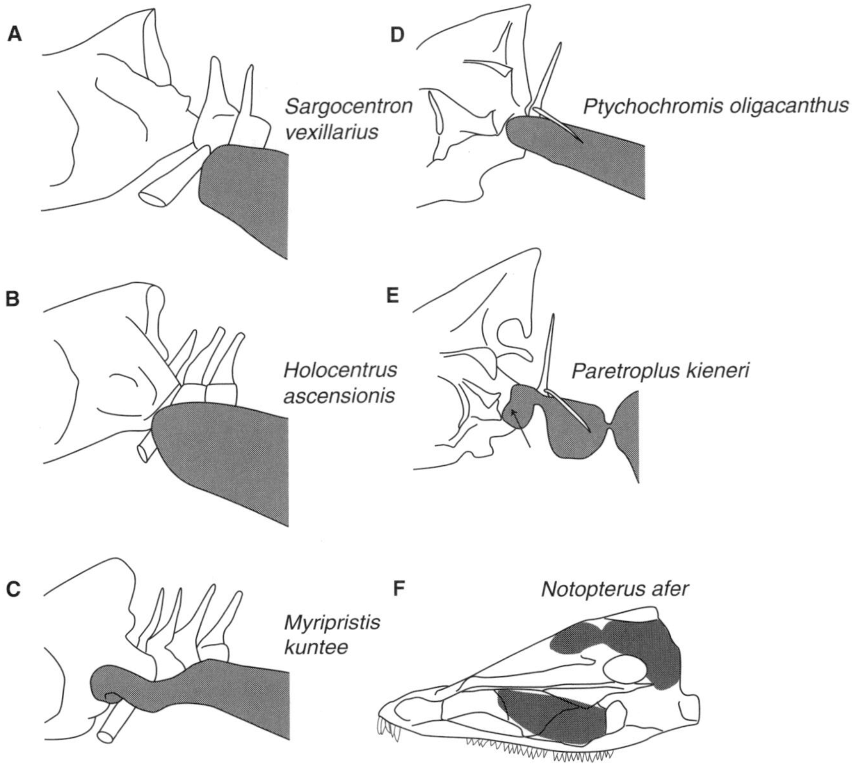

FIGURE 4.1. Examples of rostral swimbladder modifications. Swimbladders are indicated by shaded areas in **A–E**. In Holocentrids (**A–C**), rostral extensions of the swimbladder vary from short extensions that do not contact the cranium (**A**) to intimate assocations with the saccular capsule (**C**). In specialized cichlids (**D, E**), some taxa have rostral swimbladder extensions that contact the skull (**D**), and some taxa have a rostral swimbladder extension that enters the cranium (arrow in **E**) and expands into an intracranial auditory bulla immediately adjacent to perilymphatic spaces (not pictured). In notopteroids (**F**), cranial bullae are present (shaded areas), but these are generally separated from perilymphatic spaces by thick bones (see text). (**A–C** redrawn after Nelson [1955], with permission; **D, E** redrawn after Stiassny et al. [2001], with permission; **F** redrawn after Greenwood [1963], with permission.)

to catalog some of the lesser known evolutionary specializations. Because the Weberian apparatus, found in otophysan fishes, is the most well known hearing specialization, the evolutionary history and possible origins of this linkage to the inner ear are discussed in some detail.

1.1 Types of Stimuli Available to the Inner Ear

To properly appreciate the specializations that alter inner ear function in fishes, it is necessary to understand the stimulus field that surrounds a moving or sound-producing object in an aqueous medium (see also Webb, Montgomery, and Mogdans, Chapter 5; Sand and Bleckmann, Chapter 6). This is a dauntingly complex task, but has been handled in detail by Kalmijn (1988, 1989) and Rogers and Cox (1988). When an object moves through water, the medium in front of the object is pushed aside and water is drawn into the low-pressure region created by the trailing edge of the object. If the object vibrates with enough force, it will create slight compressions of the medium, and as these rarify, they will propagate outward as the familiar sound wave (a propagating oscillation of pressure). Close to the source, however, the bulk flow of water created by the advancing surface of the source dwarfs these pressure fluctuations, and the energy of the propagating wave is weak in comparison. These bulk movements attenuate rapidly with distance, so the relationship soon inverts, and the hydrodynamic flow becomes negligible relative to the pressure fluctuations of the propagating wave. Close to the source, the stimulus field is dominated by hydrodynamic flow and large particle motions. Farther from the source, it is dominated by fluctuations in ambient pressure. These two regions are referred to as the near- and farfields, respectively. In principle, all of these forces extend outward from the source infinitely, so the pressure wave and the hydrodynamic flow are present in both the near- and farfield, but the energy of the incompressible flow attenuates so rapidly with distance that it quickly becomes negligible when considered in the context of sensory thresholds and ambient background noise levels. The boundary between the near- and farfields has been described in measures of source wavelengths; hence long-wavelength (low-frequency) sources have a relatively more extensive nearfield. Further, given the speed at which most biological systems move or create disturbances, low-frequency sources have potentially greater relevance to the lives of most fishes, particularly if pressure cannot be detected. Thus the auditory and hydrodynamic scene important to most fishes is dominated by large hydrodynamic effects surrounding low-frequency sources and limited acoustic effects at greater distances. High-frequency sources can be detected only if the pressure field can be sensed directly. Unfortunately for them, most fishes are only capable of detecting bulk hydrodynamic flows and steep pressure gradients and are unable to sense pressure directly (see Section 1.2).

Many derived taxa, on the other hand, have evolved the ability to exploit the acoustic-pressure fields surrounding higher frequency sources. The ability to detect pressure provides fishes with the potential to detect weaker sources at all distances, and to detect equal energy sources at greater distances, both because of increased sensitivity and less rapid attenuation of pressure with increasing distance from the source. Pressure sensitivity also allows fish to detect higher frequency sources that produce less intense or more spatially limited hydrodynamic effects. Individually, hydrodynamic and acoustic fields both present challenges for source localization (see later), but if both types of energy can be sensed and compared, it is possible that fishes could use this comparison to compute source location and distance. Finally, because shallow waters prevent the propagation of low-frequency sounds (Rogers and Cox 1988), fishes living in the shallows may gain an advantage if they can extend the frequency range they use for prey or predator detection and communication. As we show, the ability to sense pressure has evolved repeatedly within teleosts, which makes a strong argument for adaptive value of sound detection.

1.2 Primitive Transduction Mechanisms

The fundamental sensor of the ear and lateral line system is a directionally sensitive hair cell (Jørgensen 1989). This is an excitable cell whose membrane potential changes in response to deflection of an apical bundle of a single cilium (kinocilium) and multiple microvilli (stereovilli). The physical arrangement of this ciliary bundle gives the cell directional response properties such that the changes in membrane potential are proportional to the displacement along a single axis (Hudspeth and Corey 1977). This basic displacement sensitive unit is exploited for several sensory modalities, each distinguished by the mechanical apparatus interposed between the hair cell bundle and the external milieu (the "linkages of the vertebrate ear"). In its most familiar (but certainly *not* primitive) arrangement, the hair cells rest on a flexible membrane within the mammalian cochlea. Fluctuations in ambient pressure establish a pressure differential between the outer and middle ears, causing the tympanic membrane to oscillate with the changes in pressure. These oscillations are transmitted to the ear by a series of bony levers that terminate on a flexible window in the labyrinth. The push–pull movements of the ossicles set the perilymph into motion, ultimately resulting in displacements of the basilar membrane upon which the hair cells sit. These displacements create mechanical shearing and/or hydrodynamic forces that displace the apical bundles of the hair cells (Yost 1994). This familiar story of pressure-to-motion transduction is the most well known means of diversifying hair cell function, but there are many other similar transductions performed by the vertebrate octavolateralis systems (Table 4.1).

The earliest vertebrates had no such complicated transduction mechanism, nor do most extant anamniotic vertebrates. In these animals, the apical bundles of inner ear hair cells are linked to a calcareous mass, the otolith. The bundles are enmeshed or terminate just beneath a tectorial membrane that envelops the otolith. Fishes have three otolithic macular organs: the utriculus, sacculus, and lagena. A fourth macular sensor, the macula neglecta, is typically present as well, although it is not overlain by an otolith and its function remains obscure in most, if not all, instances. As with terrestrial vestibular systems, the otolithic organs are responsive to linear accelerations due to the inertial difference between the dense otolith and body tissues. Most fishes are of approximately the same density as the water in which they live, so that when in a hydrodynamic field, the fish's body moves with the motions of the medium. The phase difference in displacement of the dense otolith relative to the fish's body (the underlying macula) displaces the ciliary bundles and modulates membrane potentials, giving rise to nerve impulses in the octaval cranial nerve. Thus for fishes, at least primitively, audition (i.e., inner ear function) is the detection of motion and the hydrodynamic field surrounding sound-producing objects, rather than the

TABLE 4.1. Summary of teleost fish families with morphological specializations capable of enhancing hearing.

	Family	Transduction mechanism	Enhancement
		Osteoglossomorpha	
Notopteroidei	Hiodontidae	Swimbladder extension (5)	Bandwidth frequency resolution?
	Notopteridae	Swimbladder extension (5)	Bandwidth resolution
	Mormyridae	Otic bullae (6)	Sensitivity; bandwidth
		Elopomorpha	
	Megalopidae	Swimbladder extension→ extracranial bullae (4)	???
	Elopidae	Swimbladder extension (4)	???
	Albulidae	Swimbladder extension (4)	???
		Clupeomorpha	
Clupeiformes		Prootic and pterotic bullae (3)	Sensitivity; bandwidth
		Prootic and pterotic bullae; recessus lateralis (3)	Pressure-sensitive lateral line?
			Self-induced motion sensing?
		Ostariophysi	
Gonorynchiformes		Protoweberian coupling? (2)	???
Otophysi	All members	Weberian apparatus (1)	Sensitivity; bandwidth Frequency resolution? Source localization?
		Acanthomorpha	
Paracanthopterygii	Gadidae	Swimbladder extension (?)	Bandwidth; sensitivity? Source localization?
Beryciformes	Holocentridae	Swimbladder extension→ otic bullae (8)	Sensitivity; bandwidth; frequency resolution
Perciformes	Anabantoides*	Suprabranchial chamber (9)	Sensitivity; bandwidth
	Carangidae	Auditory bullae (9)	???
	Centropomidae	Swimbladder extension (9)	???
	Chaetodontidae	Laterophysic connection (9)	Pressure-sensitive lateral line? Source localization?
	Cichlidae (multiple subgroups)	Swimbladder extension→ otic bullae (9)	???

TABLE 4.1. (Continued).

Family	Transduction mechanism	Enhancement
Ephippidae	Swimbladder extension (9)	???
Gerridae	Swimbladder extension (9)	???
Haemulidae	Swimbladder extension (9)	Pressure sensitivity
Kuhlidae	Swimbladder extension (9)	???
Lactanidae	Swimbladder extension (9)	???
Moronidae	Swimbladder extension (9)	???
Mullidae	Transverse lateral line linkage (9)	Self-induced motion sensing?
Nematistiidae	Swimbladder extension (9)	???
Percichthydae	Swimbladder extension (9)	???
Polyprionidae	Swimbladder extension (9)	???
Priacanthidae	Swimbladder extension (9)	???
Scombridae	Swimbladder extension (9)	???
Scaenidae	Swimbladder extension (9)	???
Sinipercidae	Swimbladder extension (9)	???

If a particular functional enhancement has been demonstrated experimentally, it is listed in the final column; inferred, but not demonstrated functional enhancements are qualified with question marks. Within the Megalopidae, Cichlidae, and Holocentridae, individual species may be arranged in a morphocline (indicated by arrows) of increasingly longer rostral swim bladder extensions that may expand into bullae in or near the cranial cavity. Numbers in parentheses refer to position of characters plotted on Fig. 4.1. Anabantoidei is marked with an asterisk to note that only two families within this perciform suborder have been examined for hearing specialization, but most families in the suborder share a similar morphology. See text for references.

acoustic pressure (sound) itself (Kalmijn 1989). Given sufficient intensity, the particle displacements that compose the propagating sound wave (in the farfield) may also accelerate the fish's body, but without additional transduction mechanisms, audition is restricted to this inertial mode and the sound wave itself is not detectable as a pressure fluctuation.

The particle motions underlying the farfield pressure fluctuations may often be too weak to engage the inertial sense, and as described in the preceding text, there appears to be a clear selective advantage to the ability to sense pressure, per se. In fishes, this is most commonly achieved by linking the inertial inner ear to a gas-filled cavity like the swimbladder. As described by Boyle's law, these gas-filled chambers change volume in response to ambient pressure and the corresponding motion of the chamber wall becomes the proximate source for the displacement sensitive inner ear. As we show in the following sections, the history of the vertebrates includes repeated exploitation of the phenomenon Boyle observed.

1.3 The Problem of Sound Source Localization

The primary function of any complex sensory system is to represent the structure of the outside world. For auditory systems, this task requires the detection and analysis of sound producing and reflecting sources, and their relationship to ambient background noise. Fay and Popper (2000) have argued that such auditory scene analysis (Bregman 1990) is the primary function of audition and that the physical constraints and limitations on scene analysis have been the driving selective forces in the evolution of hearing. Directional hearing and sound source localization have both received a great deal of attention, and spatial hearing abilities and mechanisms are probably the most important and conceptually challenging aspects of research in fish hearing. There is a large theoretical and experimental literature on this subject and a detailed review is beyond the scope of this chapter (see Schuijf 1981; Popper and Fay 1997; Fay and Popper 2000; Sand and Bleckmann, Chapter 6).

The mechanism of source localization is a thorny subject for a variety of reasons. First, although the fish auditory system is capable of coding the axis of particle motion in three dimensions (Sand 1974; Fay 1984; Lu et al. 1996; Fay and Edds-Walton 1997), this vector alternatively points toward and away from the direction of the source, leading to an ambiguity of 180°. Second, interaural cues cannot be used for localization as they are in aerial hearing because fish flesh is relatively acoustically transparent (no shadowing), and the greater speed of sound (~4.5 times the speed in air) minimizes interaural differences in arrival times. Lastly, as we describe later, the detection of the pressure component surrounding an acoustic source is typically achieved by coupling the mechanosensory inner ear to an internal gas-filled cavity, the proximate source to the ear is actually inside the fish's body and may not provide information about the location of the original source. One theoretical model suggests that fish could compute the direction of source propagation, however, by comparing the phase of particle motion with the phase of the pressure signal (Schuijf and Buwalda 1975; Schuijf 1981). As an overly simplistic example, an approaching source on the left and a receding source on the right produce the same axis of particle motion, and cannot be distinguished based on that information alone. However, an approaching source from the right would produce an increase in pressure in phase with particle motion to the left, and the receding source just the opposite. Comparison of the timing of the maximal particle displacement and pressure can resolve the 180° ambiguity inherent in the particle motion, despite the fundamentally nonspatial nature of the pressure sense. There is good evidence that pressure sensitive fish (cod) make such a comparison in discriminating sources from opposing directions (Schuijf and Buwalda 1975; Buwalda et al. 1983), but this is not yet a proven mechanism of source localization.

Unfortunately, resolving the axis of acoustic particle motion does not automatically lead to knowledge of source location. The hydrodynamic nearfield is spatially complex and the vector of particle motion only points toward the source for monopole sources or in limited regions of the field surrounding dipole or higher order sources (Kalmijn 1989). Without knowledge of the nature of the source and simultaneous sampling from multiple positions, knowledge of the vector of particle motion may not be sufficient to compute source location. The existing behavioral data (Popper et al. 1973; Chapman and Sand 1974; Schuijf and Siemelink 1974; Schuijf and Buwalda 1975; Hawkins and Sand 1977; Buwalda et al. 1983; Schuijf and Hawkins 1983) demonstrate that fish are able to use audition for some spatial analysis of the auditory scene. It is far from clear, however, that fish can compute the location of a source directly, or if they must use general information about the direction of the source to follow algorithms that lead them to it (Kalmijn 1997). Still, it appears that comparisons of pressure and particle motion may be important, so the detection of pressure is a specialization that potentially enhances spatial hearing.

The comparison of pressure and displacement may also be important in the disambiguation of intensity and source distance. In the farfield of a propagating sound source, the pressure and particle velocity are in phase and decline at the same rate with distance. Close to the source, however, particle motion is much greater in magnitude than pressure and declines more rapidly (Siler 1969; Kalmijn 1988). Thus the pressure to velocity ratio increases at a constant rate with distance from the source, independent of source intensity. Schuijf and Hawkins (1983) argued that fish could compute this ratio by comparison of pressure and displacement sensitivity and use this computation to discriminate source distances without ambiguity from source intensity. As with the discrimination of source location, sensitivity to pressure is potentially an enhancement of a fish's ability to determine the distance of a sound source.

2. Specializations that Enhance Hearing

Most known hearing specializations involve modification of the swimbladder. Fluctuations in ambient pressure cause the swimbladder to oscillate in volume, and the wall of the swimbladder pulsates. To provide hearing enhancement, this motion must be transmitted to the inner ear. Earlier authors argued that swimbladder oscillations could be transmitted directly through the animal's body tissues to the ear (Poggendorf 1952; van Bergeijk 1964; Alexander 1962; Blaxter 1981). This view was strengthened by Chapman and Sand (1974) finding that a small balloon placed underneath a fish without a swimbladder (plaice: Pleuronectidae) increased sensitivity and high-frequency hearing abilities. Studies of cod (a non-otophysan) also suggested unmodified swimbladders enhance hearing (e.g., Sand and Enger 1973), but many gadids also have modifications of the swimbladder, including rostral extensions (horns) that bring the bladder wall into proximity of the ear (see later). The auditory role of an unmodified swimbladder, however, is still mostly a matter of speculation and recent experimental evidence shows that deflation of unmodified swimbladders does not change hearing sensitivity in either *Opsanus tau,* the oyster toadfish or *Trichogaster trichopterus*, the blue gourami (Yan et al. 2000), or in *Padogobius martensii* (Lugli et al. 2003). Interestingly however, Lugli et al. (2003) did show that *P. martensii*, which has an unmodified swimbladder, was 5–10 dB more sensitive than *Gobius nigricans* (the arno goby), which does not have a swimbladder at all. The role of "unspecialized" swimbladders in hearing is probably minimal, although more experimental studies of unspecialized fishes are still needed, also focusing on the possibility of sound conduction by the spinal cord or other body tissues. If unspecialized swimbladders can function to enhance hearing, we should consider the possibility that some aspect of the evolution of lungs and swimbladders may have been affected by evolutionary pressures related to hearing.

2.1 Rostral Extensions of the Swimbladder

The simplest hearing enhancement is a forward extension or translation of the swimbladder toward the otic capsule (Fig. 4.1). By extending the swimbladder toward the cranium, the oscillations of the swimbladder wall may be transmitted to the otolithic organs. Although the holocentrids (i.e., squirrelfish) are the best known example, rostral swimbladder extensions are present in many teleost taxa, suggesting that this type of hearing specialization has evolved repeatedly within fishes (see Table 4.1 and Fig. 4.2).

2.1.1 Holocentridae

In squirrelfishes, Nelson (1955) described a trend of increasingly specialized swimbladders within in the family Holocentridae. The genera *Sargocentron* (formerly *Adioryx*) and *Flammeo* have unspecialized swimbladders, while some species of *Holocentrus* have paired rostral extensions that approach the skull (Fig. 4.2). The Myripristinae have rostral swimbladder extensions that abut the caudal wall of the braincase (occiput) and enlarge to fill a concavity in the occiput and inner ear capsule (also described briefly by Starks 1908). Experimental studies have shown that sensory performance does indeed correlate with this morphological trend: *Sargocentron* has an essentially primitive sense of hearing, with low sensitivity to acoustic pressure, and is unable to hear above 800 Hz. *Holocentrus*, with swimbladder extensions that approach, but do not contact the skull, can hear sound up to around 1000 Hz and has better sensitivity at all frequencies than *Sargocentron* (Tavolga and Wodinsky 1963; Coombs and Popper 1979). Finally, *Myripristis*, with an intimate connection between the swimbladder extension and the auditory capsule, can respond to sounds as high as 3000 Hz and is about 10–25 dB more sensitive than *Sargocentron* at comparable frequencies (Coombs and Popper 1979). Thus this family shows hearing enhancements in both sensitivity and hearing bandwidth, particularly within in the Myripristinae.

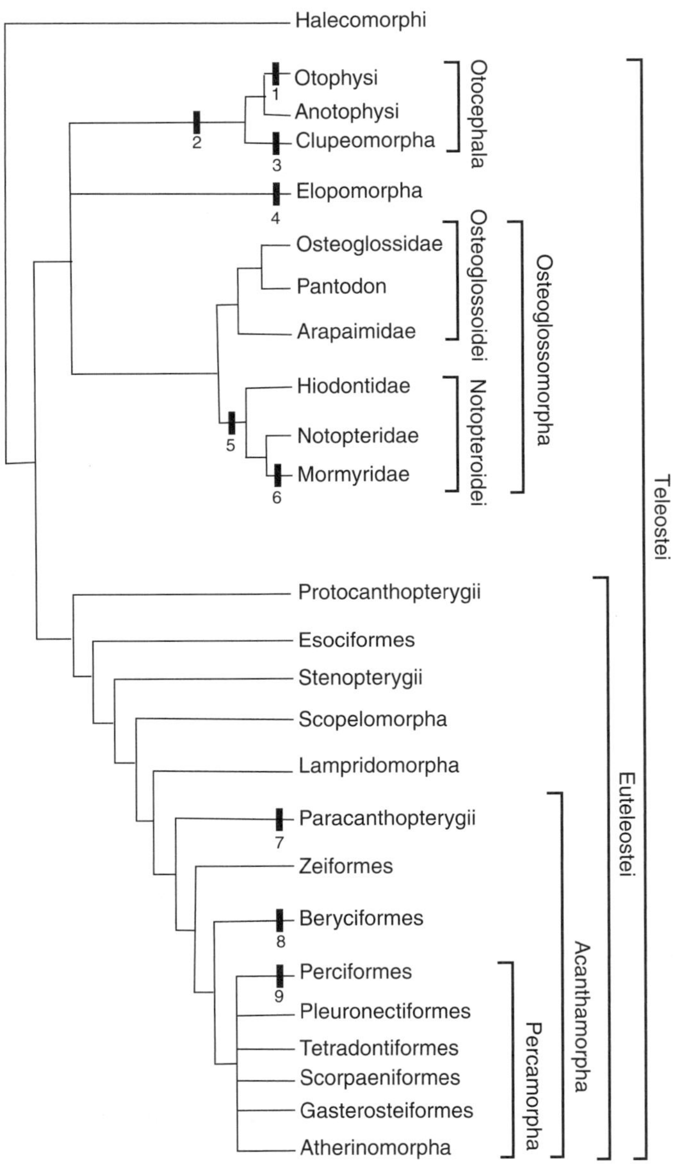

FIGURE 4.2. The distribution of hearing enhancing specializations plotted on a conventional summary of teleost relationships (based on Greenwood et al. 1966; Johnson and Patterson 1996; and Lecointre and Nelson 1996). Numbers below the bars for each node indicate the distribution of the following specializations: *1*, Weberian apparatus; *2*, modifications of the rostral vertebral elements, swimbladder morphology, and labyrinth shape and position; *3*, prootic and pterotic bullae, recessus lateralis, subcerebral canal; *4*, swimbladder extension (multiple instances?), extracranial bullae; *5*, swimbladder extension; *6*, otic bullae; *7*, swimbladder extension; *8*, swimbladder extension, otic bullae; *9*, swimbladder extension (multiple instances), subcerebral canal, laterophysic connection.

Interestingly, Popper et al. (1973) showed that two species of *Myripristis* can discriminate between sound sources presented from different locations, and they appear to have some awareness of distance that is independent of the pressure level of the source. Subjects in their experiments would only orient to sources that were within 2 m, although control experiments and comparisons with Tavolga and Wodinsky (1963) and Coombs and Popper (1979) demonstrate that this limitation was not a function of sensitivity (based on intensity thresholds, this species would be capable of detecting the sounds at much greater distances). The fish also discriminated the location of sources presented at 1.6 m, even if the sound was attenuated to match the pressure level the subject would receive from the training stimulus presented more than 10 m away. This suggests that the analysis of source location is based on more than pressure alone, such as a comparison of pressure and particle velocity. Unfortunately *Sargocentron* was not tested in a similar experiment, but this comparison (with a species that is insensitive to pressure) would be a good test case of mechanisms of source localization based on comparisons of pressure and particle velocity.

2.1.2 Gadidae

The only other fish with rostral swimbladder horns that have been studied physiologically is the Atlantic cod (*Gadus morhua*). In this species, as in many other gadids, the swimbladder is modified, with rostral extensions that approach the skull and a thickened tunica externa. The particulars of shape and swimbladder morphology vary among gadid species (Hawkins 1986). In some cases, these swimbladder modifications may be relevant to sound production, as some gadids produce sounds by the drumming of their swimbladder (Hawkins 1986). In G*adus morrhua*, however, the rostral protuberances impart some pressure sensitivity and increased auditory abilities. Cod are sensitive to both pressure and particle motion, depending on source frequency. In behavioral tests, cod do not respond to 50-Hz stimuli unless the source is close enough to present high particle velocity motions, regardless of the pressure (Chapman and Hawkins 1973). Their sensitivity to low-frequency sources declines in proportion to increasing distance. Responses to sources of 100 Hz and above, on the other hand, are generally independent of distance and decline with decreasing pressure values. Cod appear to be able to use their pressure sensitivity to detect sources up to around 600 Hz (Chapman and Hawkins 1973; Offut 1974). This pressure sensitivity is dependent on the swimbladder; deflation of the swimbladder decreases electrophysiological measures of sensitivity (Sand and Enger 1973) for sources above 100 Hz. Swimbladder deflation has its greatest effects between 200 and 400 Hz, and responses to sounds above 400 Hz are drastically reduced without an intact swimbladder. In contrast to holocentrids and otophysans, swimbladder specializations do not provide a large increase in hearing bandwidth, perhaps increasing the upper limit only from 100 Hz to 600 Hz (Chapman and Hawkins 1973; Sand and Enger 1973). In general, however, it is clear that the rostral extension of the swimbladder in cod provides the ability to detect pressure, increasing overall auditory sensitivity, particularly above 100 Hz.

Cod have also been the subjects for the most extensive series of behavioral experiments on spatial hearing to date (Chapman and Johnstone 1974; Schuijf and Siemelink 1974; Schuijf and Buwalda 1975; Hawkins and Sand 1977; Buwalda et al. 1983; Schuijf and Hawkins 1983). These studies have provided the best evidence that the resolution of the 180° ambiguity (Schuijf and Buwalda 1975; Buwalda et al. 1983) and the disambiguation of intensity and source distance (Schuijf and Hawkins 1983) depend on comparison of the phase relationships between pressure and velocity. Because the cod's ability to sense pressure is dependent (or enhanced) by specialization of the swimbladder, enhanced directional hearing may have been one of the selective advantages leading to this specialization.

2.1.3 Cichlidae

Swimbladder specializations are found in several groups of cichlids (Cichocki 1976; Sparks 2001; Stiassny et al. 2001). The most well known case is in the Malagasy and Asian taxa Etroplinae and Ptychrominae. In *Paratilapia*, the putative sister taxon to Etroplinae + Ptychrominae (Stiassny et al. 2001), the swimbladder has rostral diverticula that approach the skull but do not contact the occiput (Fig. 4.1). In the Ptychrominae, these extensions do contact the skull and generally fill large concavities in the exoccipital bones. In the Etroplinae (e.g., *Paretroplus*), the swimbladder has a thickened and tough tunica externa and extends through a large foramen in the exoccipital and expands into a large bulla in direct contact with the labyrinth (Sparks 2001). The morphological variation in this group is highly reminiscent of the variation among holocentrid subfamilies described by Nelson (1955) (Section 2.1.2), and it would be surprising if it did not include similar functional changes. Similar relationships between the swimbladder and the occiput and/or otic capsule are also seen in some South American cichlids (Cichocki 1976), suggesting repeated convergent evolution of swimbladder specializations within the Cichlidae.

The hearing capabilities of a few cichlids, the oscar (*Astronotus ocellatus*), tilapia (*Tilapia macrocephala*), and the Princess of Burundi (*Neolamprologus birchardi*) have been studied experimentally, but these fishes all have unmodified swimbladders and apparently no hearing specializations. The oscar is sensitive within a limited bandwidth, below 900 Hz (Yan and Popper 1992), is sensitive to particle acceleration rather than pressure (Lu et al. 1996) and has unexceptional intensity discriminatory abilities (Yan and Popper 1993). *Neolamprologus* shows physiological responses to sounds as high as 2 kHz, but sensitivity decreases greatly above 250 Hz (Ladich and Wysocki 2003). In terms of stimulus pressure, *Neolamprologus* is at least 20 dB less sensitive than goldfish, and this difference is even greater at high frequencies (~60 dB at 1000 Hz). *Tilapia* is similar to the other two cichlid species in sensitivity and may also have poor frequency resolution: Tavolga (1974) showed that detection of 500-Hz tones is not impaired selectively with the frequency of background noise. That is, all frequency background noise has similar masking effects on the signal tone, suggesting that the auditory system has broadly tuned frequency filters. Clearly, comparisons of sensitivity and frequency response with more the specialized species (e.g., *Paretroplus*) should be the subject of future studies.

Cichids also present a good test case for the effects of unspecialized swimbladders in hearing capabilities. Riverine cichlids from West Africa, such as species within the genus *Steatocranus*, have reduced swimbladders or lack a swimbladder altogether (*S. tinanti, S. carassius and S. gibbiceps*, C.B.B., unpublished observations). Any differences in hearing ability between these species and *Astronotus* might be due to the absence or reduction of an otherwise unspecialized swimbladder.

2.1.4 Notopteroidei

Rostral swimbladder extensions also occur within in the Notopteroidei (Osteoglossomorpha). The entire suborder is characterized by a rostral extension of the swimbladder that approaches the inner ear. In the basal genus *Hiodon*, rostral extensions of the swimbladder enter the cranium and terminate at thin membranous openings separating the bladder from perilymphatic space, which is associated with the utricle (Greenwood 1963). In the Notopteridae (Fig. 4.1), rather large extensions of the anterior chamber of the swimbladder extend into the skull and lie lateral to the large saccular capsules (Coombs and Popper 1982a). Finally, in the Mormyridae, anterior diverticulae of the swimbladder create an air-filled bulla that invades the otic capsule, and their connections with the swimbladder are lost during ontogeny. These otic bullae are nestled within the semicircular canals and are intimately connected to the dorsolateral surface of the saccule (von Heusinger 1826; von Frisch 1938; Stipetič 1939).

There are limited data on the hearing abilities of arawana, *Osteoglossum bicirrhosum* (Coombs and Popper 1980, 1982b), but a complete audiogram has not been published. Using a simultaneous masking paradigm (in which a conditioned tone is played at the same time as a masking signal of overlapping frequencies), Coombs and Popper (1982b) showed that the psychophysical tuning curve of this species is generally similar to those of the more specialized *Notopterus*, but less sharply tuned. In *Notopterus*, which has an intimate relationship between the swimbladder and the ear, the effectiveness of a masker is strongly frequency dependent. That is, as the frequency of the masker differs from the signal (both 300 and 500 Hz were tested), the fish's ability to detect the signal improves. Its ability to detect a 300-Hz tone is most impaired in the presence of a 300-Hz masker, and the impairment declines as the frequency of the masker differs from 300 Hz. The tuning curve of *O. bicirrhosum* is less sharply tuned, i.e., masker effectiveness declines more gradually as the frequency seperation between signal and masker increases. *Gnathonemus petersii,* the elephantnose mormyrid (with a more advanced specialization than *Notopterus*), has an even sharper tuning curves than *Notopterus* (McCormick and Popper 1984). This suggests that the swimbladder–ear associations in Notopteroidei are specializations that enhance the frequency selectivity of the auditory system.

Judged on the audiogram alone, however, the hearing enhancement in *Notopterus* is modest. It is capable of detecting sounds between 100 and 1000 Hz, with a sharp decline in sensitivity above the most sensitive band of 500 Hz (Coombs and Popper 1982a). It is less sensitive than mormyrids, otophysans, or *Myripristus* (see Section 2.1.1) at all frequencies, and is very similar to *Sargocentron* in hearing sensitivity, despite the fact that *Notopterus* has an intimate swimbladder–inner ear connection and *Sargocentron* does not. Coombs and Popper (1982a) suggest that their measurements may not accurately reflect the true sensitivity of *Notopterus* because there was a large variability in responses and their behavioral measure may have probed two different sensory channels (pressure and displacement) simultaneously. That is, they may have been measuring two thresholds, a more sensitive pressure threshold and less acute displacement sensitivity. Without more conclusive evidence that this was in fact the case, they were forced to present the averages of an apparently bimodal data set, thus underestimating the sensitivity of *Notopterus* audition. Still, only a few of the individual thresholds they measured were within the sensitivity of goldfish or *Myripristis*, so it is possible that the swimbladder specialization of *Notopterus* does not greatly enhance sensitivity or bandwidth. It does, however, appear to provide *Notopterus* with a greater frequency selectivity than *Osteoglossum* (Coombs and Popper 1982b), and its possible role in source localization has not been tested.

The hearing enhancement provided by the otic bullae in Mormyridae is clearer. Behavioral audiograms of *Gnathonemus petersii* (McCormick and Popper 1984; Fletcher and Crawford 2001) and *Pollimyrus adspersus* (Fletcher and Crawford 2001), and physiological measures in *Brienomyrus brachyistius* (Yan and Curtsinger 2000) have measured auditory abilities in these species. All three species have a relatively wide frequency bandwidth, with demonstrated sensitivity between 100 and 3000 Hz (Yan and Curtsinger (2000) showed responses in *B. brachyistius* up to 4000 Hz). Mormyrids are less sensitive (~10 dB) than goldfish at frequencies below 1000 Hz (McCormick and Popper 1984). Sound sensitivity in mormyrids is clearly enhanced by the presence of otic bullae, however. When these bullae were deflated (Yan and Curtsinger 2000) or filled with saline (Fletcher and Crawford 2001), detection thresholds increased at all frequencies tested, by 5–15 dB in *B. brachyistius* and 10–40 dB in *P. adspersus* and *G. petersii*. Interestingly, although the audiograms became more uniform with regard to frequency, the animals still displayed auditory sensitivity throughout the intact frequency range. In the case of *Pollimyrus*, at least, this sensitive hearing is part of a complex acoustic communication system (Crawford et al. 1997), but the function of such acute hearing in other species is not known. Cain (1995) has shown that *G. petersii* makes mistakes navigating familiar through passages if the water depth is changed, suggesting that these fish use hydrostatic pressures in the perception and memory of their spatial surroundings (Cain et al. 1994). It is possible that the auditory bullae could play a role in the detection of hydrostatic pressure (but see Blaxter 1978).

2.1.5 Elopomorpha

Among the elopomorph fishes, species in the family Megalopidae (tarpons) have precoelomic extensions of the swimbladder that enter bone-encased periotic chambers and expand into three pairs of bullae that lie external to the lateral skull wall. These bullae are strictly extracranial, and make no direct contact with the inner ear or perilymphatic spaces (Greenwood 1970), but they are in close proximity nonetheless. In the tarpon *Megalops*, the most anterior pair of diverticula extend into a groove in the prootic bone, underneath the anterior portion of the saccule. These diverticula extend medially and nearly meet at the midline but are separated by a thin bony wall. There are two pairs of diverticula that extend from the more posterior portion of each periotic bulla. The ventromedial pair extends under the basioccipital and fuse in the midline. This transverse gas-filled space underlies the posterior portion of the saccule. Despite this spatial proximity to the saccule, all of these gas-filled cavities are separated from the otic organs by thick cranial bones and never contact peri- or endolymphatic spaces (Greenwood 1970).

The genus *Tarpon* exhibits a similar arrangement, but the medial walls of the periotic chamber have a thinned flexible portion in the basioccipital and (sometimes) the prootic bones (Greenwood 1970). This thinning could allow for transmission of vibrations from the periotic bullae to the underlying sacculi. In both *Tarpon* and *Megalops*, the gas-filled spaces are not in direct contact with perilymphatic spaces of the inner ear, so in the absence of experimental evidence, it is uncertain if the rostral swimbladder extensions have effects on hearing capabilities. As in other taxa possessing rostral swimbladder extensions without direct contact with the ear (e.g., *Holocentrus* or *Gadus*), it is likely that the mere proximity of a gas-filled cavity can alter inner ear function, but experimental verification is needed. The difference in intervening bone thickness between *Megalops* and *Tarpon* is also likely to result in differences in hearing capability between these genera.

Although only the Megalopidae possess intricate periotic bullae, some species within the family Elopidae also have rostral extensions of the swimbladder that approach the ear. For example, in *Elops lacerta* (West African ladyfish), two narrow finger-like projections extend forward from the anterior part of the swimbladder to a point just behind the saccular swelling on the basioccipital. Each projection lies in a shallow groove in the basioccipital. A peritoneal sheath covers each diverticulum but the tissue is not thicker or distinguishable from the peritoneal covering of the swimbladder as it is in the Megalopidae (Greenwood 1970). In other species, such as the bonefish, *Albula vulpes* (Albulidae), two shorter anterior extensions protrude from the swimbladder, but do not contact the skull (Greenwood 1970).

Unfortunately, there are no experimental data on hearing abilities in any elopomorph species. The anatomical data suggest hearing enhancements in several taxa, with limited enhancements in the Elopidae and successively greater enhancements to hearing within Megalopidae.

2.1.6 Other Swimbladder Specializations

The preceding sections review only those taxa for which we have either detailed anatomical descriptions or experimental evidence of hearing specializations.

It is likely that these examples represent only a subset of teleost taxa that possess hearing specializations. Within the Percomorpha, many taxa have rostral swimbladder extensions and may have independently evolved enhanced hearing, as shown in Table 4.1. In some instances, e.g., Haemulidae (grunts), descriptions of rostral swimbladder extensions (Johnson 1980) can be related to demonstrations that some species of grunts are sensitive to the pressure component of 400-Hz sound stimuli (Cahn et al. 1969). In other cases, morphological observations have not yet been augmented by behavioral or physiological measures of sensory performance, including the families Nematistiidae (Starks 1908); Sparidae (Dijkgraaf 1952, cited in Schellart and Popper 1992); Centropomidae (*Lates*: Katayama 1959), Ephippidae (Herre and Montalban 1927), Gerridae (Green 1971), Kuhliidae (Gosline, 1966), Lactariidae (Leis 1994), Moronidae (*Lateolabrax*: Katayama, 1959), Percicthyidae (MacDonald, 1978), Polyprionidae (*Stereolepis*: Katayama, 1959), Priacanthidae (Starnes 1988), Sciaenidae (Chao 1986; Sasaki 1989; Ramcharitar 2002), Scombridae (*Gasterochisma*: Collete and Nauen 1983) and Sillanginidae (McKay 1985), Sinipercidae (Katayama 1959). Swimbladder extensions have a limited distribution within many of these groups, so regardless of uncertainties in percamorph relationships, each listing in Table 4.1 represents a potentially independent evolution of hearing enhancement.

2.2 Auditory Bullae in Clupeomorpha

In the Clupeomorpha bilaterally paired diverticula of the swimbladder penetrate the exoccipital and expand into gas-filled chambers in the lateral walls of the braincase. Each of these extensions expands into two bullae within the prootic and the pterotic bones respectively (Grande 1985), but retain a pneumatic connection with each other and with the body of the swimbladder throughout life. Such connections are likely important in regulating pressure within these bullae and/or sensing hydrostatic pressure during large changes in depth (Denton and Blaxter 1976). The prootic bullae housed within the prootic bones and intimately associated with the utricle are diagnostic of Clupeomorpha (Grande and de Pinna; Fig. 4.3). The prootic bulla is most certainly an auditory specialization. The second air filled bulla within the pterootic bones has received less attention from auditory physiologists. It is a lateral expansion of the cranial diverticulum (pneumatic duct) of the swimbladder, situated caudal and lateral to the prootic bullae, within the loop of the horizontal semicircular canal, but tightly bound by the pterotic bones (O'Connell 1955). It is not intimately associated with the macular organs of the inner ear or the recessus lateralis (see Section 2.5.1), but some kind of functional linkage with the prootic bullae can not be excluded. The pterotic bullae are absent in several sprat genera (*Sprattus*, *Clupeonella*, † *Spratticeps*[1], and the Pacific menhaden, *Ethmidium*). All of these are considered advanced Clupeomorphs (Grande 1985). Since the morphological and functional

[1]The dagger symbol (†) is used to denote extinct taxa.

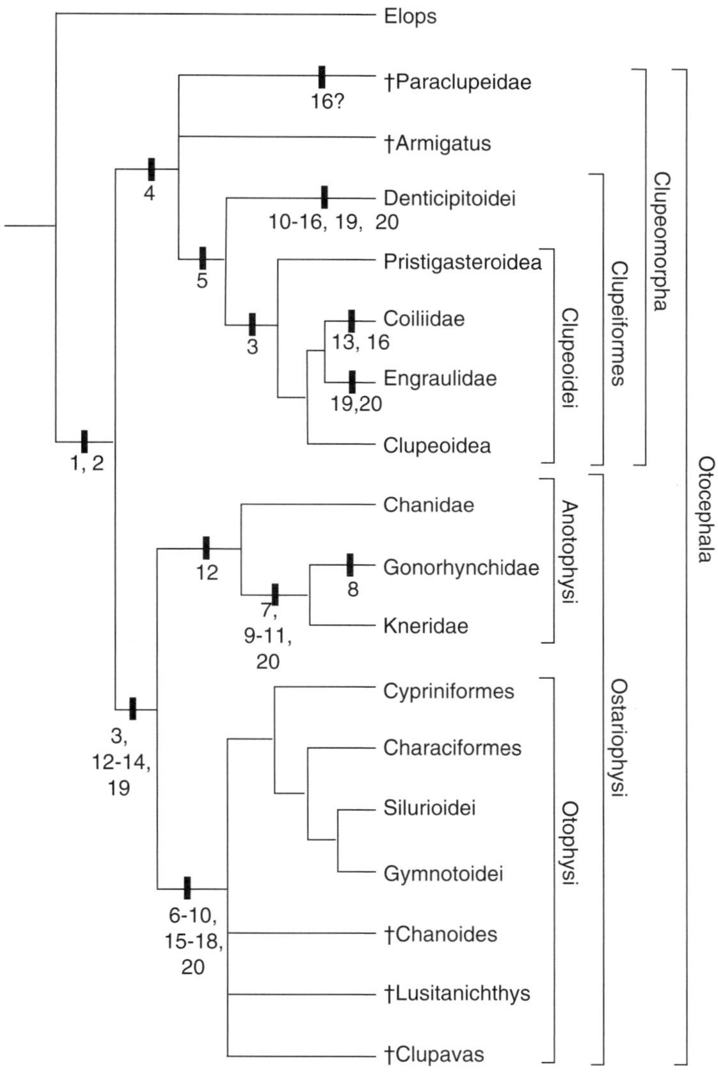

FIGURE 4.3. Character plot from Grande and de Pinna (2004) of anterior vertebral/otophysic characters shared between and among Clupeomorph and Ostariophysan taxa. Numbers below the bars at each note indicate the distribution of the following specializations: *1*, saccular and lagenar otoliths in a posterior and median position (Otocephala); *2*, posterior elongation of labyrinth (Otocephala); *3*, swimbladder with silvery peritoneal tunic covering a least the anterior portion (Clupeoidei + Ostariophysi); *4*, presence of prootic and pterotic bullae (Clupeomorpha); *5*, presence of a recessus lateralis (Clupeiformes); *6*, presence of a sinus impar (Otophysi); *7*, anterior supraneurals forming a joint with associated neural arches: Gonorynchoidei (i.e., Gonorynchidae and Kneriidae + Otophysi); *8*, anterior supraneurals in contact with each other either directly or through cartilage (Gonorynchidae + Otophysi); *9*, expansion of anterior supraneurals (Gonorynchoidei + Otophysi); *10*, expansion and contact of anteriormost neural arches

studies most familiar to hearing scientists (e.g., Best and Gray 1980) describe species without pterotic bullae, such as *Sprattus*, its potential role in hearing remains unknown.

The prootic chamber is subdivided by the bulla membrane (Fig. 4.4), with gas on one side of the membrane and perilymph on the other (the following description is based on the anatomical studies of O'Connell 1955; Denton and Blaxter 1976; and Best and Gray 1980). The utricular recess sits atop this perilymphatic space dorsally. The bulla membrane is also connected to the macular organs of the utricle by a thin elastic thread. There are three distinct macular organs on the floor and ventral portions of the walls of the utricular recess (anterior, medial, and posterior maculae). The otolith sits over the medial and posterior maculae and is connected to the cupula of the anterior macula rostrally (O'Connell 1955). The medial macula is attached to the anterior and posterior walls of the utricular chamber by elastic membranes, with perilymph below and endolymph above. The perilymphatic chamber itself is bipartite, with a large recess below the posterior portion of the utricle and a fenestration (the prootic fenestra) connecting it to the dorsal portion of the prootic bulla (that is, dorsal to the bulla membrane). The perilymphatic chamber beneath utricle communicates with a much larger sinus that terminates laterally in an elastic membrane that forms a medial wall of part of the lateral line canal system (the recessus lateralis; see Section 2.5.1), and surrounds the sacculus medially. The perilymphatic chambers on either side of the cranium are connected medially by a subcerebral canal (Denton and Blaxter 1976). The elastic thread connecting the bulla membrane to the floor of the utricle passes through the prootic fenestra and attaches to the rostral pole of the medial macula. Deflections of the bulla membrane induce motion of the perilymph, deflecting the medial and anterior macular surfaces, which leads, ultimately, to hair cell stimulation (Denton and Blaxter 1976; Best and Gray 1980). The elastic membranes supporting the macula are much stiffer ($\sim 200\times$) than the elastic thread connecting the medial macula

FIGURE 4.3. (Continued) with one another (Pristigasteroidea + Gonorynchoidei + Otophysi); *11*, anteriormost neural arch abuts the back of the skull via the exoccipitals (Pristigasteroidea + Gonorynchoidei); *12*, anterior 1–3, vertebral centra shorter that posterior ones (Pristigasteroidea + Ostariophysi); *13*, first pleural rib attaches to the peritoneal tunic of swimbladder (Pristigasteroidea + Coiliidae); *14*, second pleural rib attached to peritoneal tunic of swimbladder (Pristigasteroidea + Ostariophysi); *15*, presence of tripus (sensu Fink and Fink 1981), consisting of a modified and expanded parapophysis and modified rib of the third centrum which connects to the swimbladder (Otophysi + Pristigasteroidea); *16*, modification and expansion of the parapophysis of third centrum (Pristigasteroidea + Coiliidae + Otophysi); *17*, presence of scaphium (Otophysi); *18*, presence of intercalarium (Otophysi); *19*, constriction of swimbladder into two chambers (Pristigasteroidea + Engrauloidea + Ostariophysi); *20*, fusion of the first two parapophyses to centra (Gonorynchoidei + Otophysi + Pristigasteroides + Engrauloidea). Cladistic relationship follows Grande (1985), Fink and Fink (1981), Grande and Poyato-Ariza (1999) and Di Diario (2002).

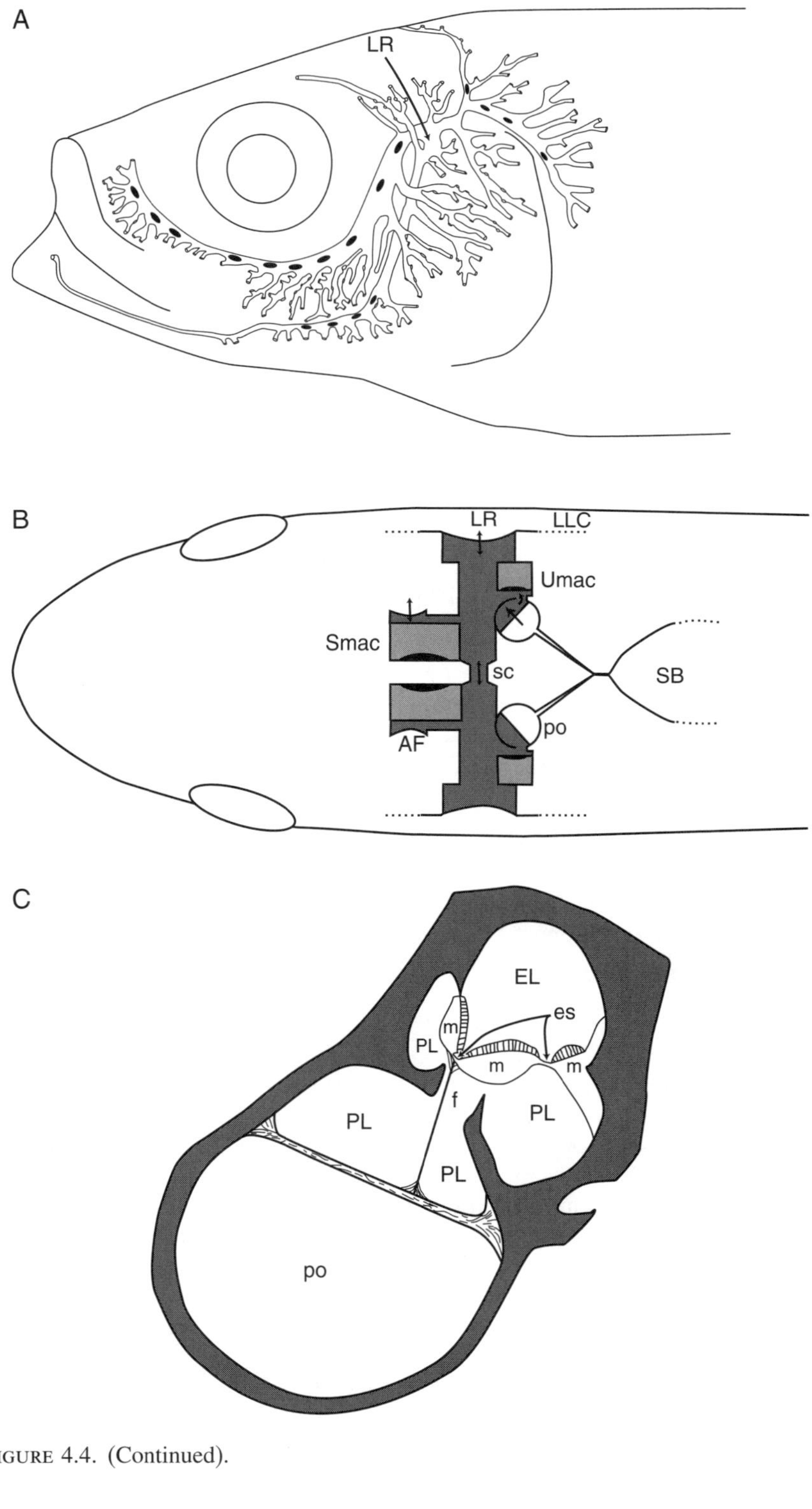

FIGURE 4.4. (Continued).

to the bulla membrane, leading Best and Gray (1980) to speculate that if the thread is functionally important, it may be only at very low frequencies, perhaps by deforming the macula rather than displacing it. This could allow the utricle to play a role in baroreception and perception of depth as well as auditory functions. Best and Gray (1980) also noted that both the medial and anterior macula vary in stiffness along the anterioposterior and ventrodorsal axes, respectively. This variation in stiffness could have similarities to the cochlea, mechanically filtering hair cell stimulation by frequency.

The compliant surfaces in this system, the floor of the utricle, the bulla membrane, and the membrane bounding the recessus lateralis all have relatively flat frequency responses between less than 1 and 1000 Hz (Denton et al. 1979). This agrees well with the measured auditory sensitivity of herring (Enger 1967), which is relatively equally sensitive to low-frequency sounds (up to around 750 Hz) and declines sharply in auditory sensitivity above 1000 Hz (but they are still capable of detecting 1600- and 3200-Hz stimuli (Mann and Lu 1998; and see Section 2.2.1, on ultrasonic hearing). It is likely, however, that this complex of specializations is important in more ways than extending frequency bandwidth or enhancing sensitivity. It would appear that clupeomorphs have an excellent system with which to compare pressure and displacement components of a sound source (Denton et al. 1979), possibly providing unambigious cues for source localization.

2.2.1 Ultrasonic Sensitivity

One of the most interesting recent findings in fish hearing is the rigorous documentation of ultrasonic sensitivity in selected species. Although fisheries scientists have long suspected such sensitivity based on responses to echosounders, dolphin repelling pingers, and acoustic sounders used to guide fish away from hydroelectric facilities (e.g., Dunning et al. 1992; Nestler et al. 1992; Ross et al. 1996), most bioacousticians were skeptical and rarely tested fish hearing with stimuli above a few kilohertz. An early report that goldfish can

FIGURE 4.4. (Continued) Otophysic and laterophysic connections in clupeomorphs. (**A**) Lateral view of the head of *Sprattus sprattus*, showing the lateral line canals and the location of the recessus lateralis (LR). Neuromasts are indicated by dark ovals. (Redrawn with permission after Gray 1984.) (**B**) Schematic representation of the perilymphatic and air-filled cavities in clupeomorphs. Arrows indicate the potential paths of fluid motion. Light gray shading depicts endolymphatic spaces and darker gray depicts perilymphatic spaces. (Redrawn after Gray and Denton [1979], with permission.) AF, Auditory foramen; LLC, lateral line canal; LR, recessus lateralis; SB, swimbladder; po, prootic bulla; sc, subcerebral canal; Smac, saccular macula; Umac, utricular macula. (**C**) Frontal section through the prootic bulla of *Sprattus sprattus*, showing the relationship between air-filled spaces (po) and the utricle. (Redrawn after Denton and Gray [1979], with permission of Nature Publishing Group.) es, Elastic suspensorium; EL, endolymphatic space; f, prootic fenestra; m, utricular maculae; PL, perilymphatic space.

detect sounds above 50 kHz (Offut 1968) has mostly been ignored or discounted due to the lack of control and calibration of the stimulus. More recently, carefully controlled studies have shown conditioned and neurophysiological responses to stimuli in the ultrasonic range in the cod (Astrup and Møhl 1993) and in two species of clupeids (Mann and Lu 1998; Mann et al. 2001). This very high frequency sensitivity must still be considered an extreme specialization and not a general ability of most teleost species. Mann et al. (2001) showed that goldfish are not capable of detecting sounds above a few kilohertz, using the same equipment used to document ultrasonic sensitivity in shad. The ability of shad and cod to detect ultrasound (the former is at least an order of magnitude more sensitive) probably allows these species to detect the echolocation pulses of odontocetes and make appropriate escape responses (Mann and Lu 1998; Astrup 1999).

This is a fascinating evolutionary story, with familiar themes of predator–prey interactions, ecological adaptation, and evolutionary innovation (Astrup 1999); more importantly in the present context, it is still unclear exactly how such sensitivity is mediated. Other than the swimbladder extensions described in the preceding text, *Gadus morhua* is morphologically unspecialized, and similar morphologically to other presumably ultrasound-insensitive fishes. Among the Clupeomorpha, all of whom have the auditory bullae described above, only two species have been shown to possess ultrasonic sensitivity (*Alosa sapidisimma,* American shad, and *Brevoortia patronus,* gulf menhaden), and several other species do not (*Anchoa mitchilli,* bay anchovie, *Harengula jaguana*, scaled herring, and *Sardinella aurit,* round sardinella; Mann et al. 2001). Although one might speculate that part of the utricular organ may be specialized to detect ultrasonic pulsations of the bulla, or that the neuromasts near the recessus lateralis may be somehow stimulated by ultrasound (discussed by Astrup 1999 and Mann et al. 2001), the two species with ultrasonic sensitivity do not have any well documented differences in either of these features from their close relatives who are not sensitive to high-frequency sounds.

Higgs et al. (2004) have examined the developmental time course of ultrasound sensitivity and inner ear morphology. Although they could not provide quantitative correlations, it appeared that ultrasound-sensitivity developed with a similar time course to the thinning of the elastic connections supporting the medial macula of the utricle. They also note that other ultrasonically sensitive species have an apparently thinner membranous connection supporting this macula than species that are not sensitive to ultrasound. Quantitative measurements of the mechanical responses of this macula to ultrasonic stimulation and physiological recordings from the utricular nerve will be needed to confirm this intriguing hypothesis. Recently, Plachta et al. (2004) recorded physiological responses to ultrasonic stimuli in neurons in the auditory brainstem. Most ultrasonically sensitive units were not sensitive to sonic stimuli (i.e., below 10 kHz), suggesting that clupeids have a specialized processing pathway for ultrasonic sound detection. Understanding the mechanisms and the evolutionary history of ultrasonic sensitivity is one of the outstanding problems to be faced by the current generation of researchers in fish hearing.

2.3 Suprabranchial Chambers in Anabantoidei

Some teleosts have evolved auditory specializations involving air-filled chambers other than the swimbladder. The most well known of these is the suprabranchial chambers of gouramis (Perciformes: Anabantoidei). These fishes possess a dorsal outpocketing of the branchial roof, which protrudes into the cranial floor. This chamber, or labyrinth, is isolated from the pharyngeal space by muscular valves and lined with a richly vascularized respiratory epithelium (Liem 1963). These animals swallow air and hold it in the labyrinth, which allows them to survive in hypoxic and anoxic waters. This air-filled space is apposed to a membranous window in the skull floor that also provides these fishes with enhanced hearing sensitivity (Schneider 1941; Yan 1998). Although nearly all anabantoids have a labyrinth organ, only species in the families Helostomatidae (Saidel and Popper 1987) and Belontidae (Ladich and Yan 1998) have been tested for hearing abilities. Gouramis are sensitive to sounds up to 4500 Hz, and probably capable of detecting the pressure component of sound fields (but this was not directly tested). When measured by similar techniques, auditory sensitivity in gouramis is either similar to the catfish *Ictalurus* and the goldfish *Carassius* (Saidel and Popper 1987) or somewhat (~10–20 dB) less sensitive (Ladich and Yan 1998; Yan 1998) at low frequencies. Pressure sensitivity in gouramis declines much more rapidly above several hundred hertz (~400 in Saidel and Popper 1987, 1000 in Yan 1998) than in goldfish (or cyprinids generally, see Ladich 1999). The high-frequency limit may be related to the size of the suprabranchial chamber and its resonance properties. Pygmy gouramis (*Trichopsis pumila*), the smallest species tested, had better sensitivity than larger species above 3 kHz.

Yan (1998) showed that gouramis' high auditory sensitivity arises from the gas-filled superbranchial chamber. Removal of the air in the labyrinth resulted in a decrease of auditory sensitivity by 5–30 dB, depending on the species and stimulus frequency (Yan 1998). Although the effects were frequency specific and tended to be greater at high stimulus frequencies, auditory responses were still measured throughout the original wide frequency band after gas removal (albeit with greatly reduced sensitivity). The greatest effects were seen within the most sensitive frequency band for each particular species, typically around 500–700 Hz. Interestingly, in a later study, Yan et al. (2000) demonstrated that removal of the gas in the swimbladder (leaving the suprabranchial chamber intact) had no effect on hearing thresholds in the blue gourami (*Trichogaster trichopterus*).

As with *Pollimyrus*, many gouramis are active vocalizers (e.g., *Trichopsis* is the genus of "croaking gouramis"), and sound detection is part of a complex communication system. But all gouramis tested, including species that do not vocalize (*Macropodus* and *Trichogaster*), have excellent hearing, in some cases better than vocalizing species. In other cases, the predominant energy of the vocalizations does not match the most sensitive hearing band. Ladich (2000, see also Ladich and Yan 1998) interpreted these findings to mean communication was not the driving force behind the evolution of enhanced hearing. As discussed later, the most well known examples of hearing specialists, the otophysans, are often mute (Ladich 2000).

2.4 Mechanical Linkage Between the Ear and the Swimbladder: The Weberian Apparatus

The most well known hearing specialization, the Weberian apparatus, is a mechanical linkage that transmits motion of the swimbladder wall directly to a perilymphatic space, the sinus impar of the inner ear. As demonstrated by von Frisch (1938), this chain of modified skeletal elements transmits volume changes of the swimbladder to the ear and generally enables otophysans to hear frequencies up to several thousand Hertz and increases sensitivity throughout the frequency range (Fay 1988). Originally described by Weber (1820), the apparatus is diagnostic of the ostariophysan subgroup Otophysi (e.g., carps, minnows, suckers, knifefishes, and catfishes, Rosen and Greenwood 1970). The Weberian apparatus consists of a series of modified anterior centra, neural arches, supraneurals, and pleural ribs that lie in a linear sequence and connect a modified swimbladder (see later) to the sinus impar of the inner ear. Three pairs of ossicles (Figs. 4.5 and 4.6) are held in position on each side of the vertebral column by interossicular ligaments (Alexander 1962). The tip of the tripus articulates with the tunica externa of the camera aerea Weberiana (i.e., anterior surface of the anterior chamber of the swimbladder), and is connected rostrally to the intercalarium by the ligamentum tripae. Vibrations of the camera area Weberiana are transmitted to the intercalarium by the tripus. Motion of intercalarium is transmitted to the scaphium via the ligamentum scaphae. The concha scaphium forms the posterolateral wall of the sinus impar perilymphaticus. Displacements of the scaphium set the perilymphatic fluid in motion, ultimately causing otolithic displacements in the nearby saccule. Traditionally, the claustrum is included as a fourth weberian ossicle, but it is not linked to the other ossicles by ligaments and may not be involved in sound transmission (Grande and de Pinna 2004).

The evidence that the Weberian apparatus enhances hearing is considerable. Audiograms have been constructed for goldfish, carp, a number of other cyprinids and several catfishes (see Fay 1988 for summary and references). With one notable exception (see Section 2.4.1), these species all demonstrate wide bandwidth hearing and are sensitive to sounds from <100 Hz to a few thousand Hertz. Where it has been determined, the thresholds, particularly over 150 Hz, are proportional to the pressure component of the sound stimulus (e.g., Fay and Popper 1975). It has been known since von Fritsch (1938) that the morphology of the swimbladder is important for the acute sense of hearing in otophysans, and Poggendorf (1952) showed that extirpation of the tripus impaired hearing in catfishes. More recently, Yan et al. (2000) showed that deflation of the swimbladder in goldfish decreased sensitivity by as much as

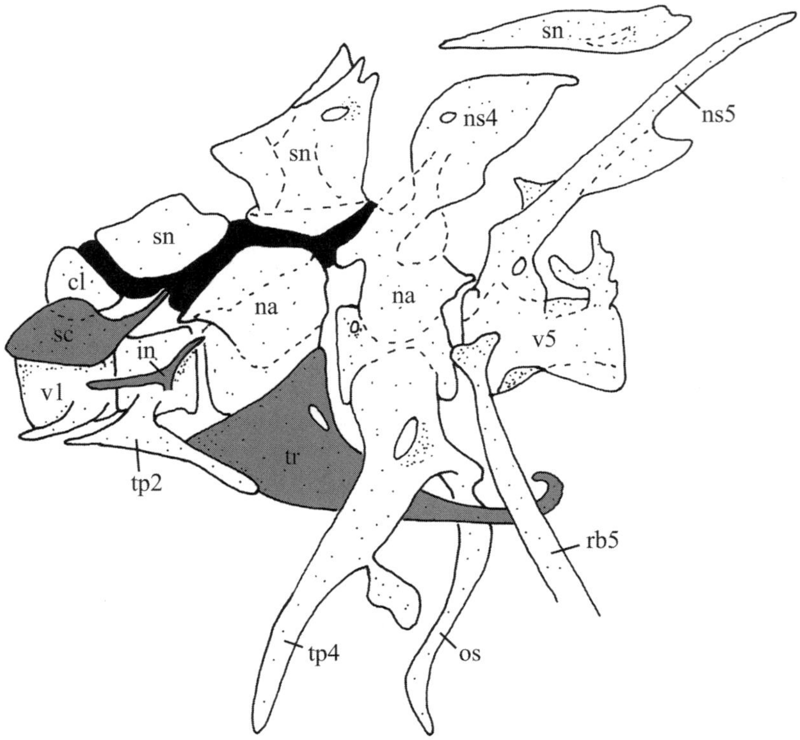

FIGURE 4.5. Weberian apparatus of *Danio rerio* (redrawn after Grande and Young [2004], with permission), LUF.082314, 30 mm SL. Anterior to the left. Stippling indicates bone; black shading indicates cartilage. Weberian elements are shaded gray. Boc, basicoccipital; cl, claustrum; exo, exoccipital; in, intercalarium; na, neural arch; ns, neural spines; os, os suspensorium; pa, parapophysis; rb, pleural rib; sc, scaphium; soc, supraoccipital; sn, supraneural; tp, transverse processes; tr, tripus; v, vertebra (centra).

60 dB. Extirpation of the tripus produced a smaller decrease in sensitivity, ranging from 7.3 to 32.5 dB (Ladich and Wysocki 2003). The decrease in sensitivity is highly frequency dependent, increasing linearly (greater deficit) with increasing frequency (in contrast to the results of Yan et al. [2000], where the deficit was roughly constant below 1500 Hz). Further, responses to 4 kHz stimuli were completely abolished by tripus extirpation (swimbladder deflation reduced sensitivity by ~40 dB). Ladich and Wysocki (2003) also tested a cichlid (*Neolamprologus brichardi*, a putative hearing "generalist") in the same apparatus, with results qualitatively similar to those of another unspecialized cichlid, the Oscar (*Astronotus*) described in Section 1.3.1 (Yan and Popper 1992). Tripus-extirpated goldfish were still approximately 4–25 dB more sensitive than *N. birchardi*, and the differences were greatest above 500 Hz.

Although the Weberian apparatus increases auditory sensitivity, and its effects are greatest at high frequencies, the mere presence of a swimbladder may be

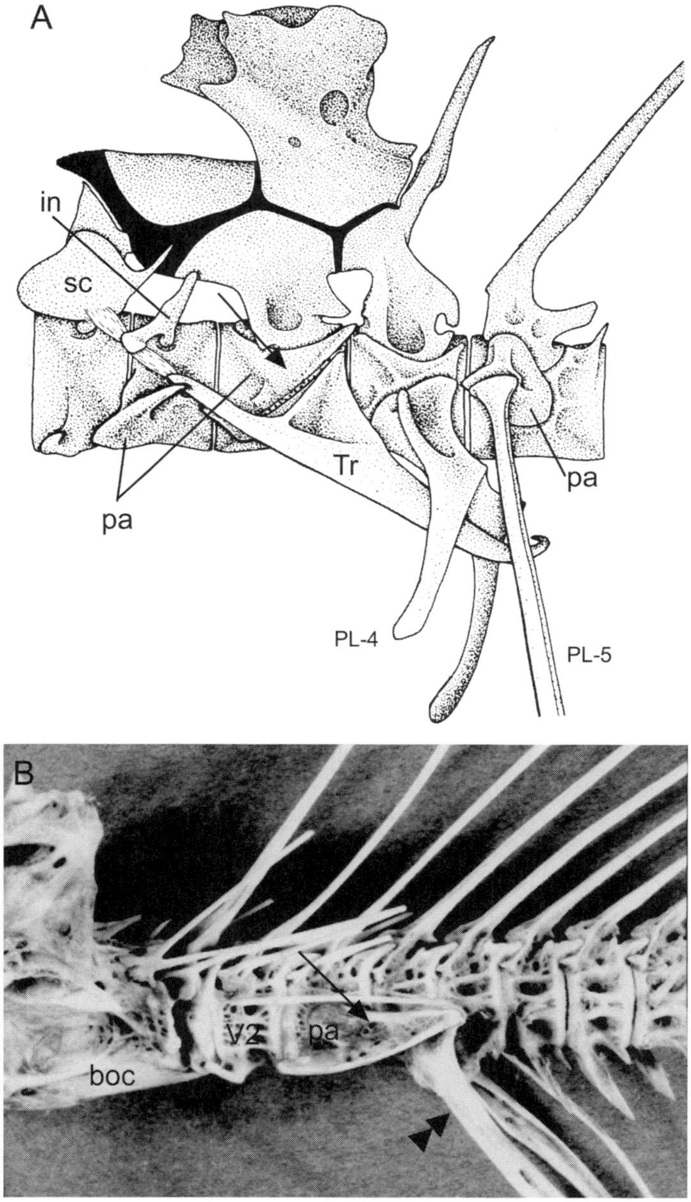

FIGURE 4.6. (A) Illustration of the Weberian apparatus of *Opsariichthys*, MCZ 32375. (Modified from Fink and Fink [1981], with permission.) Note that the parapophysis of the veterbra three (arrow) is not fused to the tripus. (**B**) Photograph of the anterior vertebral column of *Pellona castelnaeana*, AMNH 9319SD(S) showing the expanded parapophysis (arrow) on vertebra three. The broad rostralmost rib (arrowheads) is attached to the posterior surface of this expanded parapophysis. (Modified from Grande and de Pinna [2004], with permission.) Anterior directed to the left. Abbreviations as in Fig. 4.5.

extremely important. The existing experimental data suggest that the Weberian ossicles proper are not the only mechanical linkage between the gas bladder and the ear. Tripus extirpation reduces sensitivity by as much as 32 dB, but removal of gas from the swimbladder reduces sensitivity more, by as much as 60 dB (Yan et al. 2000; Ladich and Wysocki 2003).

The behavioral and ecological importance of the Weberian apparatus is not at all clear. Although many otophysan species (particularly catfishes) make sounds, many other species may be mute, including very large groups (e.g., Gymnotiformes). Ladich (2000) argues that vocal communication systems are not the driving force behind the evolution of hearing enhancements. Indeed, many vocalizing species have rather insensitive hearing (e.g., the arno goby, *Gobius nigricans*, Lugli et al. 2003), although hearing may be enhanced within particular frequency bands. The best hypothesis for the original selective advantage of enhanced hearing is probably the most general one: an improved image of the auditory world in general (Bregman 1990; Fay and Popper 2000). Whatever selective forces were at work at the origin of the Otophysi, enhanced hearing may have been a key innovation for this most successful and diverse group of freshwater fishes.

2.4.1 Variation in Weberian Apparatus Structure and Function

The frequency-dependent effects of tripus extirpation suggest that changes in Weberian ossicle morphology, for instance in the robustness or gracility of each ossicle or the strength of the connections between them, could have a tremendous effect on the range and sensitivity of hearing across otophysans. Considerable variation in the adult Weberian apparatus morphology has been reported among otophysan subgroups (Fink and Fink 1981). Within cypriniforms, there is little obvious variation in the Weberian apparatus morphology, but variation within Siluriphysi (i.e., catfishes and knifefishes), on the other hand, is extraordinary. For example, the ascending and articular processes of the scaphium are lost in loricariids and callichthyids (Chardon et al. 2003). According to Coburn and Grubach (1998), the scaphium is missing as a separate element altogether in callichthyids. In many catfishes and gymnotiforms the intercalarium is often reduced to an intraligamentous ossification (Fink and Fink 1981), and lost altogether in loricariids. In many catfishes, the interossicular ligament that connects the tripus and the scaphium is shortened so that the tripus physically contacts the concha scaphii. The tripus and the scaphium essentially become one element (Chardon et al. 2003). The claustrum is secondarily lost in gymnotoids (Fink and Fink 1981). Chardon and Vandewalle (1997) suggested that the reduction in the size and number of Weberian ossicles may serve to increase the efficiency of high-frequency sound transmission from the swimbladder to the inner ear by decreasing the inertia of the ossicles. In the absence of experimental evidence of hearing abilities, the functional significance of this variation remains unclear.

Ladich (1999) provides the only systematic survey of hearing abilities among a limited number of the very diverse Otophysi. His studies were conducted in relation to differences in vocalizations rather than Weberian apparatus morphology, but he demonstrated that all otophysans do not hear equally well. No clear correlations between hearing abilities and either taxonomic relationships or Weberian apparatus morphology were seen in these hearing differences, but some generalizations can be made. Cyprinids have moderately sensitive hearing, with best sensitivity around 800 Hz, and a rather dramatic decline in sensitivity above 2000 Hz. As expected from the modest diversity in Weberian apparatus morphology in this group (see preceding text), the two distantly related cyprinids were similar in hearing abilities. The one species of characid tested (*Serrasalmus nattereir,* piranha) also had a similar audiogram. The only gymnotiform tested, *Eigenmannia sp. cf. virescens* (glass knifefish) demonstrated qualitatively similar results, although the gymnotiform was approximately 5–10 dB less sensitive than *Serrasalmus* or cyprinids at all frequencies tested (up to 5000 Hz). Hearing sensitivity in two other gymnotiforms, *Gymnotus carapo* (banded knifefish) and *Hypopomus* sp., has also been measured by Kramer et al. (1981). *Gymnotus* was substantially more sensitive than *Hypopomus*, with a higher frequency of best sensitivity and a broader range of sensitivity (possibly as high as 5 kHz).

Catfishes are, in general, more sensitive at most frequencies and display a more uniform sensitivity below 3000 Hz, but they are also the most diverse group of test subjects, in terms of auditory function (Ladich 1999) and morphology. *Corydoras paleatus* (peppered corydora), an armored callichthyid, had the poorest sensitivity, averaging 25 dB less sensitive than the most sensitive species tested, *Platydoras costatus,* or Raphael catfish (Doradidae). Hearing sensitivity in *Corydoras* declined dramatically above 1000 Hz and a similar, but less severe, fall-off in sensitivity was also seen in another doradid, *Agamyxis pectinifrons* (whitebanded catfish). The two doradids tested differed greatly: *Platydoras* was the most sensitive species tested and *Agamyxis* was only slightly more sensitive than *Corydoras*, the least sensitive catfish tested. These latter two species were approximately equal in sensitivity to the cyprinids at low frequencies (<300 Hz) and somewhat less sensitive than cyprinids at middle and high frequencies (Ladich 1999).

Popper and Tavolga's earlier study (1981) on the marine catfish, *Arius felis* (Ariidae) also showed the difficulty in generalizing among otophysans. This species has an unusually large utricle, but Popper and Tavolga (1981) did not observe anything unusual about its Weberian apparatus. Unlike most other catfishes (except perhaps *Corydoras*), *Arius* was fairly insensitive to high-frequency sounds and did not respond to sounds above 1000 Hz. On the contrary, this species has its best sensitivity around 200 Hz, some 20 dB more sensitive than either *Ictalurus* or *Carassius* in this low-frequency range. Based in part on Tavolga (1976), Popper and Tavolga (1981) suggested that this species uses low-frequency vocalizations as a primitive form of echolocation, using its great sensitivity in this bandwidth to detect the acoustic shadows created by objects within its emitted sound field. The specializations of the utricle, both ultrastructurally and mechanically may explain this enhanced sensitivity. This does not explain why this species is so insensitive to higher frequency sounds, despite the presence of a Weberian apparatus. Clearly, "specialization" does not always mean greater frequency bandwidth.

2.4.2 Origins of the Weberian Apparatus and the Relationship Between Otophysi and Clupeomorpha

Hypotheses involving the evolution and development of the Weberian apparatus have been proposed by many authors (e.g., Rosen and Greenwood Rosen 1970; Gayet and Chardon 1987; Chardon and Vandewalle 1997; de Pinna and Grande 2003). Debates among ichthyologists have stemmed from the reexamination and description of basal ostariophysans (e.g., † *Ramallichthys*), otophysans († *Chanoides*), and clupeomorphs (e.g., † *Armigatus* and pristigasteroids). These debates have centered on the relationships of ostariophysans and clupeomorphs, but they have also led to suggestions that the parts of the Weberian apparatus have evolved in several stages (Fink and Fink 1996; Chardon and Vandewalle 1997; Chardon et al. 2003; de Pinna and Grande 2003; Grande and de Pinna 2004; Grande and Young 2004). Results from these studies point to a mosaic evolution of the Weberian apparatus and perhaps enhanced hearing in general in this group (Fig. 4.3).

As detailed in Grande and de Pinna (2004), all otocephalans (Clupeomorpha + Ostariophysi) have modifications of the labyrinth (medial migration of saccular and lagenar otoliths and a posterior elongation of perilymphatic spaces) that may represent a preadaptation for linkages with posterior swimbladder and mechanical elements. Modifications of the swimbladder are common in this group as well. All ostariophysans and some clupeomorphs (Clupeoidei) have a silvery peritoneal tunic covering the anterior portion of the swimbladder; and all ostariophysans and some clupeomorphs (Pristigateroidea and Engraulidae) have a constriction that divides the swimbladder into anterior and posterior chambers.

Modifications of the first and second pleural ribs, including attachments to the peritoneal tunic of the swimbladder (echoing the configuration of a tripus), are common in pristigasteroids, coliids and anotophysans. Modifications of the first several vertebrae, including articulations of adjacent supraneural and neural arches and direct apposition with the cranium (Fig. 4.7), are found repeatedly (e.g., Gonorynchiformes, Pristegasteroidea). For instance, in the American coastal pellona, *Pellona harroweri*, the first three vertebrae are foreshortened, the anterior neural arches are expanded and articulate with each other posteriorly, and the first neural arch articulates with the exoccipitals (Fig. 4.7). Grande and de Pinna (2004) showed that the parapophysis of the third centrum in pristigasteroids, coilids, and otophysans is strikingly similar and possibly homologous across all three groups. As seen in Fig. 4.6, the parapophysis in pristigasteroids and coilids is elongated and oriented at an angle relative to the axis of the vertebral column in lateral view, similar to the cyprinid *Opsariichthys* (freshwater minnows). In addition the first pleural rib in pristigasteroids is flattened and expanded proximally and attaches to the posterior margin of parapophysis by a flat movable articulation, much like the rib-like portion of the tripus in *Opsariichthys* (Grande and de Pinna 2004).

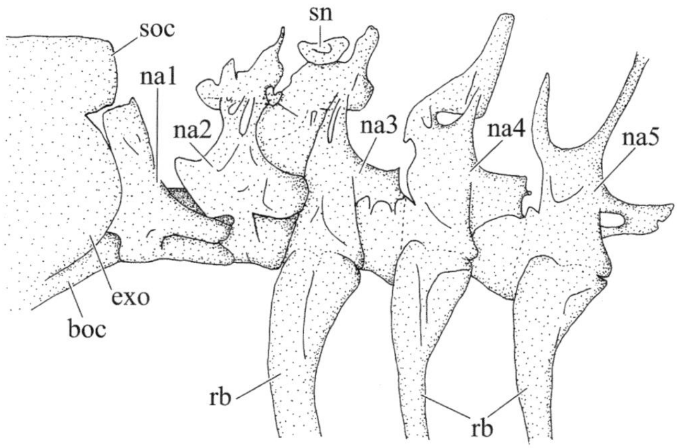

FIGURE 4.7. Anterior vertebrae, associated neural elements, and posterior skull region of the gonorhynchiform *Phractolaemus ansorgei* FMNH 63938, directed to left. (From Grande and Poyato-Ariza [1999], with permission.) Note the broad and untapered rostralmost rib, attached to the third centrum, and the relatively massive abutments of the neural arches, centra, and the occiput. Abbreviations as in Fig. 4.5.

A primitive Weberian apparatus morphology is found in †*Lusitanichthys characiformis* (Gayet 1981, 1985), from the marine Cenomanian deposits of Portugal, and †*Chanoides macropoma* (Patterson 1984) from a marine Eocene locality of Monte Bolca. In †*Chanoides*, the anterior centra are not reduced in size as they are in extant otophysans. The scaphium and intercalarium are more generalized in shape and retain foramina for dorsal and ventral spinal nerve roots. The tripus is smaller than that found in extant otophysans (Patterson 1984). And according to Fink and Fink (1996) the tripus of †*Chanoides* is more rib-like in comparison to extant otophysans, and thus more primitive. It indeed seems that the Weberian apparatus has changed over time, and has probably become more efficient in transmitting sound vibrations as suggested by Chardon and Vandewalle (1997).

Possible homologies between clupeoids and ostariophysans may reflect a common evolutionary history, which is also supported by molecular data (Lê et al. 1993; Lecointre 1995) and caudal fin characters (Lecointre and Nelson 1996). Eschmeyer (1966) first suggested that the Clupeomorpha and Ostariophysi may have evolved from the same herring-like ancestor (which would imply that ostariophysans may be a derived group of clupeomorphs). Currently, the preferred hypothesis is that clupeomorphs and a monophyletic Ostariophysi are sister-taxa (Johnson and Patterson 1996; Lecointre and Nelson 1996). As the two lineages diverged, the clupeomorphs developed their unique acoustico–lateralis system while the ostariophysans perfected the Weberian apparatus.

If, as is very likely the case, Ostariophysi and Cluepomorpha share a common history, it should not be surprising to find elements or precursors of the Weberian apparatus in some clupeomorphs. As pointed out by Lecointre and Nelson (1996,

p. 203) "if clupeomorphs and ostariophysans are related, then too, are their otophysic connections." Gayet and Chardon (1987) proposed that gonorynchiforms exhibit an intermediate or primitive form of the otophysan Weberian apparatus. This hypothesis was rejected by Grande and Poyato-Ariza (1999) however, because many of these are characters also found among clupeomorphs. The gonorynchiform morphology (Fig. 4.7) may be closer to the ancestral protootophysic linkage, and the keys to understanding the Weberian apparatus itself may be in studies of more advanced otocephalans that have a mosiac of individual weberian features, such as pristigasteroids (Grande and de Pinna 2004). This also suggests that experimental studies of gonorynchiform and pristigasteroid hearing may be used to establish the basal hearing abilities of the Otocephala and trace the evolution of hearing enhancements in both Clupeiformes and Otophysi. The difference in hearing deficits caused by tripus extirpation (Ladich and Wysocki 2003) and swimbladder deflation (Yan et al. 2000) are especially interesting in this context, and suggest that anterior vertebral modifications other than direct linkages may tie the swimbladder to inner ear function.

A final reconstruction of the history of hearing enhancement in these groups requires a sound phylogenetic footing. Did both the Weberian apparatus and the clupeiform otophysic link evolve from a gonorynchiform-like morphology, or did the Weberian apparatus evolve more directly from something like the clupeiform condition? To answer these questions, we will need to confirm that the superorders Ostariophysi and Clupeomorpha are monophyletic. Do clupeiforms (possibly clupeomorphs) and otophysans form a monophyletic group, with Gonorynchiformes forming the sister-group to the clupeiform + otophysan assemblage? If so, this relationship would explain why so many otophysic characters appear among clupeiforms. It would also strengthen the argument that the Weberian apparatus evolved from the clupeomorph otophysic system (Chardon et al. 2003). Morphological and molecular data concerning the relationships among these taxa are, unfortunately, not congruent. While Ishiguro et al. (2003) and Saitoh et al. (2003) place gonorynchiforms with clupeomorphs, Lecointre (1995) argues for the monophyly of Ostariophysi (i.e., Otophysi + Gonorynchiformes). Clearly additional phylogenetic studies of cluepocephalans are necessary for a better understanding of the origins of the Weberian apparatus. As pointed out by Grande and Young (2004), these phylogenetic studies must also provide a better understanding of the morphology of basal fossil clupeomorphs such as † *Diplomystus* and † *Ellimichthys* whose relationships to the more advanced clupeiforms are poorly known.

Despite the tantalizing commonalities described in the preceding text, de Pinna and Grande (2003) and Grande and de Pinna (2004) argued that such scenarios must be based on a well-supported phylogenetic framework. But within such a framework, the evolution of particular characters (e.g., tripus and claustrum) point to a Weberian apparatus that has evolved in pieces over time. Any future studies of hearing abilities in pristigasteroids or gonorynchiformes also requires a sound phylogenetic framework to trace the history of hearing abilities. A complete understanding of the evolutionary history of the Weberian apparatus and enhanced hearing awaits better phylogenetic systematics of the Otocephala.

2.5 Laterophysic Connections: Associations Between the Swimbladder and the Lateral Line Canal System

Although the relationship between the lateral line and sound detection has been a debated subject, there are clearly at least three instances of specialization that could provide the lateral line with some sensitivity to pressure (see Webb, Montgomery, and Mogdans, Chapter 5). For a thorough review of lateral line function and biophysics, the reader is referred to the excellent reviews by Kalmijn (1988, 1989), Coombs and Montgomery (1999), Webb, Montgomery, and Mogdans (Chapter 5), and Sand and Bleckmann (Chapter 6).

2.5.1 Clupeomorpha: The Recessus Lateralis

In the Clupeiformes the cranial lateral line canals converge just caudal to the eye forming the recessus lateralis (Fig. 4.4). The medial wall of the recessus lateralis is a flexible membrane separating it from a perilymphatic space in the inner ear. This is the same perilymphatic space that is connected to the prootic bulla and is continuous with the perilymph on the utricular side of the bullar membrane, through the prootic fenestra (Section 2.2; O'Connell 1955; Denton and Blaxter 1976). Like the sound transduction mechanism described in Section 2.2, the prootic membrane fluctuates with the volume of gas in the prootic bulla (in response to ambient pressure changes), causing the perilymph to move through the prootic fenestra and flex the recessus lateralis membrane just as it does the flexible support of the utricular macula. Flexion of the recessus lateralis membrane will cause fluid motion within the lateral line canals that radiate from the recessus lateralis (Denton et al. 1979), deflecting the cupula of nearby neuromasts, rendering them sensitive to both pressure changes and particle accelerations in the ambient environment (Gray 1984).

What might be the function of this pressure sensitivity and the dual excitation of neuromasts by both pressure and fluid motion of the ear and lateral line system? As discussed in the preceding text, phase comparisons of the displacement- and pressure-related components of a sound source is a crucial feature of many models of underwater sound source localization (e.g., Schuijf and Buwalda 1980). These comparisons could also be used by the lateral line canal system in a spatial sense, as opposed to the temporal comparison of pressure/velocity phase described in the preceding text. The fluid motions that result from the movement of the recessus lateralis membrane decline in intensity with increasing distance along the canal from the membrane. The intensity and slope of this gradient is independent of source position and is proportional to stimulus pressure, which attenuates gradually with distance from the source (Gray 1984). On the other hand, typical fluid motions within the canal, induced by the particle motion outside the canal, are highly dependent on source location, distance, and orientation (Denton and Gray 1983). The changing patterns of lateral line excitation as a fish moves relative to a source (see Webb, Montgomery, and Mogdans, Chapter 5; Sand and Bleckmann, Chapter 6) might provide the basis for calculation of source location and distance (Denton et al. 1979; Gray and Denton 1979). Gray and colleagues (Denton et al. 1979; Blaxter and Hoss 1981; Gray 1984; Gray and Denton 1991) have suggested that this mechanism may be important in determining distance between neighbors within a school.

The subcerebral canal connecting the two perilymphatic spaces on either side of the head may also be important for recessus lateralis function (Denton and Gray 1993). Pressure changes from an external source will be nearly identical on both sides of the head (particularly at low frequencies). Thus both prootic bullae will respond identically and there will be no net flow across the subcerebral canal. Pressure gradients generated by the fish's own movements on the other hand, present opposing polarities on either side of the head, and the resulting flow in the subcerebral canal will maintain a 180° phase difference in recessus lateralis movements. Canal fluids will be displaced away from the recessus lateralis on one side of the head while they will converge upon the recessus lateralis on the opposite side. Such differences could be used by the fish to distinguish stimulation caused by external sources from that caused by its own movements and to monitor its own motions to optimize swimming efficiency (Denton and Gray 1993).

Very little is known about the potential functional importance of variation in recessus lateralis morphology. A recessus lateralis is not present in †*Armigatus* (Patterson 1967) or paraclupeids (e.g., †*Diplomystus*). As reported by Greenwood (1968), the recessus lateralis in *Denticeps* (denticle herring) is incomplete and different from clupeoids in that the supra- and infraorbital laterosensory canals empty into the recessus through the same opening in the pterotic via the dermosphenoic. This morphology represents an advanced stage in the evolution of this complex, but an evolutionary pattern that is different from the one in clupeoids (Greenwood 1968).

2.5.1.1 A Similar Specialization in Mullidae?

Interestingly, Denton and Gray (1993) also described a canal linking the cranial lateral lines on either side of the head in *Mullus surmuletus* (striped red mullet: Percamorpha: Mullidae) and determined that flow through this canal is similar to that in the subcerebral canal in clupeids. This canal might also be used by mullets to monitor their own swimming motions. The use of the lateral line to optimize swimming efficiency, as suggested by Denton and Gray (1993) and Lighthill (1993) is an intriguing possibility, but it is still empirically unfounded (Webb, Montgomery, and Mogdans, Chapter 5). If this is indeed a general function of the lateral line (as proposed by Coombs and Montgomery 1999), it is tempting to speculate that some fishes have evolved morphological features that enhance this function.

2.5.2 Chaetodontidae

The butterflyfishes (Chaetodontidae) are a family of tropical coral-reef fishes that are well known to most divers and reef biologists. It has only recently been discovered that these fishes are also characterized by a unique evolutionary trend whereby the swimbladder directly impinges on the lateral line system (Webb and Blum 1990). Webb (1998) termed this connection a laterophysic connection, and surveyed the family for the presence and variation in this connection (Smith et al. 2003). Basal Chaetodontids have unmodified swimbladders and no association beween the swimbladder and any octavolateralis sytem. In the genus *Chaetodon*, all species have some form of rostral projection of the swimbladder that Webb (1998) classified as either an indirect or a direct laterophysic connection, depending on the tissues that intervene between the projection and the lateral line canal. The rostral projection of the swimbladder extends towards a thinned portion of the supracleithral bone which forms the medial wall of a lateral line canal (probably the middle or postotic canal). In six *Chaetodon* subgenera, the rostral horn of the swimbladder is separated from the subcleithral window by epaxial muscle. In these fishes, the swimbladder is not in direct apposition to the lateral line canal and there is substantial variation in the extent of the rostral diverticula. In the ornate butterflyfish, *C. ornatissimus*, the diverticula are very short and separated from the supracleithral bone by both epaxial muscle and a portion of the kidneys (Webb and Smith 2000). In the remaining five subgenera, there is no intervening musculature and the lateral line canal is separated from the swimbladder only by mucoid connective tissue (Webb and Smith 2000; Smith et al. 2003).

The functional significance of the laterophysic connection is still unclear, but the variation within chaetodontids provides an excellent opportunity to investigate. There are typically neuromasts within the scales lateral to the swimbladder horn and within the canal in the supracleithrum. It seems clear that pressure fluctuations within the swimbladder horns have the potential to stimulate these neuromasts. They also approach, but do not contact the otic capsule, so it is possible that they impart some pressure sensitivity to the inner ear as well. Webb and Smith (2000) were unable to find behavioral or ecological correlations with the laterophysic variation within *Chaetodon*, but such a relationship is likely to be subtle. Physiological and behavioral experiments are needed to determine conclusively the function of this fascinating morphology.

2.6 *Loricarioid Catfishes*

Another interesting set of specializations involving the swimbladder, inner ear and lateral line system is found in loricarioid catfishes. The Loricarioidea comprise five families of fresh- or brackish-water catfishes found over most of tropical Central and South America (Astroblepidae, naked sucker-mouth catfishes; Loricariidae, armored catfishes; Scoloplacidae, spiny dwarf catfishes; Callichthyidae, armored catfishes; and Trichomycteridae, pencil catfishes). The

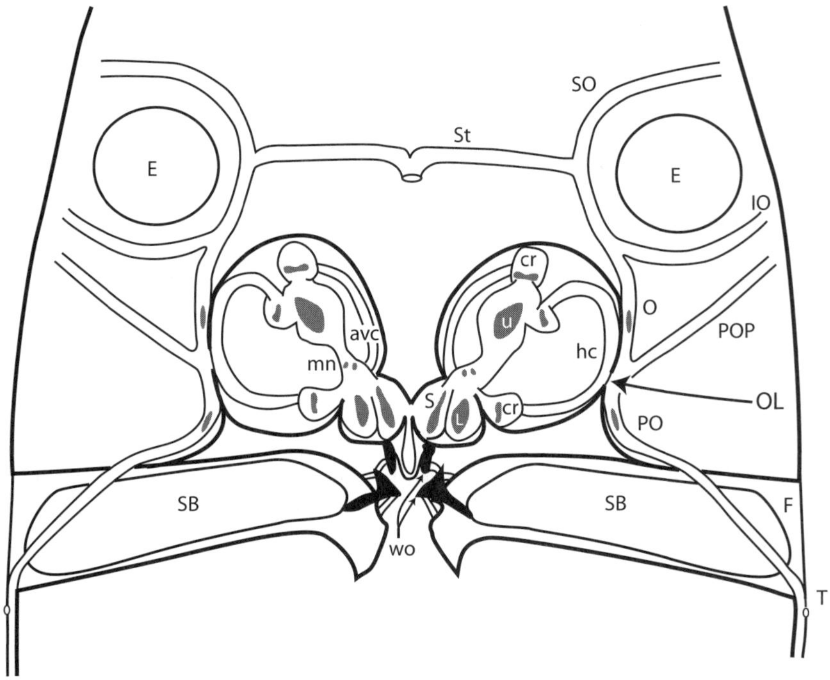

FIGURE 4.8. Ventral view of the temporal cranial region of *Ancistrus*, illustrating the unusual linkages between the swimbladder, lateral line, and inner ear systems. Heavy lines represent bony partitions and gray shapes depict inner ear and lateral line sensory organs (neuromasts). (Redrawn from Bleckmann et al. [1991], with permission of Wiley-Liss.) avc, Anterior vertical canal; cr, cristae of semicircular canals; E, eye; F, fatty tissue; hc, horizontal semicircular canal; IO, infraorbital lateral line canal; L, lagenar macula; mn, macula neglecta; OL, otolateralic connection; O, otic lateral line canal; PO, postotic lateral line canal; POP, preopercular lateral line canal; S, saccular macula; SB, swimbladder; T, trunk lateral line canal; wo, Weberian ossicles.

most well known member of this group is a loricariid, the common aquarium-trade fish Plecostomus (its formal Latin name is now *Hypostomus*). Loricarioids are negatively buoyant, bottom-dwelling fish with a very specialized swimbladder. Unlike the median unpaired organ found in most fishes, the loricarioid swimbladder is bilaterally paired (see Fig. 4.8), although in at least one genus, *Pogonopomoides*, these paired bladders are connected to a caudal unpaired section (Burgess 1989). The paired air-filled cavities are encased in bone and may be ankylosed to the temporal region of the skull (Aquino and Schaefer 2002). Each swimbladder capsule is perforated by a large lateral opening containing fatty tissue (Bleckmann et al. 1991), but the morphology of this lateral foramen is variable within the group, ranging from broad openings to small apertures or elongated tubular openings (Aquino and Schaefer 2002). The lateral opening is generally not overlain by trunk musculature but the supercleithral and/or pterotic bones may be located between the swimbladder capsule and the body wall in Trichomycteridae, Callichthyidae, and Loricariidae (Aquino and Schaefer 2002). Where there is no bone occluding the swimbladder capsule, the adipose-tissue filled space probably transmits pressure fluctuations, much like the adipose channels that enhance hearing in loaches (Kratchovil and Ladich 2000). In loricarioids, the swimbladder capsule may also fenestrated in numerous locations in some taxa (Aquino and Schaefer 2002), but it is not known if these fenestrations have a particular relationship fat-filled spaces or with the body surface.

In addition to these putative acoustic windows, each lateral swimbladder is linked to the sinus impar of the inner ear by two Weberian ossicles, which Bleckmann et al. (1991) tentatively indentified as the scaphius and tripus. The existing data on hearing abilities in this group (Ladich 1999; see Section 2.4.1) suggest that species with this derived Weberian apparatus morphology have markedly different hearing abilities from other catfishes or otophysans generally. A callichthyid catfish tested by Ladich was 20–50 dB less sensitive than other catfishes, particularly to sounds above 1000 Hz, but this may not be representative of the loricariods generally. Callichthyids are one of the taxa that have an occluded lateral opening in the swimbladder capsule (Aquino and Schaefer 2002), and it is possible that loricarioids without such occlusion (i.e., in Scoloplacidae or Astroblepidae) have more sensitive hearing than callichthyids. Further, the extent and functional significance of Weberian apparatus variation across the Loricarioidea is still poorly known and worthy of future study. The close apposition between the swimbladder capsule and caudal skull could also enhance auditory sensitivity (as it does in other specialized groups; see Section 2), reducing the importance of the Weberian linkage for auditory function throughout the group.

In some groups of loricarioids (Callichthyidae, Scoloplacidae, and Loricariidae), the lateral line system is also closely associated with the swimbladder and the inner ear (Aquino and Schaefer 2002). There is substantial variation in the relationships between the lateral line, inner ear and swimbladder within these three families (Table 4.2; Aquino and Schaefer 2002), but the most complete description comes from Bleckmann et al. (1991), who described this association in a loricariid, *Ancistrus* (see Fig. 4.8). In Loricariidae (see Aquino and Schaefer 2002), an unossified portion of the trunk lateral line canal penetrates the swimbladder chamber and courses beneath or lateral to the swimbladder itself. Caudal to the swimbladder chamber, this canal is an ossified tube coursing within the dermis. Further rostrally, the trunk canal is continued by the postotic cephalic canal, which at its junction with the preopercular canal is intimately apposed to the horizontal semicircular canal and separated by a thin membranous window. Neuromasts are present within the lateral line canal on either side of this window. This "otolateralic" connection (Aquino and Schaefer 2002) is quite unusual, and its function is not clear. Bleckmann et al. (1991) speculated that the linkage between the lateral line canals and the horizontal semicircular canal is a pressure release window from the ear to the lateral line, as the sacculus is nearly encased in bone and would require a pressure release for fluid motion in response

TABLE 4.2. Distribution of swimbladder-sensory features within the Loricarioidea.

	Trichomycteridae	Callichthyidae	Scoloplacidae	Astroblepidae	Loricariidae
Bilateral swimbladder fused to cranium	+	+	+	−	+
Lateral occlusion of swimbladder	+	+	−	−	+
Fenestrated swimbladder capsule	+	+	+	+	+
Reduced Weberian apparatus	?	?	?	?	+
Posterior lateral line nerve passes inside swimbladder capsule	−	+	+	−	+
Trunk lateral line canal passes through swimbladder capsule	−	−	−	−	+
Lateral line canal linked to semicircular canal	−	−	−	−	+

Based on Aquino and Schaefer (2002).

to Weberian ossicle vibration. This seems unlikely since the connection between the sacculus and the semicircular canals is itself very limited. The saccule and lagena are found in a medial chamber that is nearly completely distinct from the utricle and semicircular canals, joined at a membranous window in the bony separation (foramen utriculo sacculare).

A linkage between the semicircular canal and the lateral line is also problematic, as the fluid motions of the semicircular canal are induced by angular accelerations rather than compressive forces. Nonetheless, the presence of neuromasts on either side of the membranous window suggests some functional importance of this coupling. The mechanics of the unossified portion of the trunk canal flowing through the swimbladder may be more straightforward. Volume changes in the swimbladder would deform the canals, causing fluid motion within them and stimulating trunk canal neuromasts. It is possible that this requires that the more rostrally located membranous window acts as a pressure release for the lateral line into the semicircular canal, rather than vice versa. In the absence of experimental evidence of auditory or lateral line function, the enhancements provided by this peculiar morphology remain obscure.

2.7 Summary

By the early 1990s, a great deal of experimental evidence had demonstrated a tremendous diversity in the morphology of fish ears and accessory structures, hearing abilities among the 25,000 species of fishes.Yet this diversity is typically lumped into two categories of hearing "specialists" and hearing "generalists." Hearing specialists are those fishes that have some evolutionarily derived abilities, usually assumed to exhibit sensitivity to acoustic pressure and sources above a few hundred Hertz. The most-often cited example is the Otophysi, including carps, minnows, and catfishes. Species without any derived abilities are called generalists.

This convenient, explicitly nonphylogenetic means of describing the abilities of fish is useful for calling attention to the differences in hearing mechanisms. But if used as categories, rather than adjectives, these terms can also be a means of glossing over differences instead of highlighting them. To a quick or naïve reader, specialist may mean "pressure-sensitive, high frequency, high sensitivity." Generalist may mean "displacement or acceleration-sensitive, low frequency and low sensitivity." But this interpretation is at best an incompletely accurate assumption, and at worst a complete leap of faith. It is clear that many specializations do indeed widen the bandwidth of hearing and lower thresholds throughout that band. But as experimental data accumulate, it is clear that "specialist" can have a variety of advanced abilities. These could include an ability to sense pressure, which provides a greater sensitivity at all frequencies (e.g., goldfish) and a wider frequency bandwidth, or a greater sensitivity at particular frequencies with a more modest change in bandwidth (e.g., cod). Some specialized morphologies may have more subtle effects on abilities (e.g., the knifefish *Notopterus*). These might provide enhanced frequency or intensity discrimination abilities, more accurate source localization, or an improved ability to analyze the ambient acoustic scene into discrete sources. Application of the term "specialized" to some morphology should not camoflauge the lack of knowledge of the actual functional advance. In many cases, specialized fish have greater sensitivity over a wider frequency range—mediated by sensitivity to pressure—but this is almost certainly not the case in all species with derived hearing capabilities. More research is still needed to determine what other improvements have evolved over the course of vertebrate history.

This chapter has summarized the available information on the phylogenetic distribution and functional diversity of peripheral specializations that enhance octavolateralis function. Most specializations involve linking an air-filled cavity to the inner ear, giving their possessors sensitivity to pressure. Given the physics of underwater sound, this greatly expands the spatial range of the auditory systems, the types of sounds to which it is attuned, and probably enhances the ability to localize sound sources. These (and other) specializations could also provide greater frequency resolution, which, in conjunction with source localization, may improve the animal's ability to analyze the auditory scene and resolve individual sources from background noise.

Future studies will certainly examine the frequency range and sensitivity of hearing in specialized and unspecialized cichlids, otophysans with diverse Weberian apparatus morphologies (e.g., within gymnotiforms or siluriform subgroups), and clupeomorphs with and without ultrasonic hearing. Studies of hearing in gonorhynchiformes and basal clupeomorphs are also needed to establish the primitive hearing abilities for the Otophysi. The diversity *within* each of these specialized groups offers an opportunity to empirically determine the functional correlates of diversity in mechanical linkage systems. Comparisons between, for instance, catfishes with robust Weberian ossicles and those with gracile ossicles may be used to test hypotheses of morphofunctional correlations. Explicitly phylogenetic studies of auditory ability, morphological variation, and ecological and behavioral differences are desperately needed to better understand how specific specializations (e.g., those enumerated in Fig. 4.3) affect hearing and what selective pressures may have driven their evolution. The easy questions are clear (what is the sensitivity and frequency range) and the tools are now available for such studies.

More uncertain is the future refinements of meanings of the term "specialist." The field needs better experimental tools for testing spatial hearing, scene analysis, auditory discrimination and the like, and should apply its collective creativity toward testing the hypothesis that these abilities have been the specializations evolution has produced. Measures of audition using evoked potential physiology (Kenyon et al. 1998; Yan 2001) have been and will continue to be invaluable in quickly documenting the hearing abilities of large numbers of species. With little modification, these procedures may also be used to test more subtle auditory capabilities (e.g., temporal resolution: Wysocki and Ladich 2001), and such creative approaches must be pursued if we are to understand the meaning of variation in the auditory system.

As the prefatory quote by Carl Gans notes, the diversity of mechanical linkages of the octavolateralis system is a joy to behold. Within teleost fishes alone, novel morphologies capable of enhancing octavolateralis function may have evolved over twenty times! In the majority of cases, there is no experimental evidence of the function of these specializations. Considering only those cases where experimental evidence of enhanced hearing exists, the phylogenetic distribution of these traits suggests that enhanced hearing has evolved at least six times (Notopteroidei, Clupeomorpha, Otophysi, Anabantoidei, Holocentridae, Gadidae). Two other taxa have similar morphologies that have been described in detail (Cichlidae, Elopomorpha), and there is reason to believe that these specializations have evolved more than once within these families. Swimbladder extensions have been described or briefly mentioned in yet another ten taxa, and it is certainly premature to assume this survey of teleost hearing specializations is complete. Lateral line specializations that might provide pressure sensitivity have also evolved at least twice (Clupeomorpha, Chaetodonitdae), and perhaps a third time, considering Denton and Gray (1993) speculations about Mullidea.

The joy in observing this variation is that it offers us the chance to observe evolution almost directly. One can look to variation to learn how it is shaped by behavior and ecology, and how differences in morphology provide different functional abilities to each species. Where this kind of functional information is known, we have summarized it, but the main goal of this chapter has been to point out how little is known. Over 100 years ago, Bateson wrote (1894, p. 17), "In variation we look to see evolution rolling out before our eyes. In this we may fail wholly and must fail largely, but it is still the best chance left." This chapter documents some of the variations found across teleosts, but more detailed systematic surveys are still needed, particularly within cichlids and siluriphysi. It is now time to look to this variation and see the evolution of auditory function roll out before our eyes.

Acknowledgments. We are grateful to the editors of this volume for organizing this symposium and inviting us to participate and contribute this chapter. We are also most grateful to Wm. Leo Smith for fruitful discussions and for sharing his unpublished master's thesis with us. We also thank Rebecca Jones for her help in organizing Table 4.1, and Joe Schluep for helping with other figures and numerous other aspects of manuscript preparation. Dick Fay, Peter Moller, John Sparks, and Jackie Webb provided many helpful comments on earlier drafts of this chapter.

C.B.B.'s research is supported by NIH awards 1 S06 GM60654 and 1 R03 MH067808 and is helped by the infrastructure support provided by the Research Centers in Minority Institutions (RCMI) award RR-03037 from the National Center for Research Resources of the NIH. T.G.'s research is supported by grants DEB 9903533 and 0128794 from the NSF.

References

Alexander RM (1962) The structure of the Weberian apparatus in the Cyprini. Proc Zool Soc Lond 129:451–473.

Aquino AE, Schaefer SA (2002) The temporal region of the cranium of Loricarioid catfishes (Teleostei: Siluriformes): morphological diversity and phylogenetic significance. Zool Anz 241:223–244.

Astrup J (1999) Ultrasound detection in fish—a parallel to the sonar-mediated detection of bats by ultrasound-sensitive insects? Comp Biochem Physiol A 124:19–27.

Astrup J, Møhl B (1993) Detection of intense ultrasound by the Cod *Gadus morhua*. J Exp Biol 182:71–80.

Bateson W (1894) Materials for the Study of Variation Treated with Especial Regard to Discontinuity in the Origin of Species. London: Macmillan and Co.

Best ACG, Gray JAB (1980) Morphology of the utricular recess in the sprat. J Mar Biol Assoc UK 60:703–715.

Blaxter JHS (1978) Baroreception. In: Ali, MA (ed) Sensory Ecology. New York: Plenum Press, pp. 375–409.

Blaxter JHS (1981) The Swimbladder and hearing. In: Tavolga WN, Popper AN, Fay RR (eds) Hearing and Sound Communication in Fish. New York: Springer-Verlag, pp. 61–71.

Blaxter JHS, Hoss DE (1981) Startle response in herring: the effect of sound stimulus frequency, size of fish and selective interference with the acoustico-lateralis system. J Mar Biol Assoc UK 61:871–879.

Bleckmann H, Niemann U, Fritzsch B (1991) Peripheral and central aspects of the acoustic and lateral line system of a bottom dwelling catfish, *Ancistrus* sp. J Comp Neurol 314:452–466.

Bregman AS (1990) Auditory Scene Analysis. Cambridge, MA: The MIT Press.

Burgess WE (1989) An Atlas of Freshwater and Marine Catfishes. Neptune City, NJ: T.F.H. Publications.

Buwalda RJA, Schuijf A, Hawkins AD (1983) Discrimination by the cod of sounds from opposing directions. J Comp Physiol 150:175–184.

Cahn PH, Siler W, Wodinsky J (1969) Acoustico-lateralis system of fishes: tests of pressure and particle-velocity sensitivity in Grunts, *Haemulon sciurus* and *Haemulon parrai*. J Acoust Soc Am 46:1572–1578.

Cain P (1995) Navigation in familiar environments by the weakly electric elephantnose fish, *Gnathonemus petersii* L. (Mormyriformes, Teleostei). Ethology 99: 332–349.

Cain P, Gerin W, Moller P (1994) Short-range navigation of the weakly electric fish, *Gnathonemus petersii* L. (Mormyridae, Teleostei), in novel and familiar environments. Ethology 96:33–45.

Chao NL (1986) A synopsis on zoogeography of the Sciaenidae. In: Uyeno T, Arai R, Taniuchi T, Matsuura K (eds) Indo-Pacific Fish Biology: Proceedings of the Second International Conference on Indo-Pacific Fishes. Tokyo, Japan: Ichthyological Society of Japan, pp. 570–589.

Chapman CJ, Hawkins AD (1973) A field study of hearing in the cod, *Gadus morhua* L. J Comp Physiol 85:147–167.

Chapman CJ, Johnstone ADF (1974) Some auditory discrimination experiments on marine fish. J Exp Biol 61:521–528.

Chapman CJ, Sand O (1974) Field studies of hearing in two species of flatfish *Pleuronectes platessa* (L.) and *Limanda limanda* (L.) (family Pleuronectidae). Comp Biochem Physiol 47A:371–385.

Chardon M, Vandewalle P (1997) Evolutionary trends and possible origin of the Weberian apparatus. Netherlands J Zool 47:383–403.

Chardon M, Parmentier E, Vandewalle P (2003) Morphology, development and evolution of the Weberian apparatus in catfish. In: Arratia G, Kapoor BG, Chardon M, Diogo R (eds) Catfishes. Enfield, CT: Scientific Publishers, pp. 72–120.

Cichocki FP (1976) Cladistic history of cichlid fishes and reproductive strategies of the american genera *Acarichthys, Biotodoma and Geophagus*. Ph.D. Thesis,University of Michigan, Ann Arbor.

Coburn MM, Grubach PG (1998) Ontogeny of the armored catfish *Corydoras paleatus* (Siluriformes: Callichthyidae). Copeia 1998:301–311.

Collette B, Nauen C (1983) FAO Species Catalog, Vol. 2. Scombrids of the World. An Annotated and Illustrated Catalog of Tuna Mackerels, Bonitos and Related Species Known to Date. FAO Fish. Synop. No. 125: i-vii + 1–137.

Coombs S, Montgomery JC (1999) The enigmatic lateral line system. In: Fay RR, Popper AN (eds) Comparative Hearing: Fish and Amphibians. New York: Springer-Verlag, pp. 319–362.

Coombs S, Popper AN (1979) Hearing differences among Hawaiian squirrelfishes (family Holocentridae) related to differences in the peripheral auditory system. J Comp Physiol 132:203–207.

Coombs S, Popper AN (1980) Auditory sensitivity and inner ear ultrastructure in *Osteoglossum bicirrhosum*. Am Zool 20:785.

Coombs S, Popper AN (1982a) Structure and function of the auditory system in the clown knifefish, *Notopterus chitala*. J Exp Biol 97:225–239.

Coombs S, Popper AN (1982b) Comparative frequency selectivity in fishes: simultaneously and forward- masked psychophysical tuning curves. J Acoust Soc Am 71:133–141.

Crawford JD, Jacobe P, Bénech V (1997) Field studies of a strongly acoustic fish in West Africa: reproductive ecology and acoustic behavior of *Pollimyrus isidori*, Mormyridae. Behaviour 134:677–725.

Denton EJ, Blaxter JHS (1976) The mechanical relationships between the clupeid swimbladder, inner ear and lateral line. J Mar Biol Assoc UK 56:787–807.

Denton EJ, Gray JAB (1983) Mechanical factors in the excitation of clupeid lateral lines. Proc Roy Soc Lond B 218:1–26.

Denton EJ, Gray JAB (1993) Stimulation of the acoustico-lateralis system of clupeid fish by external sources and their own movements. Philos Trans Roy Soc Lond B 341:113–127.

Denton EJ, Gray JAB, Blaxter JHS (1979) The mechanics of the clupeid acoustico-lateralis system: frequency responses. J Mar Biol Assoc UK 59:27–47.

de Pinna M, Grande T (2003) Ontogeny of the accessory neural arch in pristigasteroid clupeomorphs and its bearing on the homology of the otophysan claustrum (Teleostei). Copeia 2003:838–845.

Dunning DJ, Ross QE, Geoghegan P, Reichle JJ, Menezes JK, Watson JK (1992) Alewives avoid high frequency sound. N Am J Fish Man 12:407–416.

Enger PS (1967) Hearing in herring. Comp Biochem Physiol 22:527–538.

Eschmeyer WN (1966) Re-evaluation of the position of the family Dussumieriidae in clupeoid phylogeny. Forty-sixth Annual Meeting of the Society for Ichthyhology and Herpetology, Miami, FL.

Fay RR (1984) The goldfish ear codes the axis of acoustic particle motion in three dimensions. Science 225:951–954.

Fay RR (1988) Hearing in Vertebrates: A Psychophysics Databook. Winnetka, IL: Hill-Fay Associates.

Fay RR, Edds-Walton PL (1997) Directional response properties of saccular afferents of the toadfish, *Opsanus tau*. Hear Res 111:1–21.

Fay RR, Popper AN (1975) Modes of stimulation in the teleost ear. J Exp Biol 62: 379–387.

Fay RR, Popper AN (2000) Evolution of hearing in vertebrates: the inner ears and processing. Hear Res 149:1–10.

Fink SV, Fink WL (1981) Interrelationships of ostariophysan fishes (Ostariophysi). Zool J Linn Soc 72:297–353.

Fink SV, Fink WL (1996) Interrelationships of ostariophysan fishes (Teleostei). In: Stiassny MLJ, Parenti RL, Johnson GD (eds) Interrelationships of Fishes. San Diego: Academic Press, pp. 209–249.

Fletcher LB, Crawford JD (2001) Acoustic detection by sound-producing fishes (Mormyridae): the role of gas-filled tympanic bladders. J Exp Biol 204:175–183.

Gans C (1992) An overview of the evolutionary biology of hearing. In: Webster, DB Fay, RR, Popper, AN (eds) The Evolutionary Biology of Hearing. New York: Springer-Verlag, pp. 3–13.

Gayet M (1981) Contribution à l'étude anatomique et systématique de l'ichthyofaune cénomanienne du Portugal. Deuxième partie: les ostariophysaires. Comm Serv Geol Portugal 67:173–190.

Gayet M (1985) Rôle de l'évolution de l'appareil de Weber dans la phylogénie des Ostariophysi, suggéré par un nouveau Characiforme du Cénomanien supérieur narin de Portugal. C R Séances Acad Sci, Sér 2. 300:895–898.

Gayet M, Chardon M (1987) Possible otophysic connections in some fossil and living ostariophysan fishes. Proc Euro Ichthyol Co Stockholm 1985:31–42.

Gosline WA (1966) The limits of the fish family Serranidae, with notes on other lower percoids. Proc Cal Acad Sci 33:91–112.

Grande L (1985) Recent and fossil clupeomorph fishes with materials for revision of the subgroups of clupeids. Bull Am Mus Nat Hist 181:235–307.

Grande T, de Pinna M (2004) The evolution of the Weberian apparatus: a phylogenetic perspective. In: Arratia G, Schultze H-P (eds) Mesozoic Fishes 3—Systematics, Paleoenvironments and Biodiversity. Munich: Dr. Friedrich Pfeil-Verlag, pp. 429–448.

Grande T, Poyato-Ariza FJ (1999) Phylogenetic relationships of fossil and recent gonorynchiform fishes (Teleostei: Ostariophysi). Zool J Linn Soc 125:197–238.

Grande T, Young B (2004) The ontogeny and homology of the Weberian apparatus in the zebrafish *Danio rerio* (Ostariophysi: Cypriniformes). Zool J Linn Soc 140:241–254.

Gray JAB (1984) Interaction of sound pressure and particle acceleration in the excitation of the lateral-line neuromasts of sprats. Proc Roy Soc Lond B 220:299–325.

Gray JAB, Denton EJ (1979) The mechanics of the clupeid acoustico-lateralis system: low frequency measurements. J Mar Biol Assoc UK 59:11–26.

Gray JAB, Denton EJ (1991) Fast pressure pulses and communication between fish. J Mar Biol Assoc UK 71:83–106.

Green JM (1971) Studies on the swim bladders of *Eucinostomus gula* and*E. argenterus* (Pisces: Gerreidae). Bull Mar Sci 21:567–590.

Greenwood PH (1963) The swimbladder in African Notopteridae (Pisces) and its bearing on the taxonomy of the family. Bull Br Mus Nat Hist Zool 11:377–412.

Greenwood PH (1968) The osteology and relationships of the Denticipitidae, a family of clupeomorph fishes. Bull Br Mus Nat Hist Zool 16:215–273.

Greenwood PH (1970) Skull and swim bladder connections in fishes of the family Megalopidae. Bull Br Mus Nat Hist Zool 19:121–135.

Hawkins AD (1986) Underwater sound and fish behavior. In: Pitcher, TJ (ed) Behavior of Teleost Fishes. London: Chapman and Hall, pp. 129–169.

Hawkins AD, Sand O (1977) Directional hearing in the median vertical plane by the cod. J Comp Physiol A 122:1–8.

Herre, AW and Montalban, HR (1927) The Phillipine butterflyfishes and their allies. Phillipine J. Sci 34:1–113.

Higgs DM, Plachta DTT, Rollo AK, Singheiser M, Hastings MC, Popper AN (2004) Development of ultrasound detection in American shad (*Alosa sapidissima*). J Exp Biol 207:155–163.

Hudspeth AJ, Corey DP (1977) Sensitivity, polarity, and conductance change in the response of vertebrate hair cells to controlled mechanical stimuli. Proc Natl Acad Sci USA 74:2407–2411.

Ishiguro NB, Miya M, Nishida M (2003) Basal euteleostean relationships: a mitogenomic perspective on the phylogenetic reality of the "Protocanthopterygii." Mol Phylo Evol 27:476–488.

Johnson GD (1980) The limits and relationships of the Lutjanidae and associated families. Bull Scripps Inst Ocean 24:1–114.

Johnson GD, Patterson C (1996) Relationships of lower euteleostean fishes. In: Stiassny MLJ, Parenti RL, Johnson GD (eds) Interrelationships of Fishes. San Diego: Academic Press, pp. 251–332.

Jørgensen JM (1989) Evolution of octavolateralis sensory cells. In: Coombs S, Görner P, Münz H (eds) The Mechanosensory Lateral Line: Neurobiology and Evolution. New York: Springer-Verlag, pp. 115–145.

Kalmijn AJ (1988) Hydrodynamic and acoustic field detection. In: Atema J, Fay RR, Popper AN, Tavolga WN (eds) Sensory Biology of Aquatic Animals. New York: Springer-Verlag, pp. 83–130.

Kalmijn AJ (1989) Functional evolution of lateral line and inner ear sensory systems. In: Coombs S, Görner P, Münz H (eds) The Mechanosensory Lateral Line: Neurobiology and Evolution. New York: Springer-Verlag, pp. 187–216.

Kalmijn AJ (1997) Electric and near-field acoustic detection, a comparative study. Acta Physiol Scand 161(Suppl 638):25–38.

Katayama M (1959) Studies on the serranid fishes of Japan I. Bull Fac Educ Yamaguichi U 8:103–180.

Kenyon TN, Ladich F, Yan HY (1998) A comparative study of hearing ability in fishes: The auditory brainstem response approach. J Comp Physiol A 182:307–318.

Kramer B, Tautz J, Markl H (1981) The EOD sound response in weakly electric fish. J Comp Physiol A 143:435–441.

Kratochvil H, Ladich F (2000) Auditory role of the lateral trunk channels in cobitid fishes. J Comp Physiol A 186:279–285.

Ladich F (1999) Did auditory sensitivity and vocalization evolve independently in otophysan fishes? Brain Behav Evol 53:288–304.

Ladich F (2000) Acoustic communication and the evolution of hearing in fishes. Philos Trans Roy Soc Lond B 355:1285–1288

Ladich F, Wysocki LE (2003) How does tripus extirpation affect auditory sensitivity in goldfish? Hear Res 182:119–129.

Ladich F, Yan HY (1998) Correlation between auditory sensitivity and vocalization in anabantoid fishes. J Comp Physiol A 182:737–746.

Lê HLV, Lecointre G, Perasso R (1993) A 28S rRNA-based phylogeny of the gnathostomes: first steps in the analysis of conflict and congruence with morphologically based cladograms. Mol Phylo Evol 2:31–51.

Lecointre G (1995) Molecular and morphological evidence for a Clupeomorpha-Ostariophysi sister-group relationship. Geobios 19:205–210.

Lecointre G, Nelson G (1996) Clupeomorpha, sister-group of Ostariophysi. In: Stiassny MLJ, Parenti RL, Johnson GD (eds) Interrelationships of Fishes. San Diego: Academic Press, pp. 193–207.

Leis JM (1994) Larvae, adults and relationships of the monotypic perciform fish family Lactariidae. Rec Aust Mus 46:131–143.

Liem KF (1963) The Comparative Osteology and Phylogeny of the Anabantoidei (Teleostei, Pisces). Urbana, IL: University of Illinois Press.

Lighthill J (1993) Estimates of the pressure differences across the head of a swimming clupeid fish. Philos Trans Roy Soc Lond B 341:129–140.

Lu Z, Popper AN, Fay RR (1996) Behavioral detection of acoustic particle motion by a teleost fish (*Astronotus ocellatus*): sensitivity and directionality. J Comp Physiol A 179:227–233.

Lugli M, Yan HY, Fine ML (2003) Acoustic communication in two freshwater gobies: the relationship between ambient noise, hearing thresholds and sound spectrum. J Comp Physiol A 189:309–320.
MacDonald CM (1978) Morphological and biochemical systematics of Australian freshwater and estuarine percicthyid fishes. Austral J Mar Fresh Res 29:667–698.
Mann DA, Lu Z (1998) Detection of ultrasonic tones and simulated dolphin echolocation clicks by a teleost fish, the American shad (*Alosa sapidissima*). J Acoust Soc Am 104:562–568.
Mann DA, Higgs DM, Tavolga WN, Popper AN, Souza MJ (2001) Ultrasound detection by clupeiform fishes. J Acoust Soc Am 109:3048–3054.
McCormick CA, Popper AN (1984) Auditory sensitivity and psychophysical tuning curves in the elephant nose fish, *Gnathonemus petersii*. J Comp Physiol A 155:753–761.
McKay R (1985) A revision of the fishes of the family Sillanginidae. Mem Queensland Mus 22:1–74.
Nelson EM (1955) The morphology of the swimbladder and auditory bulla in the Holocentridae. Fieldiana Zool 37:121–130.
Nestler JM, Ploskey GR, Pickens J, Menezes J, Schilt C (1992) Responses of blueback herring to high-frequency sound and implications for reducing entrainment at hydropower dams. N Am J Fish Man 12:667–683.
O'Connell CP (1955) The gas bladder and its relation to the inner ear in *Sardinops caerulea* and *Engraulis mordax*. Fish Bull Fish Wildlife Service (US) 56:505–533.
Offut GC (1968) Auditory response in the goldfish. J Aud Res 8:391–400.
Offutt GC (1974) Structures for the detection of acoustic stimuli in the Atlantic codfish, *Gadus morhua*. J Acoust Soc Am 56:665–671.
Patterson C (1967) Are the teleosts a polyphyletic group? In: Problèms Actuels de Paléontologie (Evolution des Vertebrés). Coll Int Cent Natn Res Scient 163:93–109.
Patterson C (1984) *Chanoides*, a marine Eocene otophysan fish (Teleostei: Ostariophysi). J Vert Paleo 4:430–456.
Plachta DTT, Song J, Halvorsen MB, Popper AN (2004) Neuronal encoding of ultrasonic sound by a fish. J. Neurophys 91:2590–2597.
Poggendorf D (1952) Die absoluten Hörschwellen des Zwergwelses (*Amiurus nebulosus*) und Beiträge zur Physik des Weberschen Apparates der Ostariophysen. Z Vergl Physiol 34:222–257.
Popper AN, Fay RR (1997) Evolution of the ear and hearing: Issues and questions. Brain Behav Evol 50:213–221.
Popper AN, Tavolga WN (1981) Structure and function of the ear in the marine catfish, *Arius felis*. J Comp Physiol 144:27–34.
Popper AN, Salmon M, Parvulescu A (1973) Sound localization by the Hawaiian Squirrelfishes, *Myripristis berndti* and *M. argyromus*. Anim Behav 21:86–97.
Ramcharitar J (2002) Unique ear structure of silver perch *Bairdiella chrysoura*. Bioacoustics 12:122–124.
Rogers PH, Cox M (1988) Underwater sound as a biological stimulus. In: Atema J, Fay RR, Popper AN, Tavolga WN (eds) Sensory Biology of Aquatic Animals. New York: Springer-Verlag, pp. 131–149.
Rosen DE, Greenwood PH (1970) Origin of the Weberian apparatus and the relationships of the ostariophysan and gonorynchiform fishes. Am Mus Novitates 2428:1–25.
Ross QE, Dunning DJ, Menezes JK, Kenna MJJ, Tiller GW (1996) Reducing impingement of alewives with high-frequency sound at a power plant intake on Lake Ontario. N Am J Fish Man 16:548–559.

Saidel WM, Popper AN (1987) Sound reception in two anabantid fishes. Biochem Physiol 88A:37–44.
Saitoh K, Miya M, Inoue JG, Ishiuro NB, Nishida M (2003) Mitochondrial genomics of ostariophysan fishes: perspectives on phylogeny and biogeography. J Mol Evol 56:464–472.
Sand O (1974) Directional sensitivity of microphonic potentials from the perch ear. J Exp Biol 60:881–889.
Sand O, Enger PS (1973) Evidence for an auditory function of the swimbladder in the cod. J Exp Biol 59:405–414.
Sasaki K (1989) Phylogeny of the family Sciaenidae, with notes on its zoogeography (Teleostei, Perciformes). Mem Fac Fish Hokkaido Univ 36:1–137.
Schellart NAM, Popper AN (1992) Functional aspects of the evolution of the auditory system of actinopterygian fish. In: Webster DB, Fay RR, Popper AN (eds) The Evolutionary Biology of Hearing. New York: Springer-Verlag, pp. 295–322.
Schneider H (1941) Die Bedeutung der Atemhöhle der Labyrinthfische für ihr Hörvernögen. Z Vergl Physiol 29:171–194.
Schuijf A (1981) Models of acoustic localization. In: Tavolga WN, Popper AN, Fay RR (eds) Hearing and Sound Communication in Fishes. New York: Springer-Verlag, pp. 267–310.
Schuijf A, Buwalda RJA (1975) On the mechanism of directional hearing in cod (*Gadus morhua* L.). J Comp Physiol 98:333–343.
Schuijf A, Buwalda RJA (1980) Sound localization: a major problem in fish acoustics. In: Popper AN, Fay RR (eds) Comparative Studies of Hearing in Vertebrates. New York: Springer-Verlag, pp. 43–78.
Schuijf A, Hawkins AD (1983) Acoustic distance discrimination by the cod. Nature 302:143–144.
Schuijf A, Siemelink M (1974) The ability of cod (*Gadus morhua*) to orient towards a sound source. Experientia 30:773–774.
Siler W (1969) Near- and farfields in a marine environment. J Acoust Soc Am 46: 483–484.
Smith WL, Webb JF, Blum SD (2003) The evolution of the laterophysic connection with a revised phylogeny and taxonomy of butterflyfishes (Teleostei: Chaetodontidae). Cladistics 19:287–306.
Sparks JS (2001) Phylogeny and biogeography of the Malagasy and South Asian cichlid fishes (Teleostei: Perciformes: Cichlidae), including a survey of the freshwater fishes of Madagascar. PhD Thesis, University of Michigan, Ann Arbor.
Starks EC (1908) On a communication between the air-bladder and the ear in certain spiny-rayed fishes. Science 28:613–614.
Starnes WC (1988) Revision, phylogeny and biogeographic comments on the circumtropical marine percoid fish family Priacanthidae. Bull Mar Sci 43:117–203.
Stiassny MLJ, Chakrabarty P, Loiselle PV (2001) Relationships of the Madagascan cichlid genus *Paretroplus*, with a description of a new species from the Betsiboka River drainage of northwestern Madagascar. Ichthyol Explor Freshwater 12:29–40.
Stipetič E (1939) Über das Gehörorgan der Mormyriden. Z Vergl Physiol 26:740–752.
Tavolga WN (1974) Signal/noise ratio and the critical band in fishes. J Acoust Soc Am 55:323–1333.
Tavolga WN (1976) Obstacle avoidance in the sea catfish (*Arius felis*). In: Schuijf A, Hawkins AD (eds) Sound Reception in Fish. Amsterdam: Elsevier, pp. 185–204.

Tavolga WN, Wodinsky J (1963) Auditory capacities in fishes. Pure tone thresholds in nine species of marine teleosts. Bull Am Mus Nat Hist 126:177–240.

van Bergeijk WA (1964) Directional and non-directional hearing in fish. In: Tavolga WN (ed) Marine Bio-Acoustics. Oxford: Pergamon Press, pp. 185–204.

von Frisch K (1938) The sense of hearing in fish. Nature 141:8–11.

von Heusinger CF (1826) Bemerkungen über das Gehörwerkzeug des *Mormyrus cyprinoides*, *Gastroblecus compressus* und *Pimelodus synodontis*. Arch Anat Physiol Meckel 1:324–327.

Webb JF (1998) The laterophysic connection: a unique link between the swim bladder and the lateral-line system in *Chaetodon* (Perciformes: Chaetodontidae). Copeia 1998:1032–1036.

Webb JF, Blum SD (1990) A swim bladder-lateral line connection in the butterflyfish genus *Chaetodon* (Perciformes: Chaetodontidae). Am Zool 30:98A.

Webb JF, Smith WL (2000) The laterophysic connection in chaetodontid butterflyfish: Morphological variation and speculations on sensory function. Phil Trans Roy Soc Lond B 355:1125–1129.

Weber, EH (1820) De Aure et Auditu Hominis et Animalium. Pars I. De Aure Animalium Aquatilium. Leipzig: Gerhard Fleischer.

Wysocki LE, Ladich F (2001) Can fishes resolve temporal characteristics of sounds? New insights using auditory brainstem responses. Hear Res 169:36–46.

Yan HY (1998) Auditory role of the suprabranchial chamber in gourami fish. J Comp Physiol A 183:325–333.

Yan HY (2001) A non-invasive electrophysiological study on the enhancement of hearing ability in fishes. Proc IOA 23:15–26.

Yan HY, Curtsinger WS (2000) The otic gasbladder as an ancillary auditory structure in a mormyrid fish. J Comp Physiol 186:595–602.

Yan HY, Popper AN (1992) Auditory sensitivity of the cichlid fish *Astronotus ocellatus* (Cuvier). J Comp Physiol A 171:1–5.

Yan HY, Popper AN (1993) Acoustic intensity discrimination by the cichlid fish *Astronotus ocellatus* (Cuvier). J Comp Physiol 173:347–351.

Yan HY, Fine ML, Horn NS, Colon WE (2000) Variability in the role of the gasbladder in fish audition. J Comp Physiol A 186:435–445.

Yost WA (1994) Fundamentals of Hearing. San Diego, CA: Academic Press.

5 Bioacoustics and the Lateral Line System of Fishes

JACQUELINE F. WEBB, JOHN C. MONTGOMERY, AND JOACHIM MOGDANS

1. Introduction

The complexities of the physics of underwater acoustics, the functional similarities of the hair cells in the lateral line and ear, and the presence of mechanical linkages between the ear, lateral line and swimbladder blur any hard and fast distinctions between acoustic and hydrodynamic receptors and between acoustic and hydrodynamic stimuli (see Braun and Grande, Chapter 4). Thus, any treatment of underwater sound must address the contributions of the lateral line system in the interpretation of acoustic stimuli of biological importance.

In a free-field situation (at some distance from the stimulus source, and away from physical boundaries), the acoustic field generated by a vibrating source (e.g., a monopole or dipole) is well defined mathematically (see Section 3). However, even in such a simple physical system, local flow fields close to the vibrating source (e.g., in the "nearfield") can provide information about the source that is distinct from that provided by the propagated acoustic field (e.g., in the "farfield"). In other words, the stimulus field generated by a vibrating object includes both acoustic and hydrodynamic components and movements of the acoustic source, or the water in which a fish sits, makes the distinction between these two components of the stimulus more difficult to identify. The fact that the most interesting biology happens close to a source (within a few body lengths), and in the presence of boundaries (e.g., in a rock-lined stream, or near the ocean bottom) where simple free-field conditions do not exist, make the analysis of acoustic stimuli challenging. The structural diversity of the lateral line system, including that of both the neuromasts (canal and superficial) and lateral line canals, poses challenges to the study of lateral line function in both laboratory and natural behavioral settings and should be acknowledged in any experimental analysis.

2. Anatomy and Biomechanics of the Lateral Line System

The morphology of the mechanosensory lateral line system of bony fishes has been reviewed in detail elsewhere (e.g., Coombs et al., 1988, 1992; Webb

1989b, 2000a,b; Coombs and Montgomery 1999). Here, we briefly review the structural attributes of the lateral line system, emphasizing structure–function relationships that specifically relate to the reception of both hydrodynamic and acoustic stimuli.

2.1 Structural Diversity of Lateral Line Canal Systems in Bony Fishes

A series of cranial lateral line canals are integrated into a subset of the dermal bones in the skull in bony fishes (e.g., Webb 1989b; Cubbage and Mabee 1996; Tarby and Webb 2003; Webb and Shirey 2003). Four lateral line canal patterns have been recognized: simple-narrow, branched, reduced, and widened (Fig. 5.1; reviewed by Coombs et al. 1988; Webb 1989b, 2000a,b). A narrow canal system, characterized by well-ossified canals of uniform diameter, is typical of most bony fishes, including common experimental model species (goldfish, *Carassius auratus*: Puzdrowski 1989; trout, *Onchorhynchus mykiss*: Engelmann et al. 2002; zebrafish, *Danio rerio*: Webb and Shirey 2003). Canal neuromasts are located in

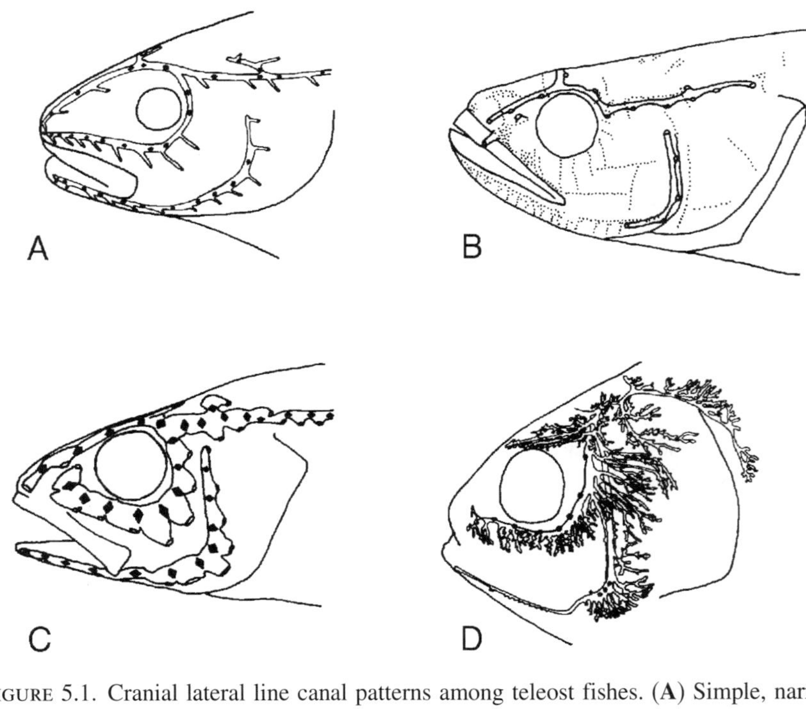

FIGURE 5.1. Cranial lateral line canal patterns among teleost fishes. (**A**) Simple, narrow canals, *Pollachius virens* (a gadid). (**B**) Reduced canals, a generalized gobiid. (**C**) Widened canals, *Percarina demidoffi* (a percid). (**D**) Branched canals, *Brevoortia tyrannus* (a clupeid). Canal and superficial neuromasts are represented by black circles or diamonds. Only canal neuromasts are illustrated in A,C,D. Canal neuromasts are not illustrated in **B**. (From Webb [1989b]. Reproduced with permission of Karger, Basel.)

predictable positions within both the cranial and trunk canals of teleost fishes; one canal neuromast is found between sequential pores of a canal and one neuromast is found in the canal in each lateral line scale in teleost fishes (Webb and Northcutt 1997).

On the trunk, eight canal patterns describe variation in the length, placement, and number of lateral line canals among teleost fishes (Fig. 5.2; reviewed by

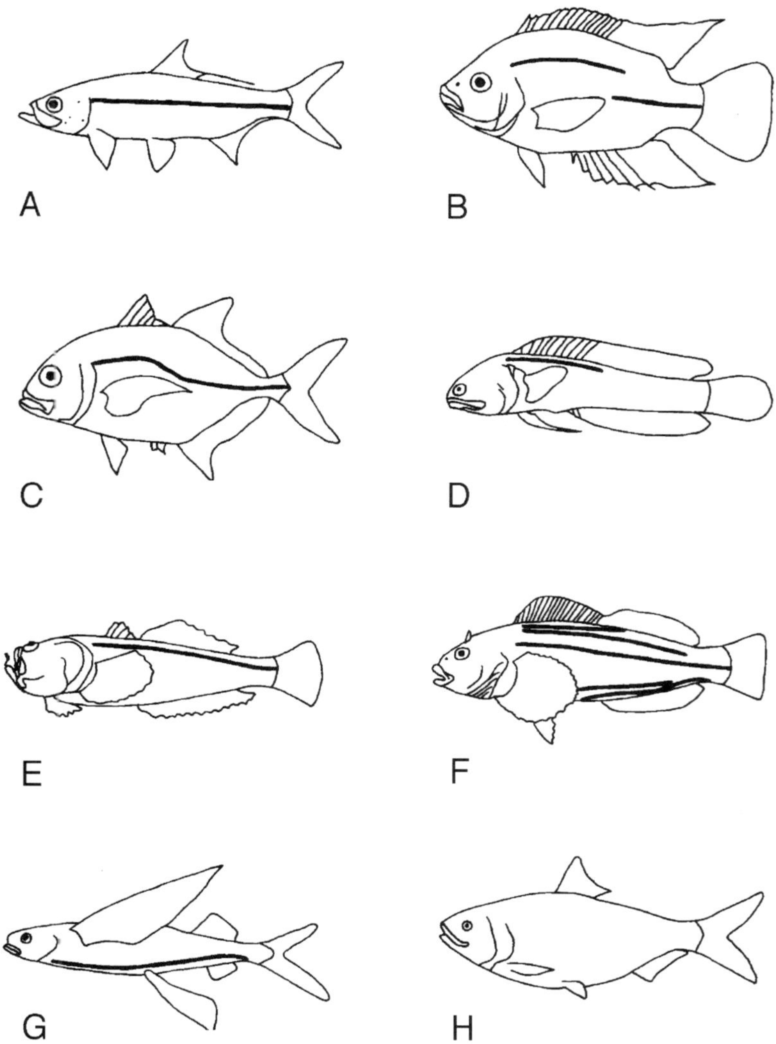

FIGURE 5.2. Trunk canal patterns among teleost fishes. (**A**) Complete (straight; a tarpon). (**B**) Disjunct (a cichlid). (**C**) Complete (arched; a jack). (**D**) Incomplete (a jawfish). (**E**) Complete (dorsal placement; a stargazer). (**F**) Multiple (a greenling). (**G**) Complete (ventral placement; flying fish). (**H**) Absent (a herring). Superficial neuromasts are not illustrated. (From Webb [1989b]. Reproduced with permission of Karger, Basel.)

Webb 1989b, 2000a,b). Typically, however, the trunk canal is generally in the form of a single tube incorporated into a series of overlapping scales on the midflank of the fish, which extend from the operculum to the caudal peduncle.

2.2 Neuromast Anatomy

Fishes generally have two kinds of neuromasts: canal neuromasts, which are located within the lateral line canals on the head and trunk, and superficial neuromasts, which occur in either linear series, clusters, or individually, on the skin of the head and trunk (Fig. 5.3). Canal neuromasts are generally up to ~400 μm in length and vary in shape (e.g., round, oval, elongate, or diamond; Webb 1989a; but see Webb and Shirey 2003). In contrast, superficial neuromasts tend to be small (~10–50 μm) and either round, or diamond shaped (Coombs et al. 1988; Webb, unpublished data).

Like the sensory epithelia (maculae) of the inner ear of fishes, neuromast receptor organs of the lateral line system are composed of a population of sensory hair cells and nonsensory support cells. Each sensory hair cell has one longer kinocilium and many stereocilia, which are graded in length and placed to one side of the kinocilium. The planar polarity of a hair cell is defined by the placement of the kinocilium relative to the stereocilia, which defines the axis of best physiological sensitivity to fluid flow. Unlike the sensory epithelia of the inner ear of fishes (in which each of four quadrants is populated by hair cells with only one orientation; see Popper and Schilt, Chapter 2), hair cells with opposing (180°) polarities are found throughout the sensory epithelium of a neuromast, creating one axis of best physiological sensitivity (see Fig. 5.3B, C). Nonsensory cells are found among the hair cells in the sensory epithelium and are generally found in a region surrounding the sensory epithelium defining the shape of the neuromast. The ciliary bundles of all of the hair cells are embedded in a gelatinous cupula, which provides the mechanical linkage between the hair cell population and the hydrodynamic environment within the lateral line canal or on the body surface. The cupula can be visualized with vital stains in living animals, but is generally lost during scanning electron microscopy and histological preparation (Webb, unpublished data), so accurate description of the cupula has been provided for only a limited number of species. Nevertheless, the minimum length of the cupula can be inferred from the maximum length of the kinocilia in a neuromast, the maximum length of the cupula of canal neuromasts can be inferred from canal diameter, and the shape of the base of the cupula of both canal and superficial neuromasts can be inferred from the shape of the neuromast.

2.3 Biomechanics of the Lateral Line System

All hair cells, whether in the inner ear or in the lateral line system, function as detectors of fluid flow. In neuromasts, hair cells are stimulated by unidirectional or oscillatory fluid flows that cause the gelatinous cupula to slide across the

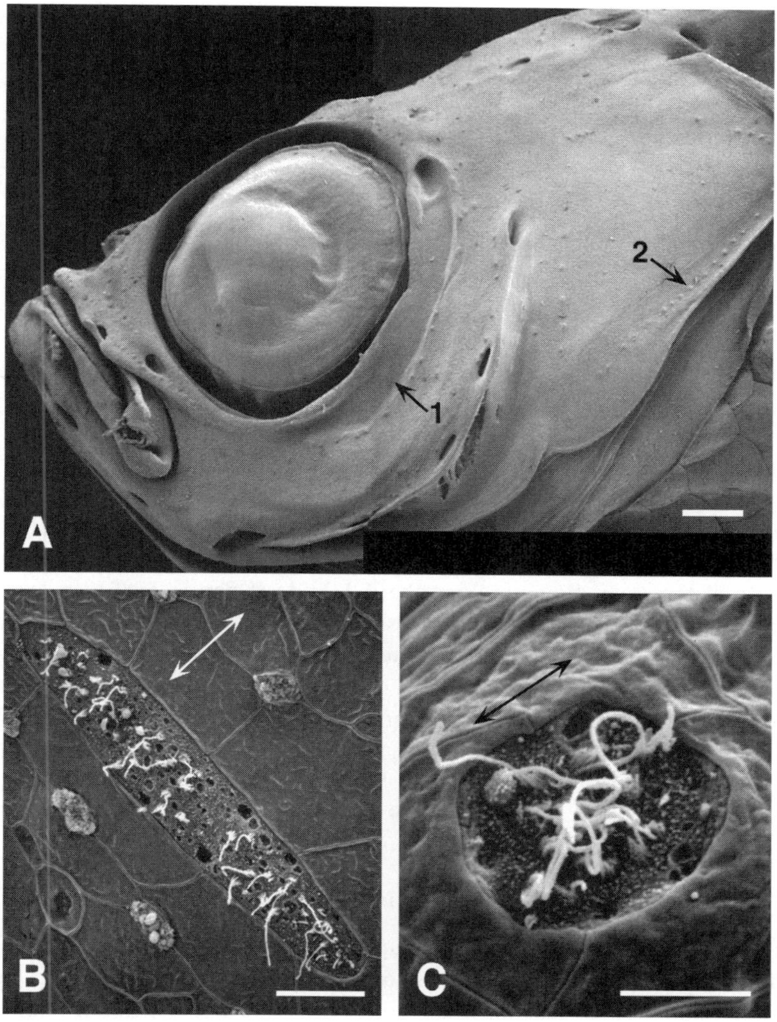

FIGURE 5.3. Two classes of neuromasts with distinctive morphologies on the head of the zebrafish, *Danio rerio*. (**A**) Lateral view of head of a 20-mm SL zebrafish. The infraorbital canal (just caudal to orbit) is still an open groove, whereas preopercular, mandibular canals and a portion of the infraorbital canal (in the lacrimal bone) are already enclosed with prominent canal pores (scale bar = 500 μm). (**B**) Presumptive canal neuromast in infraorbital canal groove at arrow 1 in **A** (scale bar = 10 μm). (**C**) Superficial neuromast in row on posterior edge of operculum at arrow 2 in **A** (scale bar = 5 μm). Double-headed arrows in **B, C** = axis of best physiological sensitivity of hair cells in neuromast. (From Webb and Shirey [2003]. Reprinted with permission of Wiley-Liss.)

sensory epithelium (van Netten and Kroese 1987), causing shearing of the ciliary bundles of all of the hair cells simultaneously. Within a certain range of amplitudes the membrane potential of a hair cell changes linearly with displacement of the ciliary bundle (e.g., in the ear, Hudspeth and Corey 1977). Displacements of less than 1 nm are sufficient to cause a neural response by hair cells of the lateral line system, and displacements greater than 100 nm cause saturation of the receptor potential in individual hair cells (Kroese and van Netten 1989).

The displacement amplitude of the cupula and ciliary bundle depend on the physical properties of the cupula–hair cell complex and the fluid forces impinging on the cupula. The stiffness of the coupling of the cupula to the sensory epithelium depends on cupula size, the pivoting stiffness of the ciliary bundles of the hair cells, and therefore on the number of hair cells in a neuromast (Denton and Gray 1988, 1989; van Netten et al. 1990). The fluid forces driving the cupula have both viscous and inertial components, which vary with the frequency-dependent thickness of the boundary layer around the cupula (Netten and Kroese 1987). The effect of these fluid forces will also vary with cupula size and shape (Netten and Kroese 1989), and with the morphology of accessory structures including canals and papillae (e.g., in *Porichthys*) surrounding a neuromast, which likely alter the hydrodynamic environment in which it functions.

Superficial neuromasts, which tend to be small, function as velocity detectors. They are driven primarily by viscous drag forces which are proportional to the velocity of the water flowing along the fish body (Kalmijn 1989). In contrast, larger canal neuromasts function as pressure gradient detectors. Fluid flow within lateral line canals occurs as a consequence of the pressure gradient between canal pore positions, which is generated by external fluid flow fields. Outside the canal, the pressure gradient is proportional to the acceleration of water particles (Denton and Gray 1982, 1983), so canal neuromasts may be considered to be acceleration detectors (Kalmijn 1989). See Section 6 for additional details on the response properties of neuromasts.

2.4 Lateral Line Function in an Ontogenetic Context

Fish body size changes dramatically (up to several orders of magnitude) from hatching through sexual maturity. Since the physics of hydrodynamics (e.g., Reynolds number) is dependent on both body size and swimming velocity, normal changes in body size, shape, and swimming capabilities that occur through the life history of a fish are likely to have important implications for the function of hydrodynamic receptors.

Neuromasts are generally present in the epithelium on both the head and trunk at hatching and are functional in larval fishes (e.g., Blaxter et al. 1983; Blaxter 1987; Metcalfe 1989; Blaxter and Fuiman 1990). Neuromast number (e.g., Vischer 1989; Metcalfe 1989; Harvey et al. 1992; Fuiman et al. 2004) and size increase, and neuromast shape often changes as a fish grows (e.g., Blaxter et al. 1983; Münz 1986, 1989; Webb 1989b; Harvey et al. 1992; Wonsettler and Webb 1997; Webb and Shirey 2003; Tarby and Webb 2003). Presumptive canal neuromasts become enclosed in lateral line canals as they develop late during the larval stage (e.g., Tarby and Webb 2003; Webb and Shirey 2003). Other neuromasts remain small and superficial, may proliferate, or may be joined by additional neuromasts that differentiate subsequent to the initial appearance

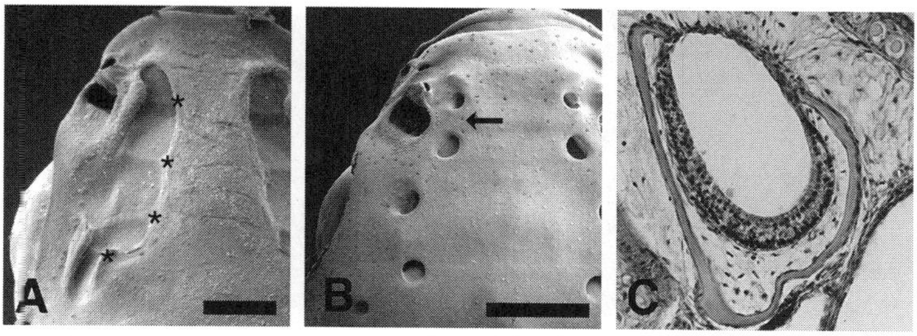

FIGURE 5.4. Ontogeny of supraorbital canal in zebrafish, *Danio rerio*. (**A**) The canal is still a groove in an individual 23 mm SL; *, location of neuromasts. (**B**) The canal has been enclosed and pores are apparent in an individual 31 mm SL. (**C**) Transverse section through canal neuromast at arrow in **B**. Scale bars = 1 mm. (From Webb and Shirey [2003]. Reprinted with permission of Wiley-Liss.)

of neuromasts (often referred to as "secondary neuromasts"). Some superficial neuromasts are found in the vicinity of the lateral line canals in adult fishes as "accessory neuromasts" (Coombs et al. 1988). They generally have hair cell orientations that are parallel to or perpendicular to the length of the canal, providing sensitivity to hydrodynamic stimuli in various axes (Fig. 5.5).

The development of the lateral line canals changes (from depression to groove and then to enclosed canal, Tarby and Webb, 2003; Webb and Shirey, 2003; Figs. 5 4 and 5.5) the hydrodynamic environment in which canal neuromasts function. Further, as canals develop and as superficial neuromasts proliferate, the relative proportions of canal and superficial neuromasts, and thus the proportion of neuromasts that act as accelerometers and velocimeters, respectively, will change. After canal morphogenesis, neuromast size and canal diameter increase (Tarby and Webb 2003), at the same time that body size and swimming speed increase. All of these factors have important implications for hydrodynamics and thus lateral line function through the lifetime of an individual. The timing of these events is likely to vary among species, and this is likely to have interesting implications for the ontogeny of lateral line function. While several behavioral studies have provided evidence that the role of the lateral line changes as a fish progresses through the larval and juvenile stages (discussed by Fuiman et al. 2004), ontogenetic changes in lateral line biomechanics and biophysics at the level of the neuromast and lateral line canal still need to be examined.

3. Lateral Line Stimuli

The lateral line system enables fishes to perceive minute water movements in a variety of behavioral contexts (see Section 4). But what physical aspects of water movements are received and analyzed by the lateral line system, and what distinguishes the function of the lateral line system from that of the auditory system?

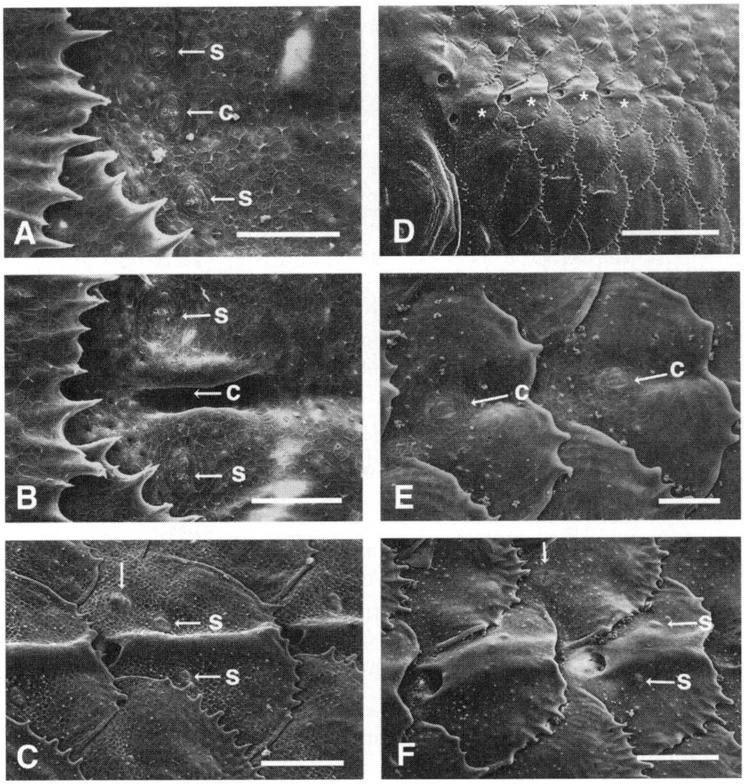

FIGURE 5.5. Formation of the lateral line canal and distribution of canal and superficial neuromasts on the trunk of two cichlids, *Lamprologus brichardi* (**A–C**) and *Archocentrus nigrofasciatus* (**D–F**). Axes of arrows indicate the axes of polarization of the hair cells in the neuromast to which it points. (**A**) In a juvenile *L. brichardi*, before formation of a canal segment, three neuromasts (presumptive canal neuromast, c; superficial neuromasts, s) sit in a vertical linear series (scale bar = 100 μm). (**B**) As the canal starts forming, the canal neuromast (c) sinks into a groove between the two superficial neuromasts (s) (scale bar = 100 μm). (**C**) After the canal has formed, the two superficial neuromasts (s) remain on the skin and other neuromasts may also arise (unlabeled arrow) with axes of polarization orthogonal to the neuromasts already present (scale bar = 200 μm). (**D**) In *A. nigrofasciatus*, the trunk canal starts forming by the sequential enclosure of a presumptive canal neuromast on each lateral line scale. Four enclosed canal segments are illustrated (*), and are followed by scales in which the canal segments have not yet formed (scale bar = 1000 μm). (**E**) Two scales in which the canal segments have not yet formed showing the position of a presumptive canal neuromast (c) in the beginnings of a horizontal depression (scale bar = 100 μm). (**F**) Two scales (from **D**) in which the canal segments have formed. Note that two superficial neuromasts (s) are found just dorsal and ventral to the canal (as in **C**), and another neuromast with orthogonal orientation is found on an adjacent scale (as in **C**) (scale bar = 250 μm). (**D, E, F** from Webb 1989a with permission.)

3.1 The Nature of Lateral Line Stimuli

Acoustic stimuli consist of two inseparable quantities: a local displacement of particles from their resting positions (local flow fields) and a simultaneous local change in sound pressure (propagated acoustic field). Both displacement and pressure spread through the medium as a wave and decrease in intensity with increasing distance from the stimulus source. In the area close to a source ("the nearfield"), the ratio of displacement to pressure is high so the displacement component predominates. With increasing distance from the acoustic source the ratio of displacement to pressure decreases so that sound pressure predominates, although particle motion is still present in the farfield. Although the distinction between "nearfield" and "farfield" is a handy concept in bioacoustics, it is important to know that the transition from nearfield to farfield is gradual and the extent of the nearfield is inversely related to stimulus frequency (for an excellent review of the physics of hydrodynamic events see Kalmijn 1988).

Fishes have approximately the same density as water and are thus moved to and fro by an impinging acoustic wave, with the water that surrounds them; they are acoustically transparent. The ears of all fishes are able to detect the displacement component of a sound wave because the calcareous otoliths in the inner ear are denser than the fish's body and lag behind the movement of the fish body when an acoustic wave passes through the fish (see Popper and Schilt, Chapter 2). This "whole body acceleration" causes relative movement of the fish's body and otoliths, which directly stimulates the hair cells of the otolithic organs ("the direct acoustic pathway"). In addition, a number of fish taxa have specializations that allow them to also detect the acoustic pressure field through an "indirect acoustic pathway." In these fishes, an incoming sound wave causes compression and rarefaction of the volume of air within the swimbladder. Oscillatory movements of the wall of the swimbladder, anterior swimbladder extensions, or modified skeletal elements (fishes with an otophysic connection) generate a secondary displacement field that is transmitted to the ear through soft tissues and fluids, thus allowing the fish to perceive sound pressure (see Popper and Schilt, Chapter 2; Fay and Edds-Walton, Chapter 3; Braun and Grande, Chapter 4).

In contrast to the inner ear, which can be stimulated either directly or indirectly, the neuromasts of the lateral line system detect only relative movement between the fish and the surrounding water. Unidirectional or oscillatory displacement stimuli are detected at each of the hundreds or, in some species, thousands of neuromast receptor organs distributed over the fish's head and trunk. As a result, the lateral line can provide a fish with information both about large-scale water flow (e.g., currents) and small-scale flow patterns (e.g., eddies and vortices). Interpretation by the lateral line system of regional differences in the pattern of water flow over the head and along the fish body can provide the fish with information about the location and nature of an acoustic source.

A substantial amount of our knowledge about lateral line function comes from studies in which a vibrating sphere is used as a dipole stimulus source (see Section 6). The physics of the flow field produced by a dipole source are well understood and can be described in mathematical terms (Kalmijn 1988, 1989). As a vibrating sphere is displaced in one direction, water molecules are pushed away, resulting in increased pressure in front of the source. This causes water to flow around the sphere into the area of decreased pressure that is generated behind the source. As the sphere is displaced in the opposite direction, this process is repeated with opposite sign. The amplitude of the flow can be described in terms of displacement (d), velocity (v), or acceleration (a) of water molecules. Acceleration is proportional to pressure gradients which are important for modeling and predicting lateral line responses (Coombs et al. 1996). Along the axis of sphere vibration, displacement decreases with increasing distance (r) from the sphere at a rate of $1/r^3$. The operating range of the lateral line system is limited to the region relatively close to the source because significant differences in flow (and pressure gradients generated in canals between pore positions) occur over short distances along the length of the fish.

Lateral line function is constrained by another physical phenomenon, the boundary layer. A boundary layer is the velocity gradient that occurs where there is relative movement at the interface between a solid and a fluid (e.g., a fish moving in water, or water flowing over the surface of a fish). In the case of water flowing over the surface of a fish, the boundary layer is defined as the region between the fish's surface, where water velocity is zero, and the distance from the surface where the water velocity reaches 99 % of the velocity beyond the influence of the fish surface (99 % of "free stream"). The thickness of the boundary layer decreases with increasing free stream velocity. In case of an oscillating (AC) stimulus (e.g., potential prey item), the thickness of the boundary layer decreases with increasing stimulus frequency. As the thickness of the boundary layer is decreased, neuromasts are subjected to higher water velocities. Thus, the effectiveness of a lateral line stimulus changes with both the velocity and frequency of water motion relative to the surface of the fish.

3.2 Natural Sources of Lateral Line Stimuli

Sinusoidal water motions produced by a vibrating sphere are convenient and important stimuli that are used to understand how neuromasts function. However, most natural lateral line stimuli are more complex than a simple sine wave. Hydrodynamic stimuli at different scales and from different sources play important roles in the lives of fishes. Water flows are generated by predators, prey, or may be generated by the fish itself. Thus, to understand fully the biological functions of the lateral line, detailed descriptions of the spatial and temporal features of naturally occurring water motions in three dimensions are still needed.

3.2.1 Inanimate Sources of Stimuli

Ocean currents, tides, river flows, wind, temperature, salinity gradients, and gravity all cause large-scale water motions. Both large- and small-scale water flows structure aquatic environments and communities in important ways. Fishes have evolved oriented locomotory behaviors to oppose passive displacement by water flows including tides and currents, or to actively utilize them to some behavioral advantage. Slow water flow may be used by the lateral line system to mediate rheotaxis (Montgomery et al. 1997; Baker and Montgomery 1999) and turbulent flows generated by obstacles in running water may be used for station holding (see Section 4.1) or for the detection of stationary objects (see Section 4.1).

3.2.2 Water Motions Produced by Animals

Planktonic and nektonic animals cannot avoid producing water movements as they feed and swim. Sessile or sedentary animals may generate water motions as a result of respiratory activity (e.g., bivalves) or by the movements of appendages used in suspension feeding or locomotion. The hydrodynamic stimuli generated by moving animals may be short, transient stimuli, long-lasting oscillatory stimuli, or a mixture of both. For instance, oscillatory water movements are generated by swimming zooplankton or fish and generate frequencies from DC (0 Hz, unidirectional flow) to approximately 45 Hz, but they rarely generate higher frequencies (Kirk 1985; Enger et al. 1989; Montgomery 1989; Satou et al. 1991). During undulatory swimming, fishes generate a trail of vortices (Blickhan et al. 1992) whose basic frequency is equal to the tail beat frequency with some high-frequency components up to at least 100 Hz (Bleckmann et al. 1991). The hydrodynamic trail generated by a swimming fish may last up to several minutes (Hanke et al. 2000; Hanke and Bleckmann 2004), which may allow predatory fish to track these trails with their lateral line systems.

3.2.3 Self-Induced Flow Fields

As fishes glide forward they generate a high-pressure bow wave in front of them, and the water that is pushed out of the way flows around the fish's body into the low-pressure area behind it, thus generating a dipole-like flow field (Hassan 1985, 1989). The low-frequency water velocity and pressure alterations generated by a swimming or gliding fish depend on swimming velocity and acceleration as well as body size and shape. As a fish approaches or passes a stationary object the flow field and the pressure distribution over the body's surface are altered, thus allowing the lateral line receptor array to detect flow distortions and provide information on the nature of that object (von Campenhausen et al. 1981; Hassan 1985, 1989, 1992a,b). When a fish glides past an object, the water velocity between an object and a fish is increased locally and this local increase travels along the fish's body as a transient event (Hassan 1985). As the distance between the object and fish decreases, the amplitude of the transient event increases, and its spatial extent decreases. As object size increases, its amplitude also increases. By measuring the spatial extent and size of such transient events, Hassan (1985) showed that the lateral line system may provide a fish with precise information about the size and distance of the object. This allows a fish to use a sort of "hydrodynamic imaging" process (Weissert and von Campenhausen 1981, reviewed by Montgomery et al. 2001) where the lateral line is employed in an "active mode," somewhat analogous to electrolocation by weakly electric fishes, although it uses quite different physical principles. There is also some evidence for the use of echolocation in the hardhead sea catfish (*Arius felis*), which are able to locate and avoid close-range obstacles by listening to the returned echos from the short, broad-band ($<$100 Hz to $\sim$1500 Hz) sound pulses that they produce (Tavolga 1976).

3.2.4 Stimuli from Surface Waves

Surface waves may be caused by terrestrial insects falling into the water or by aquatic animals contacting the water–air interface to feed or obtain oxygen. Surface waves caused by terrestrial insects last for several seconds and have displacement amplitudes less than 100 μm and an irregular time course (Lang 1980; Bleckmann 1988). Surface waves that are generated by aquatic and semiaquatic invertebrates and vertebrates that contact the water–air interface have a more regular time course and are usually shorter than insect-generated waves (Bleckmann 1988). Fish-generated surface waves have a relatively narrow-band spectrum with an upper frequency limit of 20–50 Hz, while surface waves generated by struggling insects have upper frequency limits above 50 Hz (Lang 1980).

The physical properties of surface waves differ from those generated by subsurface wave sources (e.g., Bleckmann 1993, 1994). First, the propagation velocity of surface waves is orders of magnitude slower than that of subsurface waves ($\sim$1500 cm/s). Second, the propagation velocity of surface waves is frequency dependent (e.g., for frequencies above 13 Hz, propagation velocity increases with increasing frequency). Third, surface waves attenuate with increasing distance from the wave source in a frequency-dependent manner (high frequencies attenuate more quickly than low frequencies), while attenuation of subsurface acoustic waves is frequency independent.

4. Lateral Line–Mediated Behavior

The analysis of behavior is sometimes viewed as a lesser endeavor than the theoretically and technologically more sophisticated task of defining the physical nature of acoustic or hydrodynamic stimuli, or characterizing the anatomical basis for and electrophysiological mechanisms underlying the sensory capabilities of animals. While physics can identify the possible mechanisms of detection, and anatomical and electrophysiological evidence can determine whether the animal has the capacity to utilize that information, only behavioral studies can tease apart the algorithms that animals use to navigate, feed, or communicate. Orientation to natural water flows, predatory behavior, and communication are examples of behaviors mediated by the lateral line system.

4.1 Hydrodynamics of Swimming

Stream flows experienced by fishes are rarely simple or laminar. When close to the substrate, or close to obstacles in the flow field, a fish will experience complex flow patterns that contain valuable information that can be interpreted by the lateral line system. Flow refuges (e.g., those generated by hollows, bottom contours, or obstructions, such as boulders) are prime sites where a fish can hold station at little metabolic cost yet remain close to the flow to pick up food carried in the drift. Hydrodynamic features, such as shear gradients or turbulence, may provide information to the lateral line system that will allow a fish to select and optimize the use of these locations. When a cylindrical obstacle is placed across the flow in a flume, a trout will position itself either in front of or behind the cylinder in a way that reduces the metabolic cost of swimming. In front of the cylinder, the fish is able to hold position against the flow with minimal effort. Behind the cylinder, the fish alters its body kinematics to synchronize with the shed vortices in a way that suggests it can capture energy from vortices generated by the environment to maintain station in downstream flow (Liao et al. 2003). This behavior has been termed a "Karman gait" after the name of the vortex trail that is shed by a cylinder in a flow. The lateral line system has been implicated in obstacle entrainment (Sutterlin and Waddy 1975) and involves both the superficial and canal neuromast submodalities (Montgomery et al. 2003), but the precise role of the lateral line system in organizing the Karman gait has yet to be determined.

4.2 Behavioral Orientation to Water Flows

Fishes may actively orient either positively (upstream) or negatively (downstream) to water flows. Rheotaxis is defined as the orientation to such large-scale water currents. Upstream migration is the most familiar example of positive rheotaxis, but fishes holding station in a stream also generally exhibit positive rheotaxis. In contrast, juvenile flounder (*Rhombosolea plebeia*), for instance, orient and actively move downstream (negative rheotaxis) during ebb tides, which may be a strategy to avoid being stranded on a falling tide (Benham 2001). Downstream orientation is also associated with selective tidal stream transport where a fish rides favorable tides and sits out adverse tidal streams. In both of these cases, information on tidal stream direction might appear crucial to the initiation of the behavior, but there is no direct demonstration of lateral line involvement in such behaviors (Harden Jones et al. 1979). In other cases, rheotactic behavior may be passive. For instance, salmon parr (juveniles) demonstrate station holding during the day, but then move up from the substrate and drift passively with the slow stream at night (Thorpe 1988).

If an animal is embedded within a current, rheotaxis is not possible unless the fish can detect a stationary reference frame. Such an external frame of reference is typically established by visual or physical contact with the substrate, thus suggesting that multiple sensory modalities control rheotaxic behavior (see Section 5). The role of the superficial neuromasts of the lateral line system in mediating rheotactic behavior has been demonstrated in a small number of species (Antarctic notothenioid, *Pagothenia borchgrevinki*, torrentfish, *Cheimarrichthys fosteri*, and blind cavefish *Astyanax fasciatus*: Montgomery et al. 1997; Port Jackson shark, *Heterodontus portusjacksoni*: Peach 2001; mottled sculpin, *Cottus bairdi*: Kanter and Coombs 2003). The contribution of superficial neuromasts is most clearly seen in fishes that sit on the substrate and turn to face into the flow as current speed increases. In *Pagothenia*, for instance, the threshold current speed for this behavior increases following selective ablation of the superficial neuromasts (Montgomery et al. 1997) demonstrating that superficial neuromasts provide information about water flows over the surface of a fish's body. This is presumably the case for flows generated by currents or for flows generated by the animal's own swimming movements. This raises the possibility that the lateral line is involved in the control of swimming, mediating, for instance, the dynamic modulation of fin reflexes as a function of flow speed, or the active control of boundary layer flows (Anderson et al. 2001).

Interestingly, rheotaxis is also part of the odor search strategy of fishes, extending the role of multimodal input for rheotactic behavior to include chemosensory systems. The dispersal of an odor from a point source is largely determined by the local flow conditions, and so it makes good sense for animals tracking odor plumes to process flow information. For example, the banded kokopu (*Galaxias fasciatus*) swims upstream to the source of an odor plume when in contact with the plume, and casts across the stream to relocate it if contact with the plume is lost. These fishes are still able to locate the source without their superficial neuromasts, although they take about twice as long, and follow a much more convoluted path to the source (Baker et al. 2002). Freshwater eels (*Anguilla australis* and *A. dieffenbachii*) also show an olfactory-released rheotaxis as part of their odor search strategy (Carton and Montgomery 2003).

4.3 Predatory Behavior

Animals generate water flows as the result of gill ventilation, feeding, and locomotion, which all serve as sources of stimuli for the lateral line system of piscine predators. The role of the lateral line system in feeding and predation has been studied extensively, but fishes from low-light habitats, and especially more accessible nocturnal species, have been the focus of these studies (Table 5.1). The relative importance of the lateral line system increases at night and in other low-light habitats where visual capabilities are constrained (e.g., in the deep sea, in caves, or at high latitudes, where light intensities are low or day length is short).

The overall predation strategy employed by a fish depends on the nature of the target and the combination of sensory modes employed. For pure lateral line-mediated predation, stationary (but vibrating) targets allow a saltatory approach, where the predator appears to sample the available field and then moves closer to the target in a series of steps before aiming its strike (e.g., mottled sculpin, Coombs et al. 2001). Ventilatory currents generated by stationary prey may

TABLE 5.1. Evidence for the role of the lateral line system in feeding behavior among bony fishes that feed under low light conditions (e.g., nocturnal species) or live at high latitudes where light levels and day length vary seasonally

	Common name	Behavioral evidence	Reference
Species that feed under low-light conditions			
Ostariophysi			
Gnathopogon elongatus caerulescens	Willow shiner	Live prey capture in the dark	Mukai et al. 1994
Astyanax fasciatus	Blind cavefish (banded astyanax)	Hypogean; review of feeding behavior, and nonvisual prey detection distance	Montgomery et al. 2001
Apteronotus albifrons	Black ghost knifefish	Nocturnal; modeling results suggest a role for lateral line in prey capture	Nelson et al. 2002
Salmoniformes			
Oncorhynchus mykiss	Trout (Rainbow trout)	Diurnal and nocturnal; Sensory ablation shows role for superficial and canal neuromasts in live prey capture	Montgomery et al. 2002
		Diurnal and nocturnal; Feeding rate on large benthic prey unaffected by high turbidity	Rowe et al. 2003
Gadiformes			
Theragra chalcogramma	Walleyed pollock (Alaska pollock)	Diurnal; feeding rates maintained in the dark	Ryer et al. 2002
Scorpaeniformes			
Scorpaena papillosa (=S. papillosus)	Dwarf scorpionfish (red rock cod)	Nocturnal; prey detection distance of ~10 cm in the dark	Montgomery and Hamilton 1997
Cottus bairdi	Mottled sculpin	Nocturnal; live prey capture in the dark; capture of buried prey	Hoekstra and Janssen 1985; Janssen 1990; Jones and Janssen 1992
Anoplopoma fimbria	Sablefish	Diurnal; feeding rates maintained in the dark	Ryer et al. 2002
Cyprinodontiformes			
Aplocheilus lineatus	Top minnow (Striped panchax)	Nocturnal; uses lateral line to detect surface waves generated by prey	Bleckmann et al. 1989; Vogel and Bleckmann 1997
Hyporamphus ihi	Piper (halfbeak)	Nocturnal; live prey capture in the dark	Saunders and Montgomery 1985; Montgomery and Saunders 1985

TABLE 5.1. (continued)

	Common name	Behavioral evidence	Reference
Perciformes			
Gymnocephalus cernuus	Ruffe	Nocturnal; live prey capture in the dark	Hay and Adams 1997
Cheimarrichthys fosteri	Torrent fish	Nocturnal; live prey capture in the dark	Montgomery and Milton 1993
Siniperca chuatsi	Chinese perch	Nocturnal; sensory ablation shows feeding mediated by lateral line	Liang et al. 1998
Genyagnus monopterygius	Spotted stargazer	Nocturnal; detects moving prey in the dark	Montgomery and Coombs 1998
High-latitude species			
Perciformes			
Pagothenia borchgrevinki	Antarctic fish (Bald notothen)	Functional properties of lateral line	Montgomery and Macdonald 1987
4 species	Antarctic fish	Feeding behavior in blinded fish	Janssen 1996
Other species			
Esox masquinongy	Muskellunge	Diurnal; sensory ablation shows a critical role for lateral line in final prey capture.	New et al. 2001
Batrachocottus baicalensis, Paracottus kneri	Lake Baikal sculpins	Prey detection distance in still water is greater for species with wide lateral line canals	Janssen et al. 1999
Micropogonias undulatus	Atlantic croaker	Sensory ablation shows feeding responsiveness in larvae can be influenced by lateral line	Poling and Fuiman 1997

If different than that used in the literature, scientific and/or common names in parentheses follow www.fishbase.org.

result in significant hydrodynamic flows that can be detected by the lateral line system of dwarf scorpionfish (*Scorpaena papillosus*), which use them to track and strike at their source (Montgomery et al. 1997). When fish are pursuing fast swimming planktonic prey, their lateral line system must provide sufficient information to determine and intercept the prey's locomotory trajectory. When feeding on mobile prey, the predator may encounter and then follow a turbulent wake generated by the swimming motion of the prey (Hanke et al. 2000). Such hydrodynamic trail following has been demonstrated in a nocturnal catfish (Pohlmann et al. 2001, 2004). Other evidence from rainbow trout (*Onchorhynchus mykiss*) shows that prey tracking depends on both superficial and canal neuromasts and may be accompanied by chemical trails, which may be sufficient to allow predators to track moving prey (Montgomery et al. 2002).

The topminnow, *Aplocheilus lineatus*, and the African butterflyfish, *Pantodon buchholzi*, both feed on insects that fall on the water surface. Each has a specialized cranial lateral line system that is able to determine the direction of a wave source from the intervals between the arrival times of a surface wave at different head neuromasts (Müller and Schwartz 1982; Tittel et al. 1984). Moreover, these fish can determine the distance to a wave source from the curvature of the concentric surface wave, its frequency composition, and the frequency modulation of the first seven to eight wave cycles that arrive at the fish (Hoin-Radkovski et al. 1984; Bleckmann and Käse 1987; Bleckmann, 1988, 1993; Bleckmann et al. 1989; see Sand and Bleckmann, Chapter 6).

4.4 Communication

Sound production and acoustic communication are well documented across a wide range of fish species (see Bass and Ladich, Chapter 8), but our understanding of the role of hydrodynamic signals in communication and the potential use of the lateral line in the interpretation of such signals is just in its infancy. The use of communication channels that are more hydrodynamic ("nearfield") than acoustic ("farfield") in nature has been documented in mating interactions in red salmon, *Oncorhynchus nerka* (Satou et al. 1991, 1994) and in schooling communication in herring (*Clupea harengus*) (Gray and Denton 1991). Weeg and Bass (2002) provide evidence that acoustic vocalizations by the plainfin midshipman (*Porichthys notatus*) may stimulate the lateral line system in addition to the inner ear, thus allowing multimodal interpretation of these signals. Sounds produced by chaetodontid butterflyfishes may be interpreted by the lateral line system in addition to the ear as the result of the presence of a laterophysic connection (see Section 5).

5. Multimodal Integration of Hydrodynamic and Acoustic Stimuli

Multimodal integration or functional overlap between inputs to the lateral line and inner ear is largely predicted on theoretical grounds. Braun et al. (2002) define four octavolateralis submodalities (direct and indirect acoustic stimulation of the inner ear, and hydrodynamic stimulation of canal neuromasts and of superficial neuromasts) and provide an interesting discussion about how pairs of sensory modalities can interact and guide behavior.

Consideration of the complex nature of acoustic and hydrodynamic stimuli leads to a prediction of "range fractionation" among inner ear (acoustic) and lateral line (hydrodynamic) submodalities. The otolithic organs of the ear, which respond to pressure wave stimuli only if the swimbladder provides an indirect acoustic pathway for the transmission and transduction of sound pressure), are expected to operate at greater distances from the acoustic source. Both the otolithic organs, which are directly stimulated by whole body acceleration, and the neuromast receptors of the lateral line system, provide information about hydrodynamic flows close to the source. Interestingly, in species with swimbladders stimulation of the ear via the "indirect acoustic pathway" does not provide directional information, whereas directly stimulated otolithic organs (in species lacking swimbladders) do provide directional cues, but have an inherent 180° ambiguity. The way in which some fish locate prey buried in the substrate also demonstrates a form of acoustic and lateral line fractionation (Braun et al. 2002). For instance, in the initial response to a buried source, the mottled sculpin places its mandible in contact with the substrate and then demonstrates a directed orientation of the body toward the source. The initial substrate contact is mediated via the inner ear, whereas the subsequent orientation response is mediated by the lateral line system (Janssen 1990).

Braun et al. (2002) also provide a number of examples of "complementation" between the inertial and pressure sensitive submodalities of the auditory system (direct and indirect acoustic pathways, respectively) that would allow unambiguous sound source localization. It is thought that only the lateral line can resolve source location close to the source (Coombs et al. 1996), and this prediction is supported by behavioral studies. In the orienting response of the mottled sculpin to a vibrating source, the lateral line canal system is sufficient to mediate both approach and strike behavior. Conditioning studies show that the stimulus is detected via the inner ear; however, there is no evidence that the acoustic stimulation contributes to the orientation behavior (Braun and Coombs 2000). Experimental demonstrations of multimodal overlap are problematic because selective ablation of the inner ear is quite difficult (although chemical ablation of the lateral line system has been accomplished, e.g., Karlsen and Sand, 1987; but see Janssen 2000). However, neurophysiological demonstrations of convergence of lateral line and acoustic inputs at the level of both the hindbrain and midbrain provide evidence of central integration of multimodal input (see Section 7).

The Mauthner cell mediated "C-start" escape behavior is an interesting example of how octavolateralis submodalities are integrated. In a detailed behavioral study of sensory contributions to the C-start, Canfield and Rose (1996) were able to demonstrate in goldfish and a cichlid (*Haplochromis burtoni*) that directional determination and sensory hierarchies differed between species, and that in the absence of vision, directional determination is provided by either the lateral line or the inner ear. This behavioral study did not directly reveal multimodal overlap between the lateral line and the inner ear, but a degree of overlap is implied by the physiological and anatomical evidence that shows that information from both these modalities converges on the Mauthner cell (Canfield and Eaton 1990; Zottoli et al. 1995). A similar sort of convergence is also seen in the torus semicircularis of the midbrain, and in the deeper layers of the optic tectum of several species (rainbow trout: Schellart and Kroese 1989; toadfish, *Opsanus tau*: Fay and Edds-Walton 2001; plain midshipman, *Porichthys notatus*: Weeg and Bass 2002), again implying a role for multimodal sensory integration in the determination of the spatial localization of acoustic stimulus sources.

Chaetodontid butterflyfishes of the genus *Chaetodon* have a laterophysic connection, a unique mechanical linkage of anterior swimbladder diverticula ("horns"), not with the inner ear (an otophysic connection, Schellart and Popper 1992), but with the lateral line canal in the supracleithrum at the posterior margin of the skull (Webb 1998; Smith et al. 2003; Webb et al. 2006). It is hypothesized that the laterophysic connection allows tranduction of sound pressure stimuli from the swimbladder into the lateral line canals to stimulates nearby neuromasts (Webb and Smith 2000; Smith et al. 2003), but experimental evidence is needed to test this hypothesis directly. If this hypothesis is correct, then the laterophysic connection would expand the functional repertoire of the mechanosensory lateral line system to include the reception of sound pressure, which would define yet another acousticolateralis submodality (see Braun et al. 2002). It should be noted that in *Chaetodon* spp., the anterior swimbladder horns also extend to within 1 mm of the well ossified otic capsule (Woods and Webb, unpublished data) so that such sound pressure stimuli, which cause oscillations of the volume of air within the swimbladder, are likely to also stimulate the ear. Nevertheless, the ears of both *Chaetodon* spp. and *Forcipiger* (a butterflyfish genus that lacks a laterophysic connection and swimbladder horns) demonstrate a morphology similar to other teleosts that have unremarkable hearing capabilities (Webb et al. 2005).

Scanning electron microscopy and computed tomographic (CT) imaging of the *Chaetodon* ear (Webb et al. 2005) reveals none of the morphological modifications found in the ear of species with otophysic connections (e.g., holocentrids, *Adioryx*, *Myripristis*, Popper 1977) indicating that the evolution of the laterophysic connection was not accompanied by the evolution of modifications in the size and shape of the sensory maculae of the inner ear. If sound pressure stimuli are simultaneously transmitted to the ear and the lateral line system by the swimbladder and the bilateral anterior horns, then the ear and lateral line system may work together to interpret acoustic stimuli through either "synergy" or "complementation," or as "accessory submodalities" (as defined by Braun et al. 2002). The interpretation of acoustic stimuli by these fishes is particularly important given the recent discovery that several species of *Chaetodon* as well as *Forcipiger* produce several different types of sounds (via unknown mechanisms) in social interactions in noisy coral reef environments (Tricas and Boyle 2005; Tricas et al. 2006).

6. Physiology of the Peripheral Lateral Line System

6.1 Physiology of Lateral Line Afferent Fibers

Recordings from the afferent lateral line nerves are used to characterize the response of the entire sensory transduction and encoding process in the periphery including the populations of both canal and superficial neuromasts innervated by the nerves. The details of these processes have been reviewed elsewhere (e.g., Coombs and Montgomery 1999). Thus, the emphasis here is on describing the physiological responses of the lateral line system at the level of the afferent nerves in response to stimuli presented in both still and running water, and comparing their responses with the equivalent responses of the afferent nerves of the auditory system.

6.1.1 Responses to Dipole Stimuli in Still-Water Conditions

Standard linear systems analysis makes use of sinusoidal stimuli to characterize systems properties. Borrowing the essential elements of this approach, lateral line afferent responses have been extensively characterized in their responses to sinusoidal dipole stimuli (Kroese and Schellart 1992; Montgomery and Coombs 1992; Montgomery et al. 1994). Primary afferent fibers innervating neuromasts in both fishes and amphibians respond to sine waves with sustained and strongly phase-coupled increases in discharge rate (e.g., Bleckmann and Topp 1981; Münz 1985; Zittlau et al. 1986; Elepfandt and Wiedemer 1987; Wubbels 1992). Response measures have been constructed that incorporate increases in discharge rate and phase locking. These measures have been used to examine the response as a function of stimulus frequency in order to describe the response properties of neuromasts (Fig. 5.6). These studies confirm theoretical considerations that the superficial neuromast modality should respond to the flow of water over the cupula, which is friction coupled to water flow (Denton and Gray 1983). With respect to an oscillating source, the expectation is that the gain of the response should be flat at low frequencies when gain is defined with respect to stimulus velocity. Frequency/response curves confirm this expectation. When plotted in these coordinates, superficial neuromasts show a low-pass filter characteristic with a flat response at low frequencies, and a decreasing gain at higher frequencies. The frequency at which this transition occurs varies in different species from about 10–20 Hz in some Antarctic notothenioid species (Montgomery et al. 1994), to about 30–50 Hz in the Nile tilapia, *Sarotherodon niloticus* (Münz 1985). The low-frequency characteristics of superficial neuromasts have been examined further via step changes in water velocity over the surface of the skin of the fish (Engelmann et al. 2000, 2002; Voigt et al. 2000). These experiments show that the superficial neuromasts are essentially nonadapting and provide an ongoing measure of the flow velocity over the skin of the fish. However, another explanation of the nonadapting nature of superficial neuromasts is that water flow is rarely perfectly laminar and usually contains flow fluctuations, i.e., DC changes that increase with increasing flow velocity and to which the neuromasts may respond (Chagnaud, unpublished). In agreement with this interpretation, afferent fibers typically exhibit highly fluctuating or burst-like discharge rates when exposed to unidirectional water flow (Engelmann et al. 2002).

Mechanical considerations predict that a lateral line canal should attenuate low frequency signals (Denton and Gray 1983). Thus, in comparison to the frequency–response curve for superficial neuromasts, the canal neuromast response (plotted on the same velocity coordinates) should show a band pass characteristic. These predictions have been confirmed in experimental studies in several species. The upper frequency roll-off for canal neuromasts is generally

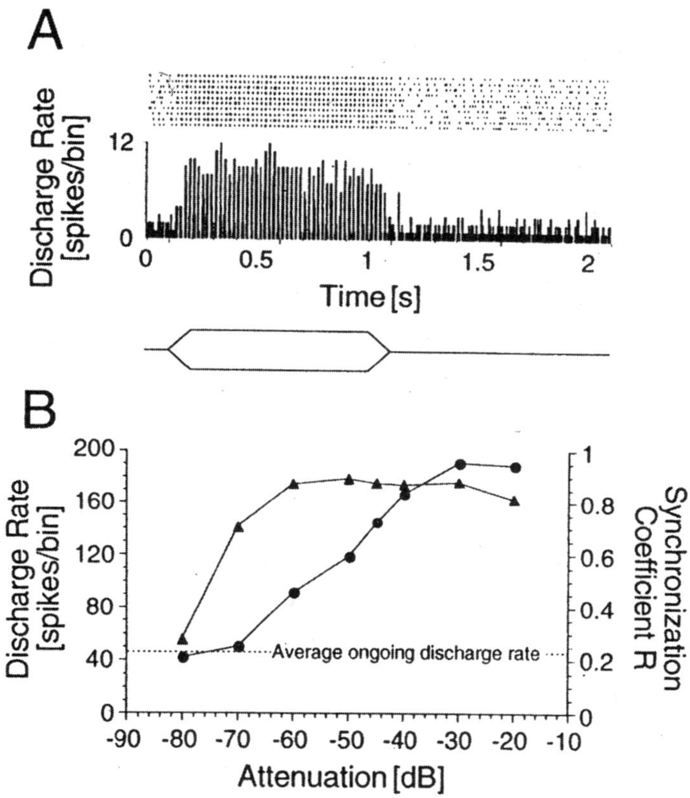

FIGURE 5.6. Responses of a primary afferent fiber in the posterior lateral line nerve of goldfish to a 50-Hz sine wave stimulus (dipole) generated by a 8 mm diameter vibrating sphere. (**A**) Raster diagram (top) and PST histograms (binwidth 2 ms) of the responses to 10 repetitions of the stimulus. Each marker in the raster diagram represents an action potential. The bottom trace represents the stimulus envelope. Displacement amplitude = 4 μm. (**B**) Input–output function of the fiber. Discharge rates (lines connecting circles re: left-hand axis) and synchronization coefficients R representing the degree of phase-locking (lines connecting triangles re: right-hand axis) are plotted as a function of level (in rel. dB). An attenuation of −20 dB corresponds to a vibration amplitude of 425 μm. (Modified after Mogdans and Bleckmann [2001], with permission of Science Publishers, Inc.)

higher than that for superficial receptors and varies from about 30–50 Hz in Antarctic species to around 100 Hz in rainbow trout (Kroese and Schellart 1992) and mottled sculpin (Coombs and Janssen 1990), and up to 200 Hz in the Nile tilapia (Münz 1985). These upper frequencies are below those typical of the auditory system of fishes (Fay and Edds-Walton 2001). The low-frequency attenuation of the response of canal neuromasts may be the result of the canal acting as a filter that reduces low-frequency noise generated by environmental water flows, or by the movements of the fish itself.

Experiments using dipole stimuli have also shown that the receptive fields of primary afferents in the posterior lateral line nerve can be predicted from the pressure gradient field generated by the vibrating sphere (Coombs et al. 1996; Coombs and Conley 1997; Fig. 5.7). The shape and width of the receptive field depend on the orientation of the axis of vibration of the sphere relative to the axis of best physiological sensitivity (hair cell polarization) of the neuromast

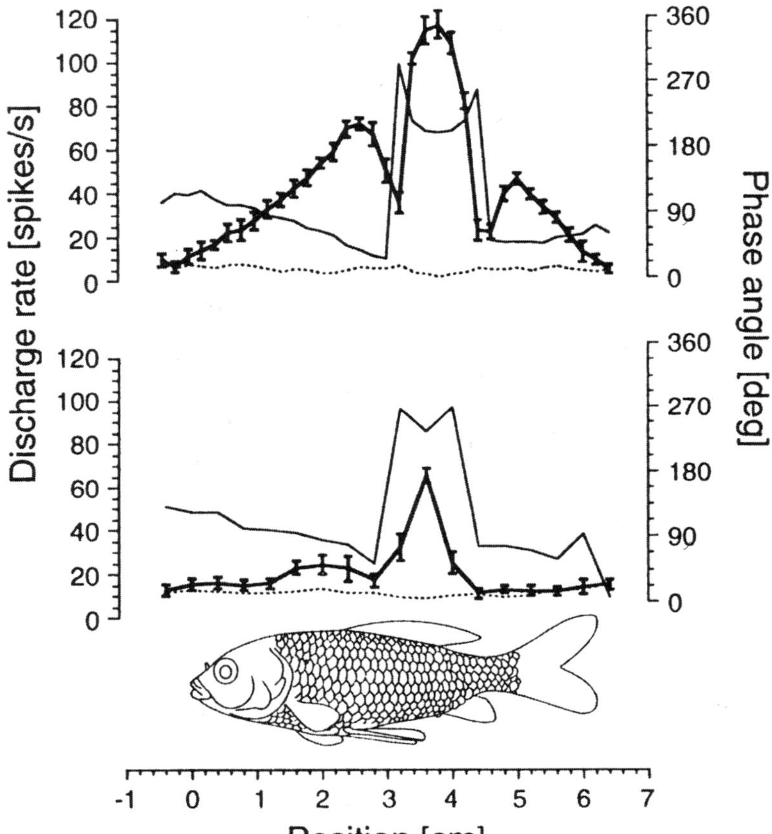

FIGURE 5.7. Receptive field organization of a primary afferent fiber at different stimulus levels in the goldfish. Average discharge rate (solid lines re: left-hand axis) and average phase angle (thin lines re: right-hand axis) are plotted as a function of the location of a sinusoidally vibrating sphere (Ø 8 mm). (**A**) Receptive field measured with a displacement amplitude of 80 μm. (**B**) Receptive field of the same fiber measured with a displacement amplitude of 20 μm. Vertical bars represent standard deviations. Dotted line represents spontaneous discharge rate recorded before each stimulus presentation. The fish symbol is drawn to match the size of the fish in the physiological experiment. Note that in terms of discharge rate the shape of the receptive field depends on stimulus amplitude, whereas the phase angle does not. (Modified after Mogdans and Bleckmann [2001], with permission of Science Publishers, Inc.)

innervated by the recorded afferent fiber (Coombs et al. 1996), the displacement amplitude of the vibrating sphere (Alnaes 1973; Caird 1978), and the distance between the sphere and the surface of the fish (Coombs et al. 1996).

6.1.2 Responses to Moving Objects in Still Water

While the analysis of responses to artificial stimuli have been very important for an understanding of lateral line function, pure sine waves emanating from stationary sources are rare in nature, if they occur at all. To find out how the peripheral lateral line system interprets more complex stimuli, small objects that are moved back and forth along the side of a fish have been used as stimulus sources (Bleckmann and Zelick 1993; Mogdans and Bleckmann 1998; Montgomery and Coombs 1998). The water motions that are caused by a moving object consist of a reproducible transient component followed by an ill-defined, long-lasting wake (Mogdans and Bleckmann 1998). The associated pressure changes and the corresponding pressure gradients that are generated by a moving object are prominent only as the object passes by the receptors, but are negligible in the object's wake. Primary afferents respond to a moving object that produces these kinds of stimuli with a characteristic discharge pattern consisting of alternating periods of excitation and inhibition (Fig. 5.8). While some fibers discharge bursts of spikes even after the object has passed the fish, other fibers do not. Fibers of the first type (type I) presumably innervate superficial neuromasts which are sensitive to water velocity, while fibers of the second type (type II) innervate canal neuromasts whose responses can be predicted from the pressure gradient generated by the moving object (Mogdans and Bleckmann 1998; Montgomery and Coombs 1998).

6.1.3 Responses to Dipole Stimuli in Running Water

Fishes are always exposed to running water because the fish and/or the water surrounding it are in motion. Researchers have begun to study lateral line function under conditions of running water in flow tanks in which a fish is exposed to a constant flow (Engelmann et al. 2000, 2002; Voigt et al. 2000; Carton and Montgomery 2002). Thresholds and slopes of flow-response functions may vary across fibers (Voigt et al. 2000; Carton and Montgomery 2002) and consequently, the number of fibers that respond to water flow increases with increasing flow velocity (Chagnaud et al. 2007). Moreover, fibers that respond to a background water flow almost always do so by increasing their discharge rates. This finding at first glance appears to be in disagreement with hair cell organization and innervation within a neuromast. As described in Section 2.2, neuromasts contain two populations of hair cells with opposite orientation and, each population is innervated by a different fiber (Görner 1963). Therefore, assuming a perfect laminar flow, about 50 % of the fibers from superficial neuromasts should respond with an increase, and the other 50 % with a decrease, in discharge rate. Natural flows usually show fluctuations (Hanke et al. 2001), and in this case afferent fibers

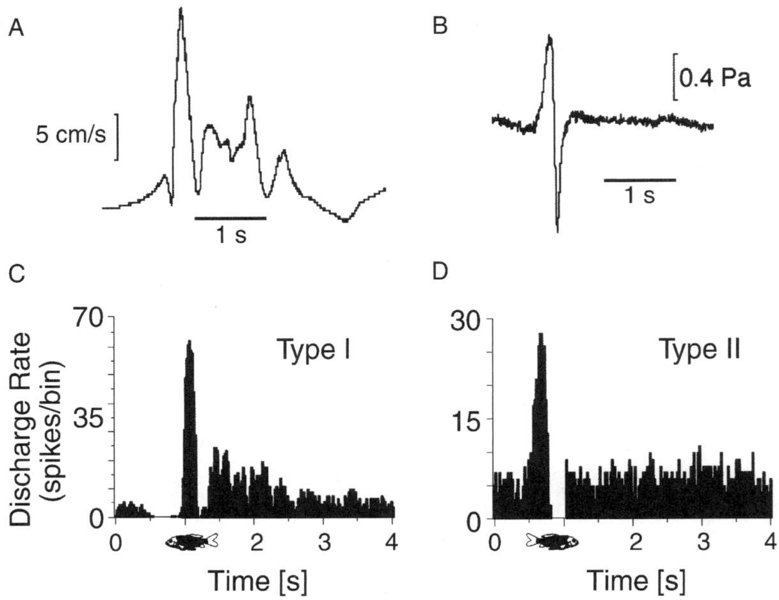

FIGURE 5.8. Peripheral lateral line responses to a moving object. (**A**) Velocity of the water motions generated by a Plexiglas rod (square cross section of 1 × 1 cm) moving with a speed of 15 cm/s. Ten consecutive measurements made with a constant-temperature anemometer were averaged. (**B**) Pressure gradient generated by the same object. Ten consecutive pressure wave measurements made with a hydrophone were averaged and the pressure gradient was calculated assuming an average distance between pores of the goldfish trunk lateral line canal of 2 mm (Coombs et al. 1996). (**C, D**) Peristimulus–time histograms (binwidth 2 ms) of the responses to the moving object of two afferent fibers in the goldfish posterior lateral line nerve. Fish symbols represent location, orientation, and size of the fish relative to the path of the moving object. Note that the neurophysiological data were not recorded simultaneously with the wave measurements. (**C**) Response of a type I unit to the object moving from anterior to posterior. The unit responded with inhibition followed by excitation and again inhibition at about the time when the object was closest to the fish. It continued to fire unpredictable bursts of spikes after the object had passed along the side of the fish. (**D**) Response of a type II unit to the object moving from posterior to anterior. The unit responded with inhibition followed by excitation but barely responded after the object had passed along the side of the fish. (From Mogdans et al. [2003] with permission.).

cannot encode the direction of bulk water flow but instead respond to the fluctuations inherent to the flow (Chagnaud, 2008). The burst-like discharge behavior that can be observed in many fibers (Engelmann et al. 2002) supports this notion.

Nevertheless, Engelmann et al. (2000) revealed two types of primary afferent fibers using unidirectional water flow of an average velocity of 10 cm/s, those that respond to water flow with an increase in discharge rate for as long as it is maintained (type I afferents), and those that do not change their discharge rate in response to water flow maintained (type II afferents). The former most likely innervated superficial neuromasts, which are located on the skin and are thus continuously stimulated by background water flow, whereas the latter presumably innervate canal neuromasts, which are unresponsive to a laminar flow because it does not create pressure differences between canal pores and less sensitive to low-frequency stimuli due to the filter properties of the canals.

The utility of the low-frequency filtration provided by the lateral line canals is elegantly demonstrated by comparing the responses of the two fiber types to a dipole stimulus in both still and running water (Engelmann et al. 2000, 2002). In still water, type I and type II afferents exhibited sustained and phase-locked responses that could hardly be distinguished from each other, demonstrating that both lateral line submodalities (superficial and canal neuromasts) respond well to a stimulus in still water. In contrast, in a water flow of 10 cm/s, the responses of type I fibers were masked, whereas the responses of type II fibers were hardly affected (Fig. 5.9). As expected, the number of fibers in which dipole-evoked responses are masked by the flow increases with decreasing signal-to-noise ratio, i.e., with decreasing ration between dipole displacement and flow velocity (Chagnaud et al. 2007). These findings demonstrate a clear functional difference between the responses of superficial and canal neuromasts. Superficial neuromasts, which are continuously exposed to the running water, become saturated in their response to the flow and thus fail to encode the oscillating stimulus, while canal neuromasts are unaffected by the flow and continue to encode the oscillating stimulus. Engelmann et al. (2002) obtained comparable results from goldfish and rainbow trout (Engelmann et al. 2002) suggesting that neuromasts are affected in a similar manner in species with different lateral line morphology that live in different hydrodynamic regimes.

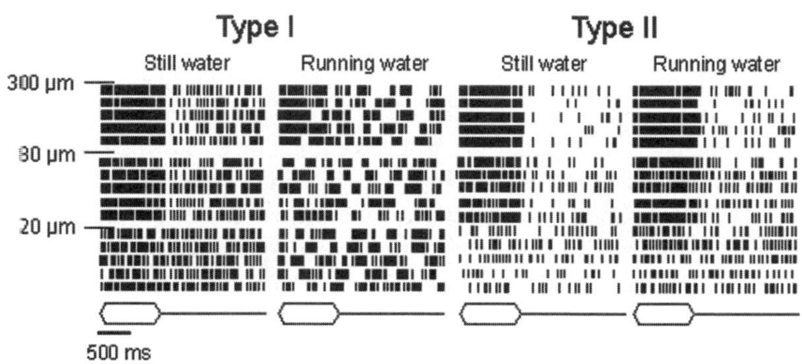

FIGURE 5.9. Responses of goldfish type I and type II posterior lateral line nerve fibers to a 50-Hz sine wave stimulus generated by a 10 mm diameter vibrating sphere. Raster diagrams of the responses to five stimulus repetitions are shown for two peak-to-peak displacements. Data were recorded in still water (no flow) and in running water (velocity 10 cm/s). Flow direction was from anterior to posterior. The flow masks the responses of the type I fiber but not that of the type II fiber. Note the burst-like discharges of the type I fiber in flow. (From Mogdans et al. [2003] with permission.)

7. Physiology of Central Lateral Line Neurons

Despite an increasing number of studies on the physiology of central lateral line neurons, our knowledge about the processing of hydrodynamic stimuli by the brain is still in its infancy. The first site of sensory integration in the ascending lateral line pathway is the medial octavolateralis nucleus of the medulla oblongata (MON, e.g., McCormick and Hernandez 1996). The neuroanatomy of this nucleus has been described in some detail (New et al. 1996). Sensory information reaches the MON via several lateral line nerves (e.g., Northcutt 1989; Puzdrowski 1989; Song and Northcutt 1991). Output neurons of the MON project to the contralateral MON, to the torus semicircularis in the midbrain (e.g., McCormick and Hernandez 1996), and then to the thalamus and telencephalon (Murakami et al. 1986; Striedter 1991). The ascending lateral line pathway runs parallel to, but is distinct from, the ascending auditory pathway (Striedter 1991).

7.1 Physiology of the Lateral Line Brainstem

A substantial amount of data concerning central lateral line processing comes from studies of MON neurons. One of the primary functions of the MON is the cancellation of hydrodynamic stimuli associated with a fish's own movements. Montgomery and Bodznick (1994) showed in the red rock cod, *Scorpaena papillosus*, that over time, MON cells learn to ignore a sine wave stimulus that is coupled to the fish's ventilatory movements. Thus, MON neurons are capable of adaptive, context-specific modification of their responses. This may allow a fish to distinguish between self-generated stimuli and those generated by other sources, and to respond readily to novel stimuli. The anatomical correlate that forms the basis for this dynamic signal conditioning is the cerebellar-like organization of the fish brainstem (Montgomery and Bodznick 1994), which is found in both the electrosensory system of fishes and the auditory system of mammals (Montgomery et al. 1995).

Comparing the responses of MON neurons to dipole stimuli with neurons in the periphery, it becomes obvious that spontaneous and evoked rates of activity are lower in the MON, and that responses exhibit greater degrees of adaptation and greater heterogeneity both in terms of the response patterns and phase-coupling (e.g., Wubbels et al. 1993; Montgomery et al. 1996; Coombs et al. 1998). Interestingly, MON units are substantially less sensitive to sine wave stimuli than are primary afferents. For instance, in the goldfish MON, about 30% of the units do not respond to a stationary dipole, even when displacement amplitudes up to 800 μm are used (Mogdans and Goenechea 2000). Such displacement amplitudes are substantially greater than those that cause rate saturation in lateral line afferents. The reason for this insensitivity is not known. However, many of the seemingly insensitive units readily responded to the water motions generated by a moving sphere. This suggests that the lateral line system is adapted for the analysis of water motions that are more complex than pure sine waves.

Receptive fields of MON units are quite heterogeneous and difficult to relate to the stimulus field of a dipole source. Some MON neurons have primary-like receptive fields that can be modeled with excitatory center/inhibitory surround and inhibitory center/excitatory surround organization of the afferent inputs to the MON (Coombs et al. 1998). The receptive fields of other neurons have a more obvious center-surround organization, and consist of areas from which a sine wave stimulus causes excitation and adjacent areas from which a sine wave stimulus causes suppression of the ongoing discharge rate (Mogdans and Kröther 2001). However, other neurons have receptive fields that consist of a single broad area from which a sine wave stimulus causes either excitation or inhibition (Coombs et al. 1998; Mogdans and Kröther 2001). Finally, the receptive fields of some MON neurons can be rather punctate (Coombs et al. 1998; Mogdans and Kröther 2001). The existing data show that lateral line neurons in the brainstem receive information from large parts of the lateral line system on the surface of the animal. However, the functional relationship between the rather complex size and shape of the receptive field and the response of most MON neurons is not clear.

Like the responses to dipole stimuli, the responses of MON units to a moving object are quite heterogeneous. When tested with the same stimulus used to study peripheral lateral line responses, many MON units responded only to the transient and reproducible part of the stimulus but did not show obvious responses to the ill-defined wake (Mogdans et al. 1997). Moreover, many unit responses were apparently independent of the direction of object motion. This is different from lateral line afferents that respond in a direction-dependent manner to a moving source due to the intrinsic directionality of the hair cells within a neuromast (Mogdans and Bleckmann 1998; Montgomery and Coombs 1998).

Studies in which recordings are made in flow tanks have shown that two types of MON neurons can be distinguished based on their neural activity in running water. Flow-sensitive MON cells change their discharge rates in response to a water flow, whereas flow-insensitive MON cells do not change their discharge rates (Kröther et al. 2002, 2004). This finding is not surprising, but it raises the possibility that flow-sensitive MON neurons receive input from flow-sensitive type I afferents (i.e., from superficial neuromast) and that flow-insensitive MON neurons receive input from flow-insensitive type II afferents (i.e., from canal neuromasts). However, the identification of only two types of MON neurons appears to be too simple. Based on their responses to a sine wave stimulus in running water, neurons in the MON of goldfish and rainbow trout were further subdivided in at least four groups (Kröther et al. 2002, 2004). Flow-sensitive type MI cells (M for MON) exhibited increased or decreased levels of activity in running water. Their responses to the vibrating sphere were masked in running water in terms of discharge rate, phase-coupling, or both discharge rate and phase-coupling. In contrast, the responses of flow-insensitive type MII units to the vibrating sphere were not masked in any respect by running water. Type MIII cells were found that, like type MII cells, were flow insensitive. However, in contrast to type MII units, the responses of type MIII units to a vibrating sphere stimulus were masked in running water. Finally, type MIV cell, like type MI units, were flow sensitive. However, unlike type MI cells, the responses of type MIV cells to a vibrating sphere stimulus were not masked in running water. This mixture of physiological cell types suggests that the functional subdivision of the lateral line periphery is not completely maintained at the level of the MON.

Kröther et al. (2004) compared the data from goldfish and rainbow trout and found that the number of MON units whose responses to a vibrating sphere are masked by running water (type MI and MII) is greater in goldfish. Across the entire population of units recorded, the degree of masking in terms of discharge rate was also greater in goldfish. These data suggest that the lateral line system of the rainbow trout is better suited to function under running water conditions than that of goldfish.

7.2 Physiology of Higher Brain Areas

The torus semicircularis of the midbrain receives input from both the auditory and the lateral line portions of the brainstem (see Fay and Edds-Walton, Chapter 3), but little is known about how it processes lateral line information. In some species, auditory and lateral line inputs project to anatomically distinct subdivisions in the torus, but there is also anatomical overlap between these two sensory systems (McCormick 1989; Schellart and Kroese 1989). For this reason, and because subsurface acoustic sources stimulate both auditory and lateral line receptors, auditory and lateral line units in the torus are difficult to distinguish. Plachta et al. (1999) studied the responses of toral lateral line units to sine wave stimuli in goldfish. They found that most units responded phasically and with poor phase-coupling to sine wave stimuli of constant amplitude. However, when the sine wave was amplitude-modulated, toral units responded strongly phase-coupled to each cycle of the modulation frequency. In many respects, responses of toral units to moving objects are similar to those of MON units (Wojtenek et al. 1998). Whereas some toral units responded with rather complex discharges to a moving object, other units exhibited a direction-dependent transient increase in discharge rate. Although there is no direct proof, it is conceivable that, as in the brainstem, the former receive input from superficial neuromasts and the latter receive input from canal neuromasts. Recent evidence suggests that the latter type is organized systematically in the torus semicircularis such that units in the anterior torus semicircularis have receptive fields in the rostral part of the fish whereas successively more caudal units have successively more caudal receptive fields (Plachta et al. 2003; Engelmann and Bleckmann 2004).

Higher brain areas beyond the torus semicircularis that are concerned with the processing of lateral line information are even less well investigated (but see Bleckmann et al. 1987, 1989b). One notable exception is the optic tectum of the African clawed toad, *Xenopus laevis*, in which lateral line units are tuned to the direction of water surface waves and are arranged topographically to form a map of surface wave directions (Claas et al. 1989). Whether tectal units in fishes encode aspects of subsurface water motions and/or whether lateral line maps exist in the tectum of fishes remains to be investigated.

8. Applied Aspects of Lateral Line Bioacoustics

There is increasing interest in the use of acoustics to control fish behavior (Popper 2002; Popper and Schilt, Chapter 2). This is focused largely on the use of sound to drive fish away from water intakes to prevent impingement, but also has potential utility in directing fish toward bypass structures at dams, particularly for down-stream passage. Advances in the basic understanding of the lateral line system and in the behavior associated with rheotaxis also have potential utility in similar situations. Probably the most important use of this information is currently in the design of culverts and fish ladders associated with upstream passage of migratory species (see Popper and Schilt, Chapter 2). An understanding of the orientation mechanisms of fishes in the flows and turbulence of natural rivers, and in reservoirs and forebays of dams provides the prospect of the development of better bypass entrances and passages (Schilt 2002).

9. Summary

The complexity of the physics of underwater acoustics requires a careful consideration of the functional contributions of the mechanosensory lateral line system and the inner ear to the interpretation of acoustic stimuli. The lateral line system, a spatial array of sensory receptor organs distributed in canals and on the skin of the head and trunk, demonstrates a great deal of structural diversity among species and in the course of the ontogeny, which suggests that the lateral line system exhibits a good deal of functional versatility in an evolutionary context and at different time points during the life history of individual species. Recent experimental work has demonstrated the role of the lateral line system in a wide range of behavioral contexts including swimming, schooling, orientation to water flows (e.g., rheotaxis), predatory behavior (including the detection of midwater and surface-borne prey), and communication. Other experimental work has demonstrated the ability of the lateral line system to detect moving objects in both still and running water. The study of central processing of lateral line input has been largely limited to analysis of the responses of peripheral lateral line nerve fibers and a primary projection center (MON, medial octavolateralis nucleus). Finally, the ability of the lateral line system to detect fluid flows near an acoustic source (in the "nearfield") overlaps with the ear's ability to detect such flows via whole body acceleration, complements the ear's ability to detect propagated sound pressure waves, and predicts the importance of multimodal integration of acoustic stimuli that arise from abiotic and biotic sources in aquatic habitats.

References

Alnaes E (1973) Two types of lateral line afferents in the eel (*Anguilla anguilla*). Acta Physiol Scand 87:535–548.

Anderson EJ, McGillis WR, Grosenbaugh MA (2001) The boundary layer of swimming fish. J Exp Biol 204:81–102.

Baker CF, Montgomery JC (1999) The sensory basis of rheotaxis in the blind Mexican cave fish, *Astyanax fasciatus*. J Comp Physiol A 184:519–527.

Baker CF, Montgomery JC, Dennis TE (2002) The sensory basis of olfactory search behaviour in the banded kokopu (*Galaxias fasciatus*) J Comp Physiol 188A: 553–560.

Benham AB (2001) Water currents and navigation by fish in shallow water habitats. MSc Thesis, University of Auckland.

Blaxter JHS (1987) Structure and development of the lateral line. Biol Rev 62:471–514.

Blaxter JHS and Fuiman LA (1990) The role of the sensory systems of herring larvae in evading predatory fishes. J Mar Biol Assoc UK 70: 413–427.

Blaxter JHS, Gray JAB, Best ACG (1983) Structure and development of the free neuromasts and lateral line system of the herring. J Mar Biol Assoc UK 63:247–260.

Bleckmann H (1988) Prey identification and prey localization in surface-feeding fish and fishing spiders. In: Atema J, Fay RR, Popper AN, Tavolga WN (eds) Sensory Biology of Aquatic Animals. New York: Springer-Verlag, pp. 619–641.

Bleckmann H (1993) Role of the lateral line and fish behavior. In: Pitcher TJ (ed) Behaviour of Teleost Fishes. London: Chapman and Hall, pp. 201–246.

Bleckmann H (1994) Reception of hydrodynamic stimuli in aquatic and semiaquatic animals. In: Rathmayer W (ed) Progress in Zoology, Vol.41. Stuttgart: Gustav Fischer, pp. 1–115.

Bleckmann H, Käse R (1987) Prey localization by surface wave-ray tracing—fish track bugs like oceanographers track storm. Experientia 43:290–293.

Bleckmann H, Topp G (1981) Surface wave sensitivity of the lateral line organs of the topminnow *Aplocheilus lineatus*. Naturwissenschaften 68:624–625.

Bleckmann H, Zelick R (1993) The responses of peripheral and central mechanosensory lateral line units of weakly electric fish to moving objects. J Comp Physiol A 172: 115–128.

Bleckmann H, Bullock TH, Jørgensen, JM (1987) The lateral line mechanoreceptive mesencephalic, diencephalic, and telencephalic regions in the thornback ray, *Platyrhinoidis triseriata* (Elasmobranchii). J Comp Physiol A 161:67–84.

Bleckmann H, Gottfried T, Blübaum-Gronau E (1989a) Lateral line system of surface-feeding fish: anatomy, physiology and behavior. In: Coombs S, Görner P, Münz H (eds) The Mechanosensory Lateral Line: Neurobiology and Evolution. New York: Springer-Verlag, pp. 501–526.

Bleckmann H, Tittel G, Blübaum-Gronau E (1989b) The lateral line system of surface-feeding fish: Anatomy, physiology, and behavior. In: Coombs S, Görner P, Münz H (eds) The Mechanosensory Lateral Line: Neurobiology and Evolution. New York: Springer-Verlag, pp. 501–526.

Bleckmann H, Breithaupt T, Blickhan R, Tautz J (1991) The time course and frequency content of hydrodynamic events caused by moving fish, frogs, and crustaceans. J Comp Physiol A 168:749–757.

Blickhan R, Krick C, Breithaupt T, Zehren D, Nachtigall W (1992) Generation of a vortex-chain in the wake of a subundulatory swimmer. Naturwissenschaften 79: 220–221.

Braun CB, Coombs S (2000) The overlapping roles of the inner ear and lateral line: the active space of dipole source detection. Phil Trans R Soc Lond 355:1115–1119.

Braun CB, Coombs S, Fay RR (2002) What is the nature of multisensory interaction between octavolateralis sub-systems? Brain Behav Evol 59:162–176.

Caird DM (1978) A simple cerebellar system: the lateral line lobe of the goldfish. J Comp Physiol A 127:61–74.

Canfield JG, Eaton RC (1990) Swim bladder acoustic pressure transduction initiates Mauthner-mediated escape. Nature 347:760–762.

Canfield JG, Rose, GJ (1996) Hierarchical sensory guidance of Mauthner-mediated escape responses in goldfish (*Carassius auratus*) and cichlids (*Haplochromis burtoni*). Brain Behav Evol 48:137–156.

Carton AG, Montgomery JC (2002) Responses of lateral line receptors to water flow in the Antarctic notothenioid, *Trematomus bernacchii*. Polar Biol 25:789–793.

Carton AG, Montgomery JC (2003) Evidence of rheotactic component in the odour search behaviour of freshwater eels. J Fish Biol 62:501–516.

Chagnaud BP, Bleckmann H and Hofmann MH (2008) Lateral line nerve fibers do not code bulk water flow direction in turbulent flow. Zoology (in press).

Chagnaud BP, Hofmann MH, Mogdans J (2007) Responses to dipole stimuli of anterior lateral line nerve fibers in goldfish, *Carassius auratus*, under still and running water conditions. J Comp Physiol A 193:249–263.

Claas B, Münz H, Zittlau KE (1989) Direction coding in central parts of the lateral line system. In: Coombs S, Görner P, Münz H (eds) The Mechanosensory Lateral Line: Neurobiology and Evolution. New York: Springer-Verlag, pp. 409–419.

Coombs S, Conley RA (1997) Dipole source localization by mottled sculpin II. The role of lateral line excitation patterns. J Comp Physiol A 180:401–416.

Coombs S, Janssen J (1990) Behavioral and neurophysiological assessment of lateral line sensitivity in the mottled sculpin, *Cottus bairdi*. J Comp Physiol 167A:557–567.

Coombs S, Montgomery JC (1999) The enigmatic lateral line system. In: Popper AN and Fay RR (eds) Comparative Hearing: Fishes and Amphibians. New York: Springer-Verlag, pp. 319–362.

Coombs S, Janssen J, Webb JF (1988) Diversity of lateral line systems: phylogenetic and functional considerations. In: Atema, J, Fay, RR, Popper AN, Tavolga WN (eds) Sensory Biology of Aquatic Animals. New York: Springer-Verlag, pp. 553–593.

Coombs S, Hastings M, Finneran J (1996) Modeling and measuring lateral line excitation patterns to changing dipole source locations. J Comp Physiol A 178:359–371.

Coombs S, Mogdans J, Halstead M, Montgomery J (1998) Transformation of peripheral inputs by the first-order lateral line brainstem nucleus. J Comp Physiol A 182:609–626.

Coombs S, Braun CB, Donovan B (2001) Orienting response of Lake Michigan mottled sculpin is mediated by canal neuromasts. J Exp Biol 204:337–348.

Cubbage CC, Mabee PM (1996) Development of the cranium and paired fins in the zebrafish *Danio rerio* (Ostariophysi, Cyprinidae). J Morphol 229:121–160.

Denton EJ, Gray JAB (1982) The rigidity of fish and patterns of lateral line stimulation. Nature 297:679–681.

Denton EJ, Gray JAB (1983) Mechanical factors in the excitation of clupeid lateral lines. Proc Roy Soc Lond B 218:1–26.

Denton EJ, Gray JAB (1988) Mechanical factors in the excitation of the lateral lines of fishes. In: Atema J, Fay RR, Popper AN, Tavolga WN (eds) Sensory Biology of Aquatic Animals. New York: Springer-Verlag, pp. 595–617.

Denton EJ, Gray JAB (1989) Some observations on the forces acting on neuromasts in fish lateral line canals. In: Coombs S, Görner P, Münz H (eds) The Mechanosensory Lateral Line. Neurobiology and Evolution. New York: Springer-Verlag, pp. 229–246.

Elepfandt A, Wiedemer L (1987) Lateral-line responses to water surface waves in the clawed frog *Xenopus laevis*. J Comp Phyiol A 160:667–682.

Engelmann J, Bleckmann H (2004) Coding of lateral line stimuli in the goldfish midbrain in still- and running water. Zoology 7:135–51.

Engelmann J, Hanke W, Mogdans J, Bleckmann H (2000) Hydrodynamic stimuli and the fish lateral line. Nature 408:51–52.

Engelmann J, Hanke W, Bleckmann H (2002) Lateral line reception in still- and running water. J Comp Physiol A 188:513–526.

Enger PS, Kalmijn AJ, Sand O (1989) Behavioral investigations of the functions of the lateral line and inner ear in predation. In: Coombs S, Görner P, Münz H (eds) The Mechanosensory Lateral Line: Neurobiology and Evolution. New York: Springer-Verlag, pp. 575–587.

Fay RR, Edds-Walton PL (2001) Bimodal units in the torus semicircularis of the toadfish (*Opsanus tau*). Biol Bull 201:280–281.

Fuiman LA, Higgs DM, Poling KR (2004) Changing structure and function of the ear and lateral line system of fishes during development. Am Fish Soc Symp 40: 117–144.

Görner P (1963) Untersuchungen zur Morphologie und Elektrophysiologie des Seitenlinienorgans vom Krallenfrosch (*Xenopus laevis* Daudin). Z Vergl Physiol 47:316–338.

Gray JAB, Denton EJ (1991) Fast pressure pulses and communication between fish. J Mar Biol Assoc UK 71:83–106.

Hanke W, Bleckmann H (2004) The hydrodynamic trails of *Lepomis gibbosus* (Centrarchidae), *Colomesus psittacus* (Tetraodontidae), and *Thysochromis ansorgii* (Cichlidae) investigated with scanning particle image velocimetry. J Exp Biol 207:1585–1596.

Hanke W, Brücker C, Bleckmann H (2000) The ageing of low-frequency water disturbances caused by swimming goldfish and its possible relevance to prey detection. J Exp Biol 203:1193–1200.

Hanke W, Rüter A, Bleckmann H (2002) Verbesserte Strömungsmessungen zur Untersuchung natürlich vorkommender hydrodynamischer Reize. Deutsche Gesellschaft für Laseranemometrie. GALA e. V., Rostock.

Harden Jones, F, Arnold GP, Greer-Walker M, Scholes P (1979) Selective tidal stream transport and the migration of plaice (*Pleuronectes platessa* L.) in the southern North Sea. J Conseil Int Explor Mer 38:331–337.

Harvey R, Blaxter JHS, Hoyt RD (1992) Development of superficial and lateral line neuromasts in larvae and juveniles of plaice (*Pleuronectes platessa*) and sole (Solea *solea*). J Mar Biol Assoc UK 72:651–668.

Hassan ES (1985) A suggested role for secondary flow in the stimulation of the cochlear hair cell. Biol Cyber 53:109–119.

Hassan ES (1989) Hydrodynamic imaging of the surroundings by the lateral line of the blind cave fish *Anoptichthys jordani*. In: Coombs S, Görner P, Münz H (eds) The Mechanosensory Lateral Line: Neurobiology and Evolution. New York: Springer-Verlag, pp. 217–228.

Hassan ES (1992a) Mathematical description of the stimuli to the lateral line system of fish, derived from a three-dimensional flow field analysis. I. The case of moving in open water and of gliding towards a plane surface. Biol Cyber 66:443–452.

Hassan ES (1992b) Mathematical description of the stimuli to the lateral line system of fish, derived from a three-dimensional flow field analysis. II. The case of gliding alongside or above a plane surface. Biol Cyber 66:453–461.

Hay AM, Adams CE (1997) Foraging of the ruffe (*Gymnocephalus cernuus*): responses in the absence of visual cues. In: International Symposium on Biology and Management of Ruffe, Abstracts, p. 43.

Hoekstra D, Janssen J (1985) Non-visual feeding behavior of the mottled sculpin, *Cottus bairdi*, in Lake Michigan. Environ Biol Fishes 12:111–117.

Hoin-Radkovski I, Bleckmann H, Schwartz E (1984) Determination of source distance in the surface-feeding fish *Pantodon buchholzi* (Pantodontidae). Anim Behav 32:840–851.

Hudspeth AJ, Corey DP (1977) Sensitivity, polarity, and conductance change in the response of vertebrate hair cells to controlled mechanical stimuli. Proc Natl Acad Sci USA 74:2407–2411.

Janssen J (1990) Localization of substrate vibrations by the mottled sculpin (*Cottus bairdi*) Copeia 1990:349–355.

Janssen J (1996) Use of the lateral line and tactile senses in feeding in four Antarctic nototheniid fishes. Environ Biol Fish 47:51–64.

Janssen J (2000) Toxicity of cobalt: implications for lateral line studies. J Comp Phys A. 186:957–960.

Janssen J, Sideleva V, Biga H (1999) Use of the lateral line for feeding in two Lake Baikal sculpins. J Fish Biol 54:404–416.

Jones WR, Janssen J (1992) Lateral line development and feeding behavior in the mottled sculpin *Cottus bairdi* (Scorpaeniformes: Cottidae) Copeia. 1992:485–492.

Kalmijn AJ (1988) Hydrodynamic and acoustic field detection. In: Atema J, Fay RR, Popper AN, Tavolga WN (eds) Sensory Biology of Aquatic Animals. New York: Springer-Verlag, pp. 83–130.

Kalmijn AJ (1989) Functional evolution of lateral line and inner ear sensory systems. In: Coombs S, Görner P, Münz H (eds) The Mechanosensory Lateral Line: Neurobiology and Evolution. New York: Springer-Verlag, pp. 187–216.

Kanter M, Coombs S (2003) Rheotaxis and prey detection in uniform currents by Lake Michigan mottled sculpin (*Cottus bairdi*). J Exp Biol 206:59–60.

Karlsen HE, Sand O (1987) Selective and reversible blocking of the lateral line in freshwater fish. J Exp Biol 133:249–262.

Kirk KL (1985) Water flows produced by *Daphnia* and *Diaptomus*: implications for prey selection by mechanosensory predators. Limnol Oceanogr 30:679–686.

Kroese ABA, Schellart NAM (1992) Velocity- and acceleration-sensitive units in the trunk lateral line of the trout. J. Neurophysiol 68:2212–2221.

Kroese ABA, van Netten SM (1989) Sensory transduction in lateral line hair cells. In: Coombs S, Görner P, Münz H (eds) The Mechanosensory Lateral Line: Neurobiology and Evolution. New York: Springer-Verlag, pp. 265–284.

Kröther A, Mogdans J, Bleckmann H (2002) Brainstem lateral line responses to sinusoidal wave stimuli in still and running water. J Exp Biol 205:1471–1484.

Kröther A, Bleckmann H, Mogdans J (2004) Effects of running water on brainstem lateral line responses in trout, *Oncorhynchus mykiss*, to sinusoidal wave stimuli. J Comp Physiol A 190:437–448.

Lang HH (1980) Surface wave discrimination between prey and nonprey by the back swimmer *Notonecta glauca* L. (Hemiptera, Heteroptera). Behav Ecol Sociobiol 6: 233–246.

Liang XF, Liu JK, Huang BY (1998) The role of sense organs in the feeding behaviour of Chinese perch. J Fish Biol 52:1058–1067.

Liao JC, Beal DN, Lauder GV, Triantafyllou MS (2003) The Kármán gait: novel body kinematics of rainbow trout swimming in a vortex street. J Exp Biol 206, 1059–1073.

McCormick CA (1989) Central lateral line mechanosensory pathways in bony fish. In: Coombs S, Görner P, Münz H (eds) The Mechanosensory Lateral Line: Neurobiology and Evolution. New York: Springer-Verlag, pp. 341–364.

McCormick CA, Hernandez DV (1996) Connections of the octaval and lateral line nuclei of the medulla in the goldfish, including the cytoarchitecture of the secondary octaval population in goldfish and catfish. Brain Behav Evol 47:113–138.

Metcalfe WK (1989) Organization and development of the zebrafish posterior lateral line. In:Coombs S, Münz H, Görner P (eds) The Mechanosensory Lateral Line: Neurobiology and Evolution. New York: Springer-Verlag, pp. 147–160.

Mogdans J, Bleckmann H (1998) Responses of the goldfish trunk lateral line to moving objects. J Comp Physiol A 182:659–676.

Mogdans J, Bleckmann H (2001) The mechanosesory lateral line of jawed fishes. In: Kapoor BG, Hara TJ (eds) Sensory Biology of Jawed Fishes—New Insights. Enfield, NH: Science Publishers, pp. 181–213.

Mogdans J, Goenechea L (2000) Responses of medullary lateral line units in the goldfish, *Carassius auratus*, to sinusoidal and complex wave stimuli. Zoology 102:227–237.

Mogdans J, Kröther S (2001) Brainstem lateral line responses to sinusoidal wave stimuli in the goldfish, *Carassius auratus*. Zoology 104:153–166.

Mogdans J, Bleckmann H, Menger N (1997) Sensitivity of central units in the goldfish, *Carassius auratus*, to transient hydrodynamic stimuli. Brain Behav Evol 50: 261–283.

Mogdans J, Engelmann, Hanke W, Kröther S (2003) The fish lateral line: how to detect hydrodynamic stimuli. In: Barth FG, Humphrey JAC, Secomb TW (eds) Sensors and Sensing in Biology and Engineering. New York: Springer-Verlag, pp. 173–185.

Montgomery JC (1989) Lateral line detection of planktonic prey. In: Coombs S, Görner P, Münz H (eds) The Mechanosensory Lateral Line: Neurobiology and Evolution. New York: Springer-Verlag, pp. 561–574.

Montgomery JC, Bodznick D (1994) An adaptive filter that cancels self-induced noise in the electrosensory and lateral line mechanosensory systems of fish. Neurosci Lett 174:145–148.

Montgomery JC, Coombs S (1992) Physiological characterization of lateral line function in the Antarctic fish, *Trematomus bernacchii*. Brain Behav Evol 40:209–216.

Montgomery JC, Coombs S (1998) Peripheral encoding of moving sources by the lateral line system of a sit-and-wait predator. J Exp Biol 201:91–102.

Montgomery JC, Hamilton AR (1997) Sensory contributions to nocturnal prey capture in the dwarf scorpion fish (*Scorpaena papillosus*). Mar Freshwater Behav Physiol30: 209–223.

Montgomery JC, Macdonald JA (1987) Sensory tuning of lateral line receptors in Antarctic fish to the movements of planktonic prey. Science 235:195–196.

Montgomery JC, Milton RC (1993) Use of the lateral line for feeding in torrentfish (*Cheimarrichthys fosteri*). NZ J Zool 20:121–125.

Montgomery JC, Saunders AJ (1985) Functional morphology of the piper *Hyporhamphus ihi* with reference to the role of the lateral line in feeding. Proc Roy Soc B Lond 224:197–208.

Montgomery JC, Coombs S, Janssen J (1994) Form and function relationships in lateral line systems: comparative data from six species of Antarctic notothenioid fish. Brain Behav Evol 44:299–306.

Montgomery JC, Coombs S, Conley RS, Bodznick D (1995) Hindbrain sensory processing in lateral line, electrosensory, and auditory systems: a comparative overview of anatomical and functional similarities. Audit Neurosci 1:207–231.

Montgomery JC, Bodznick D, Halstead M (1996) Hindbrain signal processing in the lateral line system of the dwarf scorpionfish *Scorpaena papillosus*. J Exp Biol 199: 893–899.

Montgomery JC, Baker CF, Carton AG (1997) The lateral line can mediate rheotaxis in fish. Nature 389:960–963.

Montgomery JC, Coombs S and Baker CF (2001) The mechanosensory lateral line system of the hypogean form of *Astynax fasciatus*. Environ Biol Fishes 62:87–96.

Montgomery JC, Mcdonald F, Baker CF, Carton AG (2002) Hydrodynamic contributions to multi-modal guidance of prey capture behavior in fish. Brain Behav Evol 59: 190–198.

Montgomery JC, Mcdonald F, Baker CF, Carton AG, Ling N (2003) Sensory integration in the hydrodynamic world of rainbow trout. Proc Roy Soc Biol Sci 270 (Suppl 2): 195–197.

Mukai Y, Yoshikawa, H, Kobayashi H (1994) The relationship between the length of the cupulae of free neuromasts and feeding ability in larvae of the willow shiner *Gnathopogon elongatus caerulescens* (Teleostei, Cyprinidae). J Exp Biol 197:399–403.

Müller U, Schwartz E (1982) Influence of single neuromasts on prey-localizing behavior of surface-feeding fish *Aplocheilus lineatus*. J Comp Physiol A 149:399–408.

Münz H (1985) Single unit activity in the peripheral lateral line system of the cichlid fish *Sarotherodon niloticus* L. J Comp Physiol A 157:555–568.

Münz H (1986) What influences the development of canal and superficial neuromasts?Ann Mus R Afr Centr Sci Zool 251:85–89.

Münz H (1989) Functional organization of the lateral line periphery. In: Coombs S, Görner P, Münz H (eds) The Mechanosensory Lateral Line: Neurobiology and Evolution. New York: Springer-Verlag, pp. 285–297.

Murakami T, Fukuoka T, Ito H (1986) Telencephalic ascending acousticolateral system in a teleost (*Sebastiscus marmoratus*), with special reference to fiber connections of the nucleus preglomerulosus. J Comp Neurol 247:383–397.

Nelson ME, MacIver MA, Coombs S (2002) Modeling electrosensory and mechanosensory images during the predatory behavior of weakly electric fish. Brain Behav Evol 59:199–210.

New JG, Coombs S, McCormick CA, Oshel PE (1996) Cytoarchitecture of the medial octavolateralis nucleus in the goldfish, *Carassius auratus*. J Comp Neurol 366:534–546.

New JG, Fewkes LA, Khan AN (2001) Strike feeding behavior in the muskellunge, *Esox masquinongy*: contributions of the lateral line and visual sensory systems. J Exp Biol 204:1207–1221.

Northcutt RG (1989) The phylogenetic distribution and innervation of craniate mechanoreceptive lateral lines. In: Coombs S, Görner P, Münz H (eds) The Mechanosensory Lateral Line:Neurobiology and Evolution. New York: Springer-Verlag, pp 17–78.

Peach MB (2001) The dorso-lateral pit organs of the Port Jackson shark contribute sensory information for rheotaxis. J Fish Biol 59:696–704.

Plachta D, Mogdans J, Bleckmann H (1999) The responses of midbrain lateral line units to amplitude modulated hydrodynamic stimuli. J Comp Physiol A 185:405–417.

Plachta D, Hanke W, Bleckmann H (2003) A hydrodynamic topographic map in the midbrain of goldfish *Carassius auratus*. J Exp Biol 206:3479–3486.
Pohlmann K, Grasso FW, Briethaupt T (2001) Tracking wakes: the nocturnal predatory strategy of piscivorous catfish. Proc Natl Acad Sci USA 98:7371–7374.
Pohlmann K, Atema J, Briethaupt T (2004) The importance of the lateral line in nocturnal predation of the piscivorous catfish. J Exp Biol 207:2971–2978.
Poling KR, Fuiman LA (1997) Sensory development and concurrent behavioural changes in Atlantic croaker larvae. J Fish Biol 51:402–421.
Popper AN (1977) A scanning electron microscopic study of the sacculus and lagena in the ears of fifteen species of teleost fishes. J Morphol 153:497–418.
Popper AN (2002) An overview of the applied use of sound in fisheries and fish biology. Bioacoustics 12:303–305.
Puzdrowski RL (1989) Peripheral distribution and central projections of the lateral-line nerves in goldfish, *Carassius auratus*. Brain Behav Evol 34:110–131.
Rowe DK, Dean TL, Williams E, Smith JP (2003) Effects of turbidity on the ability of juvenile rainbow trout, *Oncorhynchus mykiss*, to feed on limnetic and benthic prey in laboratory tanks. NZ J Mar Freshwater Res 37:45–52.
Ryer CH, Lawton A, Lopez RJ, Olla BL (2002) A comparison of the functional ecology of visual vs. nonvisual foraging in two planktivorous marine fishes. Can J Fish Aquat Sci 59:1305–1314.
Satou M, Shiraishi A, Matsushima T, Okumoto N (1991) Vibrational communication during spawning behavior in the hime salmon (landlocked red salmon, *Oncorhynchus nerka*) J Comp Physiol A 168:417–428.
Satou M, Takeuchi HA, Nishii J, Tanabe M, Kitamura S, Okumoto N, Iwata M (1994) Behavioral and electrophysiological evidences that the lateral line is involved in the inter-sexual vibrational communication of the hime salmon (landlocked red salmon, *Oncorhynchus nerka*). J Comp Physiol 174 A: 539–549.
Saunders AJ, Montgomery JC (1985) Field and laboratory studies of the feeding behaviour of the piper *Hyporhamphus ihi* with reference to the role of the lateral line in feeding. Proc Roy Soc B Lond 224:209–221.
Schellart NAM, Kroese ABA (1989) Interrelationship of acousticolateral and visual systems in the teleost midbrain. In: Coombs S, Görner P, Münz H (eds) The Mechanosensory Lateral Line: Neurobiology and Evolution. New York: Springer-Verlag, pp. 421–443.
Schellart NAM, Popper AN (1992) Functional aspects of the auditory system of actinopterygian fish. In: Webster DB, Fay RR, Popper AN (eds) The Evolutionary Biology of Hearing, Springer-Verlag, New York, pp. 295–322.
Schilt CR (2002) Natural history in an unnatural environment: can we help fish to help themselves? Bioacoustics 12:310–313.
Smith WL, Webb JF, Blum SD (2003) The evolution of the laterophysic connection with a revised phylogeny and taxonomy of butterflyfishes (Teleostei: Chaetodontidae). Cladistics 19:287–306.
Song J, Northcutt RG (1991) Morphology, distribution and innervation of the lateral-line receptors of the Florida gar, *Lepisosteus platyrhincus*. Brain Behav Evol 37:10–37.
Striedter GF (1991) Auditory, electrosensory, and mechanosensory lateral line pathways through the forebrain in channel catfishes. J Comp Neurol 312:311–331.
Sutterlin AM, Waddy S (1975) Possible role of the posterior lateral line in obstacle entrainment by brook trout (*Salvelinus fontinalis*). J Fish Res Bd Canada 32:2441–2446.

Tarby ML, Webb JF (2003) Development of the supraorbital and mandibular lateral line canals in the cichlid, *Archocentrus nigrofasciatus*. J Morphol 254:44–57.

Tavolga WN (1976) Acoustic obstacle detection in the sea catfish, *Galeichthys felis* In: Schuiff A, Hawkins AD (eds) Sound Reception in Fish. Amsterdam: Elsevier, pp. 185–204.

Thorpe JE (1988) Salmon migration. Sci Prog Oxford 72:345–370.

Tittel G, Müller U, Schwartz E (1984) Determination of stimulus direction by the topminnow *Aplocheilus lineatus*. In: Varju D, Schnitzler H-U (eds) Localization and Orientation in Biology and Engineering. New York: Springer-Verlag, pp. 69–72.

Tricas TC, Boyle KS (2005) The evolution of pairing behavior, sound production and hearing in chaetodontid butterflyfishes: evidence from behavior and physiology. Brain Behav Evol 66:143.

Tricas TC, Kajiura SM, Kosaki RK (2006) Acoustic communication in territorial butterflyfish: test of the sound production hypothesis. J Exp Biol 209:4994–5004.

van Netten SM, Kroese ABA (1987) Laser interferometric measurement on the dynamic behavior of the cupula in the fish lateral line. Hear Res 29:55–61.

van Netten SM, Kroese ABA (1989) Dynamic behavior and micromechanical properties of the cupula. In: Coombs S, Görner P, Münz H (eds) The Mechanosensory Lateral Line: Neurobiology and Evolution. New York: Springer-Verlag, pp. 247–264.

van Netten SM, Kelly JP, Khanna SM (1990) Dynamic responses of the cupula in the fish lateral line are spatially nonuniform. Abstracts of the 13th Midwinter Meeting for Research into Otolaryngology, 1990, St. Petershurg, Florida, pp. 341, 342.

Vischer HA (1989) The development of lateral-line receptors in *Eigenmannia* (Teleostei, Gymnotiformes). Brain Behav Evol 33:205–222.

Vogel D, Bleckmann H (1997) Water wave discrimination in the surface-feeding fish *Aplocheilus lineatus*. J Comp Physiol 180A:671–681.

Voigt R, Carton AG, Montgomery JC (2000) Responses of anterior lateral line afferent neurones to water flow. J Exp Biol 203:2495–2502.

von Campenhausen C, Riess I, Weissert R (1981) Detection of stationary objects in the blind cave fish *Anoptichthys jordani* (Characidae). J Comp Physiol A 143:369–374.

Webb JF (1989a) Developmental constraints and evolution of the lateral line system in teleost fishes. In: Coombs S, Görner P and Münz H (eds) The Mechanosensory Lateral Line: Neurobiology and Evolution. New York: Springer-Verlag, pp. 79–98.

Webb JF (1989b) Gross morphology and evolution of the mechanosensory lateral line system in teleost fishes. Brain Behav Evol 33:34–53.

Webb JF (1998) Laterophysic connection: a unique link between the swim bladder and the lateral-line system in *Chaetodon* (Perciformes: Chaetodontidae). Copeia 1998: 1032–1036.

Webb JF (2000a) Mechanosensory lateral line: microscopic anatomy and development. In: Ostrander G (ed) Handbook of Experimental Animals—The Laboratory Fish. London: Academic Press, pp. 463–470.

Webb JF (2000b) Mechanosensory lateral line: functional morphology and neuroanatomy. In: Ostrander G (ed) Handbook of Experimental Animals—The Laboratory Fish. London: Academic Press, pp. 236–244.

Webb JF, Northcutt RG (1997) Morphology and distribution of pit organs and canal neuromasts in non-teleost bony fishes. Brain Behav Evol. 50:139–151.

Webb JF, Shirey JE (2003). Post-embryonic development of the lateral line canals and neuromasts in the zebrafish. Dev Dyn 228:370–385.

Webb JF, Smith WL (2000) The laterophysic connection in chaetodontid butterflyfish: morphological variation and speculations on sensory function. Philos Trans Roy Soc Lond B 355:1125–1129.

Webb JF, Smith WL, Herman JL, Woods CF, Ketten DF (2005) The laterophysic connection: peripheral specialization for reception of acoustic stimuli in chaetodontid butterflyfishes? Brain Behav Evol 66:143–144.

Webb JF, Smith WL, Ketten DR (2006) The laterophysic connection and swim bladder in butterflyfishes in the genus *Chaetodon* (Perciformes: Chaetodontidae). J Morphol 267:1338–1355.

Weeg MS, Bass AH (2002) Structural and functional evidence for acoustic-lateral line interactions in a vocal fish. Bioacoustics 12:161–163.

Weissert R, von Campenhausen C (1981) Discrimination between stationary objects by the blind cave fish (*Anoptichthys jordani*: Characidae). J Comp Physiol 143:375–381.

Wojtenek W, Mogdans J, Bleckmann H (1998) The responses of midbrain lateral line units of the goldfish *Carassius auratus* to moving objects. Zoology 101:69–82.

Wonsettler AL, Webb JF (1997) Morphology and development of the multiple lateral line canals on the trunk in two species of *Hexagrammos* (Scorpaeniformes: Hexagrammidae). J Morphol 233:195–214.

Wubbels RJ (1992) Afferent response of a head canal neuromast of the ruff (*Acerina cernua*) lateral line. Comp Biochem Physiol A 102A:19–26.

Wubbels RJ, Kroese ABA, Schellart, NAM (1993) Response properties of lateral line and auditory units in the medulla oblongata of the rainbow trout (*Oncorhynchus mykiss*). J Exp Biol 179:77–92.

Zittlau KE, Claas B, Münz H (1986) Directional sensitivity of lateral line units in the clawed toad *Xenopus laevis* Daudin. J Comp Physiol A 158:469–477.

Zottoli SJ, Bentley AP, Prendergast BJ, Reiff HI (1995) Comparative studies on the Mauthner cell of teleost fish in relation to sensory input. Brain Behav Evol 46:151–164.

6
Orientation to Auditory and Lateral Line Stimuli

OLAV SAND AND HORST BLECKMANN

1. Introduction

1.1 Underwater Acoustics

The reader of this chapter should have basic knowledge of underwater acoustics, and is referred to other texts for a thorough discussion of the topic in a biological context (e.g., Harris 1964; van Bergeijk 1967; Kalmijn 1988). In short, sound has a dualistic nature and consists of both a pressure component and a kinetic component. Far from the sound source, particle motions associated with the pressure fluctuations constitute the kinetic component. These motions can be expressed as particle displacement, particle velocity, or particle acceleration. The ratio between sound pressure and particle velocity is constant far from the source, and defines the acoustic impedance of the medium. In addition to generating propagating sound, a vibrating sound source produces hydrodynamic flows in its vicinity. These particle motions are independent of the elastic properties of the medium, and decay very steeply with distance from the source. Consequently, particle motions close to the source are composed of both hydrodynamic flows and motions associated with the propagating sound. The attenuation of the sound pressure and the associated particle motions follow $1/r$ under free-field conditions (in a homogeneous medium far from any boundaries), where r is the distance from the source. The hydrodynamic particle motions (which dominate close to the source) attenuate much more steeply, following $1/r^2$ for a monopole source, which pulsates in volume, and $1/r^3$ for a dipole source, which vibrates with constant volume. The distance at which hydrodynamic and pressure-associated particle motions have equal amplitude is $1/2\pi$ of the wavelength for an ideal monopole source. Closer to the source the hydrodynamic particle motions dominate, and this region is commonly termed the acoustic nearfield. The region beyond, where the sound pressure-associated particle motions dominate, is then called the acoustic farfield. The nearfield is frequency-dependent and expands towards lower frequencies, and is also more extensive for a dipole source.

In the pure farfield, the particle motions are normal to the wave front, and for a monopole the particle motions are radial to the source also in the nearfield. However, for a dipole sound source the nearfield particle motions are more complex and may have any direction relative to the vibration axis of the source (see Section 2.3 and Fig. 6.6).

1.2 Physical Considerations and Historical Background

The ears of terrestrial vertebrates are sensitive to sound pressure, which is a scalar quantity, and directional information cannot be mediated via a single pressure detector. Thus, directional hearing in terrestrial vertebrates is based on the analysis of differences in sound intensity, phase, and time of arrival at the two ears. The ratio between the speed of sound in water and air is approximately 4.5, making the differences in both phase and arrival time at the two ears correspondingly less for a fish than for a comparable animal in air. Further, sound passes through the body of a fish with negligible reflective loss, minimizing the intensity differences between the two ears, while the body of terrestrial animals may constitute an effective sound barrier causing sound shadows that maximize such intensity differences. Moreover, in many teleosts a gas-filled swimbladder acts as a single sound pressure detector by transforming sound pressure into particle motion (Braun and Grande, Chapter 4).

The physical considerations mentioned in the preceding text seem to indicate unfavorable conditions for directional hearing in fish, in accordance with the ambiguous results from the initial investigations of this ability in teleosts. It is technically more difficult to conduct controlled experiments on directional hearing than to test the general hearing sensitivity, and during the first half of the previous century only three studies specifically addressed directional hearing in fish (Reinhardt 1935; von Frisch and Dijkgraaf 1935; Kleerekoper and Chagnon 1954). The first two of these papers concluded that fish cannot localize sound sources, whereas the third reached the opposite conclusion.

In spite of the scanty experimental data available at the time, van Bergeijk published a tremendously influential theoretical paper on directional hearing in fish in 1964. His analysis (van Bergeijk 1964) was based on the following assumptions:

- Hearing in the acoustic farfield is strictly a detection of sound pressure, and is based on the swimbladder acting as a transformer between pressure and particle movement.
- Within the acoustic nearfield, the lateral line may detect the incident particle movements at considerable distance (at least many body lengths) from the source.
- Biologically significant sounds have fairly high frequencies (at least several tens of Hertz), making sound detection in the farfield a dominant aspect of hearing in fish.

Later, all these assumptions were shown to be erroneous, but based on this shaky foundation van Bergeijk formulated his theory of directional hearing in fish:

- A single pressure detector (the swimbladder) cannot be used to localize a sound source, and fish behave according to their physical limitations: They do not detect the sound direction in the farfield.
- The lateral-line system amply satisfies the minimum requirements for localization of a sound source in the nearfield. Consequently, fish are capable to localize sound sources in the nearfield only.

Among his fellow scientists working on hearing in fish, van Bergeijk was unique in his physical and mathematical approach, and van Bergeik's authority was unprecedented. He was respected to such a degree that for several years his theoretical considerations hampered further progress in this field of research. However, for scientists mainly studying shark behavior, van Bergeijk's theory did just not fit their observations, and the idea that he might be wrong slowly evolved. In particular, Nelson and Gruber (1963) and Myrberg et al. (1969) concluded, based on their field observations, that sharks can detect and orient to sounds in the acoustic farfield. These important studies showed that a swimbladder is not required to detect farfield sound of normal intensities, and that sharks may react to such sounds with directed responses. Chapman and Sand (1974) showed that flatfish, which also lack a swimbladder, are sensitive to particle motion, and that the lowest auditory thresholds are less than 0.1 nm, measured as particle displacement. Recordings from primary afferent neurons from otolith organs in the goldfish (*Carassius auratus*) revealed similar low thresholds for whole-body vibrations (Fay 1984). Such sensitivity enables fish to detect farfield sounds of normal intensities even in the absence of a swimbladder.

Schuijf et al. (1972) were the first to show that teleosts may discriminate between sounds of different directions at approximate farfield conditions. Several behavioral experiments, which are reviewed in Section 3, have confirmed and extended this observation. Thus, both sharks and teleosts can discriminate between sounds of different directions, but what are the physiological mechanisms behind this ability? Further, how are fish able to not only discriminate between sounds from different directions, but also to locate the sound sources? In the following, we discuss different models for directional hearing in fish, and to which degree these models are supported by behavioral observations. The directional fast-start escape responses (startle responses), which are elicited by high intensity acousticolateral, somatic, and visual stimuli activating reticulospinal neurons, are treated in a separate section. We also include a section on the role of lateral line organs in directed responses, although these organs are insensitive to propagating sound (see review by Sand 1984).

2. Models for Directional Hearing

2.1 Vectorial Analysis of Particle Motions

Soft fish tissue has nearly the same acoustic properties as water, and will vibrate with the same phase and amplitude as the surrounding water during exposure to sound. In contrast, the otoliths have a mass density three times larger than that of the neighboring tissue, and will lag behind the motions of the hair cells when the fish oscillates in a sound field. This creates shear movements of the sensory hairs in close contact with the otoliths (Fay and Edds-Walton, Chapter 3). Otolith organs are thus accelerometers, as outlined by de Vries (1950), and the inner ear of fish is inherently sensitive to the kinetic sound component (particle acceleration). The hair cells are directionally sensitive displacement detectors (Flock 1965; Hudspeth and Corey 1977), and in the otolith organs the hair cells are organized in patterns where the direction of the optimal sensitivity axis varies along the maculae (see review by Popper and Coombs 1982). It is therefore reasonable to suggest that the fish brain may calculate the direction of particle movements of the incident sound by vectorial weighing of the input from different regions of the sensory maculae, and *all current models of directional hearing in fish are based on this idea.*

In the acoustic farfield, the particle motions are radial to the source (normal to the wave front), simplifying the computation of sound direction. However, in the nearfield the situation may be more complex, as discussed in Section 2.3. Another complicating factor is the auditory function of a swimbladder. The advantage of a swimbladder in lowering auditory thresholds may be at the expense of acute directional hearing, because the amplified vibrations reradiating from the swimbladder may mask the incident particle motions. This issue is discussed further below.

At the time when the hypothesis of directional hearing based on vectorial weighing of inputs from different populations of hair cells emerged, it seemed difficult to provide experimental support, due to the inevitably complex sound fields in small laboratory tanks (Parvulescu 1967). However, because a fish in water is nearly acoustically transparent and vibrates with the same phase and amplitude as the surrounding medium, vibration of the fish in air simulates the kinetic sound component in water. Such an approach eliminates the problem of making directionally well defined stimuli in small tanks (Enger et al. 1973; Sand 1974). In retrospect, this seems rather obvious, but at the time it was a mental leap. Nearly all later neurophysiological studies of directional hearing in fish have employed variations of this stimulation technique. The only exception is a study by Buwalda and van der Steen (1979), who employed a standing wave controlled by pairs of opposing sound projectors to investigate the directionality of saccular microphonic potentials in the Atlantic cod (*Gadus morhua*). This stimulation technique is described in detail by Buwalda (1981).

In the experiments by Sand (1974), microphonic potentials were recorded from various positions along the saccular macula and from the lagena in perch during whole body vibration in air. For vibrations in the horizontal plane the angular response pattern was the same for all electrode positions along the sacculus, and showed a cosine function with maximal amplitude of the microphonic potentials at an angle of about 20° relative to the long axis of the fish (Fig. 6.1A). The angle between the optimal axes for the two ears was thus about 40°, which corresponds to the mean angle between the saccular otoliths in this species.

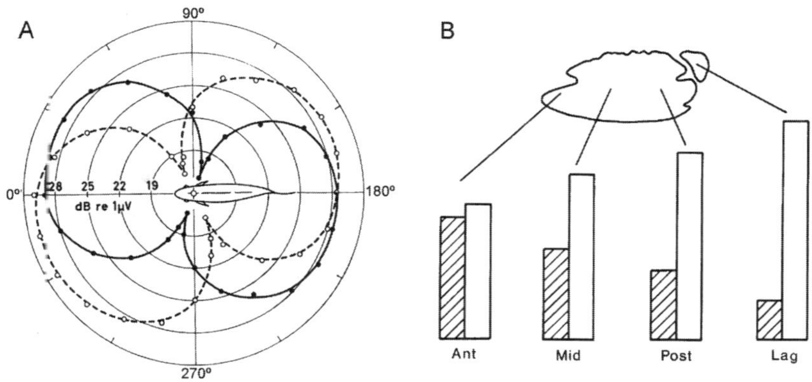

FIGURE 6.1. (**A**) Polar diagram of microphonic potentials recorded from the right (•) and left (o) sacculus in perch as a function of the horizontal vibration angle. Maximal microphonic responses were evoked by vibration directions deviating about 20° from the long axis of the fish, which corresponds to the mean angle of about 40° between the sacculi in this species. (**B**) Comparison of mean ratios between the microphonic sensitivity to vertical (open columns) and optimal horizontal (hatched columns) vibrations at 200 Hz for the anterior (Ant), mid (Mid), and posterior (Post) parts of the sacculus, and for the lagena (Lag). The lagena and posterior part of the sacculus are predominantly sensitive to vertical vibrations, whereas the anterior part of the sacculus is equally sensitive to horizontal and vertical vibrations. (From Sand 1974.)

Consequently, the saccular microphonic response is greatest when the relative otolith movements are along the main orientation axis of the sacculus in the head. Since the paired sacculi have different orientations in azimuth, the output from the right and left sacculus will be different (except for sources in the mid sagittal plane), although the incident particle motions are virtually the same at both ears. It was thus concluded that fish might determine the azimuth of a sound source by comparing the output from the two ears. Fish would then be similar to terrestrial vertebrates in utilizing interaural response differences as the basis for computation of azimuth. In agreement with this hypothesis, Schuijf and Siemelink (1974) found that the ability of Atlantic cod to discriminate between horizontal sound directions was lost after unilateral severing of the auditory nerve. Further, binaural interactions have been demonstrated by recordings from single units in both the acoustical lobes of the medulla oblongata and in the mesencephalic torus semicircularis in Atlantic cod (Horner et al. 1980), and Edds-Walton (1998) has provided anatomical evidence for binaural processing in the toadfish.

Figure 6.1B shows the relative microphonic responses to vertical and horizontal vibrations for the different recording positions in perch. The lagena and the posterior part of the sacculus were relatively more sensitive to vertical vibrations than the anterior part of the sacculus, in agreement with the pattern of saccular and lagenar hair cell orientation in this species (Enger 1976), and it was suggested that input from only one ear might provide sufficient information to estimate the elevation of a sound source. It was also concluded that the ability of the fish to determine the elevation of a sound source should be at least as good as the resolution of azimuth. The latter suggestion was later confirmed in behavioral experiments (Section 3).

The experiments by Sand (1974) provided the first electrophysiological data supporting a vectorial weighing mechanism for directional hearing in fish. However, recordings from afferent fibers from different parts of the ear give more refined information about regional differences in directional sensitivity of the sensory maculae, and reveals directly the directional information conveyed to the next level in the auditory pathway. Such experiments were first performed by Fay and Olsho (1979), who recorded responses from saccular and lagenar afferents in the hearing specialist goldfish during head vibrations in three orthogonal directions. The vibration directions of optimal sensitivity in the horizontal and sagittal planes were then calculated, and found to correspond roughly with the hair cell orientation maps. Polar diagrams of the directional sensitivity to primary auditory afferents in fish were first presented by Hawkins and Horner (1981), who recorded from the saccular and utricular branches of the auditory nerve in Atlantic cod during whole-body vibrations in the horizontal plane. The saccular units showed a narrow angular distribution of their optimal axes of vibration, with a mean optimal angle of 6° relative to the long axis of the fish (Fig. 6.2A).

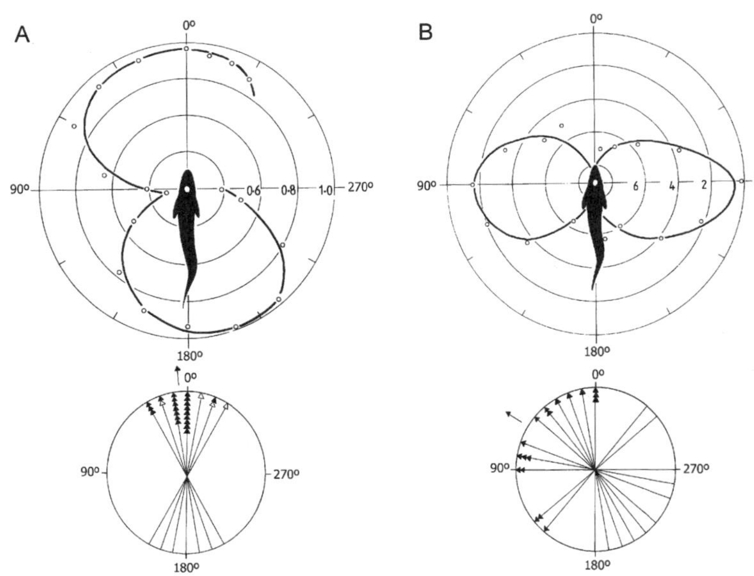

FIGURE 6.2. (**A**) Polar diagram of the spike frequency recorded from a primary afferent neuron from the left sacculus in cod as a function of the horizontal vibration angle. Lower panel: Distribution of the optimal horizontal vibration angle of afferent neurons from the left sacculus. Closed arrows represent units from the anterior ramus, and open arrows are from the posterior saccular ramus. (**B**) Comparable diagrams for primary afferent neurons from the left utriculus in cod. (From Hawkins and Horner 1981.)

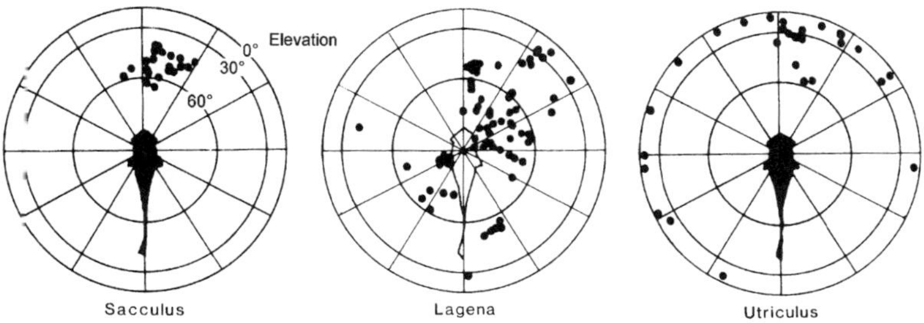

FIGURE 6.3. Distributions in spherical coordinates of the optimal vibration axes of single afferents from the right sacculus, lagena, and utriculus in the goldfish. The ear is depicted in the center of a globe, and the position of the symbols on the northern hemisphere shows the location at which the optimal axis would penetrate the surface of the globe. (From Fay 1984.)

In contrast, the utricular units displayed a wide angular distribution, with some optimal sensitivity axes even perpendicular to the long axis of the fish (Fig. 6.2B).

Soon after, Fay (1984) studied the responses in branches of the auditory nerve innervating all three otolith organs in the goldfish. The stimulation system was designed to produce whole-body accelerations along any axis in space, and the directional sensitivity of saccular, lagenar and utricular units was determined in three dimensions (Fig. 6.3). The optimal vibration axes of saccular units were tightly grouped in space, in agreement with the notion that the sacculus is mainly stimulated by reradiated swimbladder motions in otophysan species (which possess Weberian ossicles, Braun and Grande, Chapter 4). Lagenar units were more widely scattered in elevation, but with azimuth roughly grouped around 60°. The optimal axes of utricular units covered a wide range in a nearly horizontal plane, corresponding to the horizontal orientation of the utricular macula.

The most sensitive units responded to vibrations down to 0.1 nm at 140 Hz, which corresponds to the auditory particle motion thresholds previously obtained in behavioral experiments on flatfish (Chapman and Sand 1974). There were no major differences in sensitivity between units from the different branches of the auditory nerve, indicating that all otolith organs may be involved in hearing.

Following these pioneering studies, similar recordings from primary, auditory afferents have also been performed in toadfish (*Opsanus tau*; Fay and Edds-Walton 1997, 2000; Edds-Walton et al. 1999), sleeper goby (*Dormitator latifrons*; Lu and Popper 1998, 2001; Lu et al. 1998, 2003, 2004), and plainfin midshipman (*Porichthys notatus*; Weeg et al. 2002). Based on all the electrophysiologcal studies on the peripheral auditory organs in fish, the following main conclusions related to directional hearing may be drawn:

- Primary auditory afferents tend to have directional response patterns similar to the cosine directional response function of a single hair cell, indicating that each afferent contacts a population of hair cells with the same directional orientation.

- Afferents from all otolith organs are sufficiently sensitive to respond to particle motions associated with sounds of normal intensity, indicating that the brain may use information from all otolith organs in its computation of sound direction.
- Strong phase locking to the stimulus is a common feature of the neural responses in primary auditory afferents. Information about stimulus phase is thus conveyed to the central nervous system.
- The optimal axes of saccular and lagenar afferents display a wide scatter in elevation, but with azimuths grouped along axes coinciding with the physical orientation of the maculae. Optimal axes of utricular afferents are mainly in the horizontal plane, and show a wide scatter in azimuth. These patterns are compatible with the mainly vertical orientations of the saccular and lagenar maculae, and the horizontal orientation of the utricular macula in most species (with clupeids as the major exception). In otophysan species, the optimal axes of saccular units are more tightly grouped than in hearing non-specialists, in agreement with the tight link between the sacculus and the swimbladder.
- The distribution of the optimal axes of primary afferents suggests that information from one ear might suffice for computation of sound source elevation, while information from both ears might be required for computation of azimuth. Thus, the peripheral auditory apparatus of a fish appears capable of three-dimensional detection of sound direction.

If the directional information conveyed to the central nervous system via the primary afferents is utilized in the control of behavior, there should exist central auditory nuclei where the sound direction is represented in a manner appropriate for decision-making. The torus semicircularis (TS) in the mesencephalon, which is homologues to the inferior colliculus in mammals, is a likely candidate for such functions. TS receives input from major auditory areas in the medulla oblongata (McCormick and Hernandez 1996). Further, single-unit recordings from the TS in the Atlantic cod have demonstrated binaural interactions, and sound may induce both excitation and inhibition of neurons in this area (Horner et al. 1980).

The directional sensitivity of TS neurons has been studied in three species: rainbow trout (*Oncorhynchus mykiss*; Wubbels et al. 1995; Wubbels and Schellart 1997, 1998), goldfish (Ma and Fay 2002), and toadfish (Edds-Walton and Fay 2003, 2005a). The data from rainbow trout and toadfish showed that the directionality of the primary afferents was not only preserved in the TS, but various degrees of sharpening of the directional responses were also observed. The sharpening has been tentatively explained via a combination of excitatory and inhibitory inputs to the same neuron (Ma and Fay 2002; Edds-Walton and Fay 2003, 2005a). In the rainbow trout, directional sensitivity was studied only in the horizontal plane, where the optimal vibration directions of the TS units displayed a wide distribution covering any vibration angle (Fig. 6.4A). Thus, the TS in rainbow trout seems well equipped for determination of sound direction, at least in the plane studied.

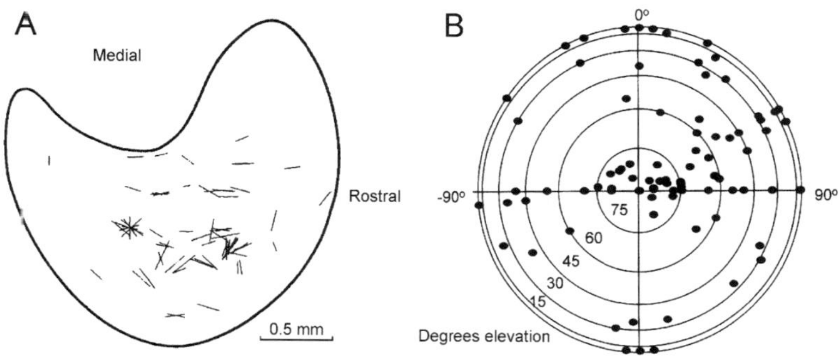

FIGURE 6.4. (**A**) Dorsal view of location of directional sensitive units in the right torus semicircularis in the rainbow trout. The center of each line represents the location of the unit (mean depth about 300 μm below the surface of TS), the length denotes its directional sharpness, and the orientation shows its optimal direction of vibration in the horizontal plane. The distribution of optimal directions covers any vibration angle. (From Wubbels and Schellart 1998.) (**B**) Distribution in spherical coordinates of the optimal vibration axes of directional single units in the left TS in toadfish. The fish is in the center of a globe, and the position of the symbols on the northern hemisphere shows the location at which the optimal axis would penetrate the surface of the globe. The optimal axes are widely distributed in space, covering any azimuth and elevation. (From Edds-Walton and Fay 2003.)

In the toadfish, the three-dimensional directional responses of TS single units have been studied in great detail (Edds-Walton and Fay 2003, 2005a). Most units were directionally sensitive, with a sharpened directional response compared to primary afferent neurons. The optimal vibration axes were arrayed widely in spherical space, covering all azimuths and elevations (Fig. 6.4B). Interestingly, some TS neurons seemed to be bimodal, responding both to whole body acceleration and to stimuli that presumably activated the lateral line system (Edds-Walton and Fay 2003, 2005b). The distribution of optimal vibration axes in toadfish TS is much wider than the comparable distribution of optimal axes of saccular afferents in this species (Edds-Walton et al. 1999). This discrepancy could be due to input from other otolith organs than the sacculus, which is commonly considered the main auditory end organ in fish, or to computations based on the directional properties of saccular afferents from both ears. In either case, the data indicate that all axes of acoustic particle motion around the fish are represented in the TS.

Recordings from the TS in the hearing specialist goldfish (Ma and Fay 2002) show a strikingly different picture than the data from the hearing nonspecialists rainbow trout and toadfish. The optimal vibration axes of directional goldfish TS units were tightly grouped in space (primarily vertical), in a similar fashion as the distribution of optimal vibration axes of saccular afferents (Fig. 6.3). The wide scatter of best axes of lagenar and utricular afferents was thus not reflected in the properties of TS neurons. The authors discussed if the lack of diverse optimal axes of TS units indicates poor directional hearing in goldfish (and other hearing specialists), or is due to inadequate sampling of higher order auditory neurons. It would be remarkable if the seemingly useful directional information conveyed by lagenar and utricular afferents is not utilized in these species.

The electrophysiological data on directional sensitivity obtained by vibrating the fish in air exclude pressure stimulation via the swimbladder. It is still not understood how the directional information in the incident particle acceleration is protected against masking by the amplified secondary particle motions radiating from the swimbladder. Stimulation of the otolith organs via the swimbladder is likely to be identical for symmetrical parts of the two ears, and also independent of sound direction. In agreement with this notion, Buwalda and van der Steen (1979) observed that whereas the saccular microphonic responses in Atlantic cod showed a cosine relationship to the axis of particle motion in a standing wave with a high ratio between particle motion and pressure, the responses became omnidirectional when this ratio was inverted. However, the incident particle motion input to the otolith organs will generate different responses in symmetrical elements of the ears (except for sources in the midsagittal plane). Consequently, information about the direction of the incident particle motion may be obtained by subtracting the responses from the two ears (common mode rejection), whereas adding the responses from the two ears will emphasize the sound pressure waveform (which are common to both ears). The existence of such mechanisms lacks experimental verification. However, parallel detection of incident particle motion and sound pressure is a prerequisite for the phase model for directional hearing, which is discussed in the next section.

2.2 The Phase Model for Directional Hearing

As outlined in the previous section, the otolith organs in fish are inertial motion detectors directly stimulated by the particle accelerations of a sound wave, and fish may use these organs to determine the three-dimensional directionality of the incident sound. However, detection of the direction of the particle motion is in itself not sufficient to determine the direction to the sound source, since the particle motion in the farfield is alternately either away from or toward the source. Consequently, there is an inherent bidirectionality or 180° ambiguity in the response of a simple particle motion detector, making it impossible to discriminate between opposing sound sources (180° apart).

This inherent 180° ambiguity is solved by the phase model for directional hearing in fish (Schuif 1975, 1976, 1981). The model is based on the fact that the direction of farfield particle motion and wave propagation coincides during acoustic compression, while these directions are opposite during rarefaction. The model was inspired by Piddington's (1972) observation that goldfish can discriminate between sounds of inverted polarity. In species with a swimbladder, the model assumes that the fish is able to simultaneously detect the direction of the incident particle movements and the sound pressure—via the swimbladder (Fig. 6.5). By decoding the phase difference between these components, fish may be able to discriminate between opposing sound sources. In the previous

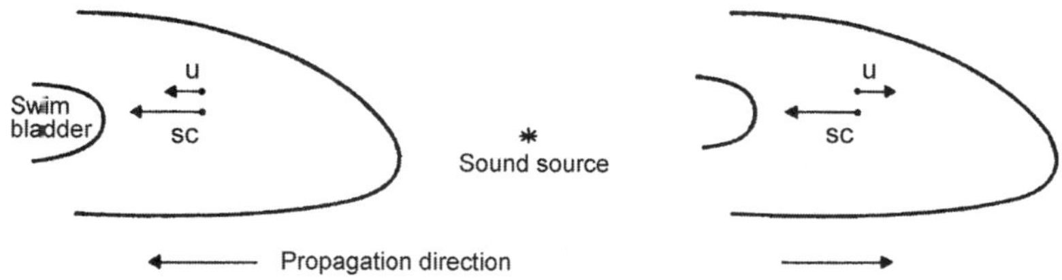

FIGURE 6.5. Principle of the phase model for solving the 180° auditory ambiguity in fish with a swimbladder. The left and right parts of the figure compare instants of maximum compression in the farfield at the same distance directly behind or in front of the source. The particle motions in the direct wave and the scattered wave emanating from the swimbladder are *u* and *sc*, respectively. If the sign of *sc* equals that of *u*, the source is in front of the fish. If the signs are different, the source is behind the fish. (From Scuijf 1975.)

section, a simple neural processing strategy for separation of acoustic pressure and particle motion was briefly described. Rogers et al. (1988) have elaborated on the phase model and proposed algorithms that the central nervous system might use to process acoustic information in order to localize the source.

The phase model was also adapted for sharks and other species lacking a swimbladder, and it was postulated that the 180° ambiguity was then resolved by comparing the phase between the direct sound and sound reaching the fish after reflection from the surface or the bottom (Schuijf 1975, 1976, 1981). Of course, this extension of the model has its limitations, and will fail if the fish is in midwater far from any reflecting surfaces. Rogers and Zeddies (Chapter 7) have suggested alternative, theoretical models that may resolve the 180° ambiguity for species both with and without a swimbladder. However, their speculations are not based on experimental data.

A prerequisite for the phase model in species possessing a swimbladder is separate encoding of sound pressure and incident particle motion, and phase comparison of these sound parameters. In the hearing specialists, such a task is clearly feasible, and the otophysan species ide (*Leuciscus idus*) is able to discriminate between opposing sound sources in the horizontal plane (Schuijf et al. 1977). Also in Atlantic cod, where an auditory function of the swimbladder has been demonstrated in both behavioral (Chapman and Hawkins 1973) and electrophysiological (Sand and Enger 1973) experiments, the phase relationship between incident particle motion and sound pressure is used to discriminate between opposing sound sources in both the horizontal (Schuijf and Buwalda 1975; Buwalda et al. 1983) and the median vertical plane (Buwalda et al. 1983).

Further, the Atlantic cod is capable of discriminating between much smaller phase differences than the 180° required for unambiguous detection of the propagation direction. The phase difference between particle motion and pressure is a function of distance to the source within the nearfield, and it was suggested that phase analysis could also enable the fish to detect the distance to a monopole source (Buwalda et al. 1983). Later behavioral experiments did in fact demonstrate that the Atlantic cod can discriminate between sound form sources at different distances (Schuijf and Hawkins 1983). It is likely that this ability is based on phase comparison, although the ratio between the amplitudes of sound pressure and particle motion is also a function of distance within the nearfield. These behavioral experiments on Atlantic cod, which will be further discussed in Section 3, mark the culmination of the phase model for directional hearing in fish, and it was concluded that: "... it is quite likely that the cod is able to estimate the true distance of a sound source in its vicinity. Combined with its three-dimensional directional hearing capabilities, this would provide the cod, an animal living essentially in a three-dimensional habitat, with a real acoustical sense of space. In this respect the auditory capacities of cod would far exceed those of most terrestrial vertebrates..." (Schuijf and Hawkins 1983, p. 144).

This view on directional hearing in fish is certainly attractive, but should be treated with caution. At the time when the phase model was introduced and tested, the prevailing assumption was that most hearing nonspecialists possessing a swimbladder are sensitive to sound pressure close to threshold, at least within the upper part of the audible frequency range, but this view has recently been challenged (Yan et al. 2000). Further, the choices of sound sources (monopoles rather than dipoles) and frequencies (too high) in the behavioral tests of the model have been rather unphysiological, as discussed in Sections 2.3 and 3. It may turn out that the main physiological relevance of the phase model is related to the fast start escape responses, which are discussed in Section 4.

In the original phase model, neural common mode rejection by subtracting the outputs from binaurally symmetrical hair cell populations is suggested as a mechanism for resolving the incident particle movements. Conversely, adding such outputs emphasizes the pressure component. Therefore, the resolution of the 180° ambiguity according to the phase model depends on a binaural mechanism. However, an alternative processing strategy may utilize a basically monaural mechanism for resolving both the direction and the distance to the sound source. This variation of the phase model was suggested by Schellart and de Munck (1987), and is termed the orbit model.

The fact that particle movements associated with the incident sound wave and the scattered wave from the swimbladder are not in phase and have, in general, different directions, leads to elliptical particle motion orbits for pure tones (Schuijf 1981; de Munck and Schellart 1987; Scellart and de Munck 1987). The orbits are unique for each source position, and the orbit model predicts that the characteristics of the orbits themselves (shape, orientation of the length axis, direction of revolution) are analyzed, rather than extracting segregated information on the incident and scattered sound waves. For example, the 180° ambiguity is resolved by detection of the direction of revolution of the elliptical orbit. Although such a mechanism is inherently monaural, binaural comparison of motion orbits provides additional information that may improve source localization.

Behavioral determination of hearing thresholds as a function of source direction failed to provide unambiguous support of the orbit model (Schellart and Buwalda 1990). Further, recordings of directional responses of midbrain auditory neurons in rainbow trout gave results more in support of the original phase model than the orbit model (Wubbels and Schellart 1997, 1998). Hence, experimental support of the orbit model is scanty.

The focus on the 180° ambiguity problem in fish audition, and the efforts to develop a single, unifying model for its solution, may have been exaggerated. Terrestrial vertebrates, which determine sound direction by comparing sound parameters at the two ears, also encounter auditory ambiguity problems. Since no interaural differences will occur in timing, phase, and intensity of sound for all potential sound sources in the median plane, terrestrial animals must handle ambiguities in both elevation and front–back. Such ambiguities are solved by various measures, including movements of the head and pinnae, visual and olfactory cues, and estimation of the most likely source location based on experience. Considering the familiar solutions to the auditory ambiguity problems in terrestrial vertebrates, it is reasonable to suggest that also fish may employ a variety of mechanisms to resolve the 180° ambiguity.

2.3 The Guidance Model for Sound Source Localization

The emerging picture based on the phase model of directional hearing is that species with a swimbladder may have a vision-like, acoustical sense of space. However, the phase model is hampered with several uncertainties, and might not enable the fish to unambiguously locate the sound source at a distance. The model is based on two main assumptions:

- The particle motions are radial to the source.
- The phase information required to solve the 180° ambiguity is provided by a swimbladder or reflecting surfaces.

The first assumption is valid throughout the acoustic field for a monopole, but is valid only in the farfield for sources of higher order. For example, within the nearfield of a dipole, the direction of the particle movements at a certain point yields no information about the location of the sound source (van Bergeijk 1964), as illustrated in Figure 6.6. This limitation of the phase model would be of little significance if most natural underwater sound sources were monopoles, or if most acoustically evoked behaviors took place in the farfield. However, apart from fish that produce sound by swimbladder pulsations, most natural underwater sound sources do not change volume, and are thus not monopoles. Moving objects, like swimming fish, are best approximated by a dipole.

Further, fish and other animals moving underwater mainly produce extremely low-frequency sound (Kalmijn 1989; Bleckmann et al. 1991). The major components of the particle accelerations caused by swimming fish are even below 10 Hz. For biological sound sources generating such low frequencies the nearfield

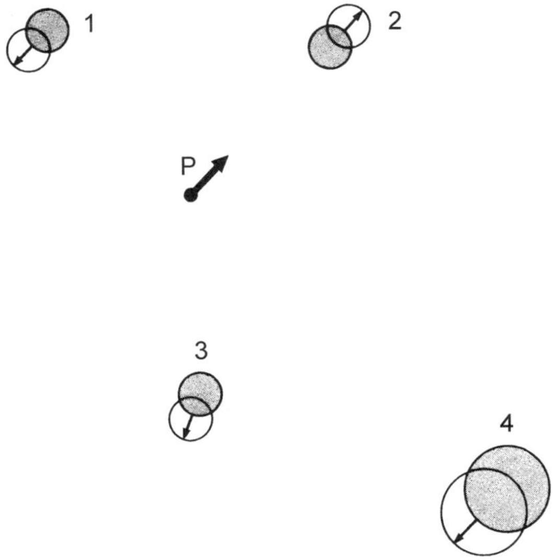

FIGURE 6.6. Ambiguity of the nearfield particle motions generated by a dipole sound source (vibrating sphere). An infinite number of potential dipoles might generate the indicated particle motions at point P, four of which are depicted. The magnitude of the vector at P is exaggerated for clarity. (From van Bergeijk 1964.)

extends beyond the audible range, and farfield detection is hardly biologically relevant. Fish are sensitive to extremely low-frequency sounds, even down to below 1 Hz, and infrasonic particle accelerations may be particularly effective in evoking behavioral responses in fish (see reviews by Sand and Karlsen 2000; Sand et al. 2001). It is also clear, in contrast to the view of van Bergeijk (1964), that the lateral line system is able to detect nearfield particle motions only up to at a distance of a few centimeters (Sand 1981, 1984; Enger et al. 1989). Within most of the nearfield, the acoustic detectors are the otolith organs, responding to whole-body acceleration of the fish. However, the lateral line can detect the vortices in a fish wake at considerable distance from the actual location of the wake generator (see Section 5.2.2).

The second assumption fails for a fish without a swimbladder in midwater, far from any reflecting surfaces. Moreover, the auditory gain provided by a swimbladder is frequency dependent, as the swimbladder pulsations exceed the free field particle motions only above a certain frequency, which will depend on both swimbladder volume and depth (Sand and Enger 1973; Sand and Hawkins 1973). Thus, the very low frequencies generated by moving objects are detected without the aid of the swimbladder at levels close to threshold (Sand and Karlsen 1986). However, pressure detection is essential for the startle responses evoked in otophysan species by low-frequency stimuli at high intensity (see Section 4).

The emerging picture from these considerations is that dipole sources producing extremely low-frequency sounds with extensive nearfields are among the biologically most important sound sources. Such sources are detected within the nearfield without the aid of the swimbladder, and the direction of the particle motions at a single spot provides no information about the location of the source. The celebrated phase model is thus inadequate to explain directional hearing under such circumstances. The phase model is not necessarily wrong. Phase analysis of pressure and particle motion may tell the fish if it is detecting farfield sound or not, and thus if the direction of particle movements can be trusted to be radial to the source. The phase model may also explain the directed startle responses to high-level stimuli, as discussed in Section 4. However, an alternative model is needed to explain localization of dipole sources within the nearfield.

Kalmijn (1989, 1997) has suggested such an alternative model, based on a previous model for orientation of elasmobranches to bioelectric fields (Kalmijn 1982). The acceleration fields in the vicinity of moving objects are governed by the same mathematical equations as the bioelectric fields produced by aquatic animals. Hence, he has proposed that fish may reach the sound source by using their otolith organs in a similar fashion as elasmobranches use their electroreceptors in directed approaches toward concealed prey. The model predicts that the fish may locate a dipole source by merely sensing the direction of the acceleration field. While proceeding, the fish only has to turn in a manner that keeps a constant angle between the body axes and the incident particle acceleration detected by the inner ear. This simple strategy will guide the fish to the source (Fig. 6.7). The algorithm also works for monopoles and for combinations of monopoles, dipoles, and higher order sources. The hypothesis applies equally well for both sharks and teleosts and all types of sound sources at any distance. This unifying *guidance model for sound source localization* suggests that fish do not actually perceive the absolute location of sound sources at a distance, but are instead guided to the source. Of course, this strategy requires a more or less continuously emitting source during the approach.

It should be stressed that the guidance model has not yet been rigorously tested in behavioral experiments. However, the next section shows that most experiments on directional hearing in fish have revealed only that fish are able

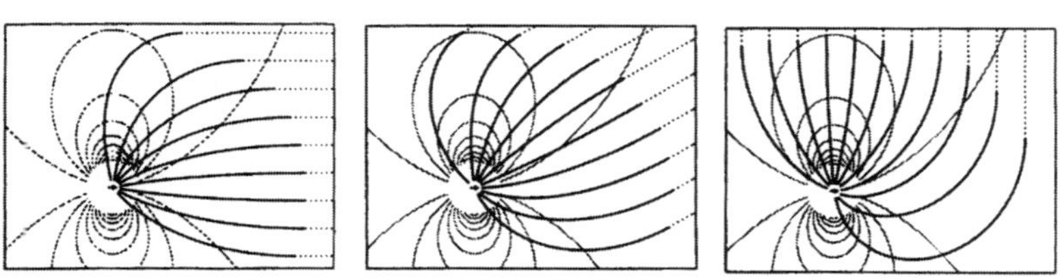

FIGURE 6.7. The guidance model for sound source localization. The shaded dipole field lines represent the acceleration field of a moving prey. The predator enters the field from three different directions along the paths indicated by dotted lines. When the local acceleration stimuli received by the otolith organs are sufficiently strong, the predator starts a guided approach. Along the approach paths indicated by heavy lines the predator maintains a constant angle between the local accelerations and its body axes. (From Kalmijn 1997.)

to discriminate between sounds from different directions. Very few, if any, experiments have demonstrated that fish are able to actually localize the sources from a distance.

3. Behavioral Studies of Directional Hearing

In the previous sections, various models for directional hearing in fish were described in considerable detail, but without confirmation by behavioral studies the validity of these models will remain uncertain. However, to design such studies is not a trivial task. The initial behavioral experiments in this field were simply aimed at demonstrating directional auditory responses in fish, but did still give conflicting results, probably due to the use of exceedingly complex stimulus fields in small tanks. Hence, the first indications of directional hearing in fish came from field observations of freely ranging sharks orienting toward sound sources, often from large distances (reviewed by Myrberg et al. 1976).

The first definite evidence of directional hearing in a teleost was provided by Schuijf et al. (1972), who trained the Ballan wrasse (*Labrus berggylta*) to discriminate between sounds emitted from either of two spatially separated sound sources. The experiments were carried out in a Norwegian fiord several meters from reflecting surfaces, and were based on conditioning with food as a reward. Within its netting cage the fish was trained to orient toward the active sound projector. Therefore, the fish was only required to detect the change in sound direction, rather than the actual *location* of the sound source. This serious limitation has also hampered most of the later behavioral studies of directional hearing in fish. The most noticeable exception is a study by Popper et al. (1973), who observed unconditioned directional orientations of Hawaiian squirrelfish (Holocentridae) toward a sound projector emitting a playback of squirrelfish alarm calls.

In the mid-1970s, the suggestion by van Bergeijk (1964) that any directional response to acoustic cues in fish must depend on the lateral line was still debated. However, Schuijf and Siemelink (1974) and Schuijf (1975) showed that Atlantic cod lost the ability to discriminate between different sound directions in the horizontal plane after severing the nerve roots innervating one of the labyrinths, although the lateral line system was still functioning. These experiments also indicated that information from both ears might be required for computation of azimuth, as originally proposed by Sand (1974).

In support of the phase model described in Section 2.2, Schuijf and Buwalda (1975) showed that Atlantic cod can discriminate traveling sound waves impinging on the head from those impinging on the tail. Furthermore, phase reversal of the acoustic pressure in the traveling wave, obtained by interference from a perpendicular standing wave generated by an opposing pair of sound projectors, caused 180° reversal of the directional response. Directional discrimination was also possible in the loop of a horizontal, standing wave, provided that appropriate pressure information in the correct phase was added.

Similar experiments were later performed by Schuijf et al. (1977) on the otophysan species ide (*Leuciscus idus*). In otophysan species, which possess Weberian ossicles that efficiently transmit sound pressure-induced pulsations of the unpaired swimbladder equally to both sacculi, it is reasonable to assume that the potential masking effects of the reradiated swimbladder motions on the particle motions in the incident wave is particularly severe. Nevertheless, even the ide displayed directional hearing and discriminated between sounds from sources 180° apart. However, except for this coarse discrimination, the ability of angular resolution was not studied further.

In all the studies by Schuif and collaborators mentioned so far, the fish was moving freely within a netting cage, and trained to approach or orient toward a particular source in a choice situation (Fig. 6.8A). During the training, the fish was rewarded at the food dispenser in line with the active sound projector. As noted, although the fish makes a correct choice during the test by displaying an oriented response toward the source, this experimental design cannot prove that the fish has in fact perceived the location of the source.

An alternative experimental strategy is to restrict the movements of the fish by a narrow confinement, and to monitor the response to a relevant stimulus (e.g., a change in the direction of a pulsed tone) by recording the heart rate. The fish is conditioned to display a reduced heart rate (bradycardia) in response to the stimulus, in anticipation of a mild electric shock applied just after the stimulus

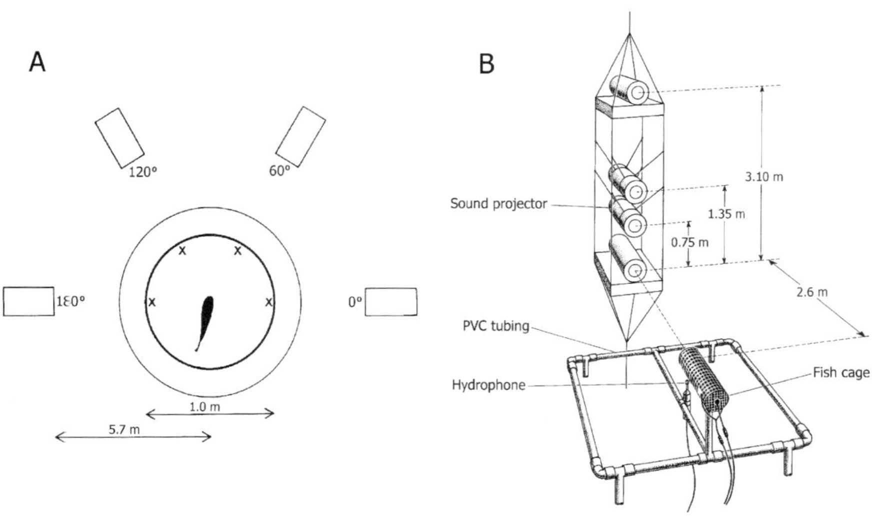

FIGURE 6.8. Examples of experimental designs for behavioral testing of directional discrimination. (**A**) The fish is free to move within a netting cage and is trained to orient toward the food dispenser (*x*) in line with the active sound transducer. (Redrawn from Schuijf 1975.) (**B**) The fish is confined in a small cage and gives a conditioned physiological response (i.e., reduced heart rate) to a change in sound direction. (From Hawkins and Sand 1977.) None of these experimental approaches can unambiguously determine if the fish can perceive the actual location of the sound source.

(Fig. 6.8B). Of course, also this method will only be able to reveal if the fish is able to discriminate between sound directions, whereas the ability to determine the position of the source is not tested.

By employing this technique, Chapman (1973) and Chapman and Johnstone (1974) studied auditory masking in Atlantic cod and haddock (*Melanogrammus aeglefinus*) to test if the fish nervous system is able to process differences in sound direction. The threshold of masked tones where recorded during emission of pure tones and noise from different projectors, and the auditory masking was reduced by about 7 dB when the angle between the sound projectors was 45° or greater (Fig. 6.9A). This result suggests that directional discrimination is well developed in these species. Chapman and Johnstone (1974) also showed that the Atlantic cod could readily be conditioned to a change in the direction of a pulsed tone switched between two equidistant sources. The limit for angular discrimination was close to 20° (Fig. 6.9B), which is in agreement with the angular threshold of about 22° estimated by Schuijf (1975), based on his choice experiments on the same species.

Further, Chapman and Johnstone (1974) reported that the threshold for discrimination between sound directions in Atlantic cod was considerably higher than the threshold for simply detecting the presence of a sound. This finding is in agreement with the notion that only the incident sound will give relevant directional information, in contrast to the amplified vibrations emanating from

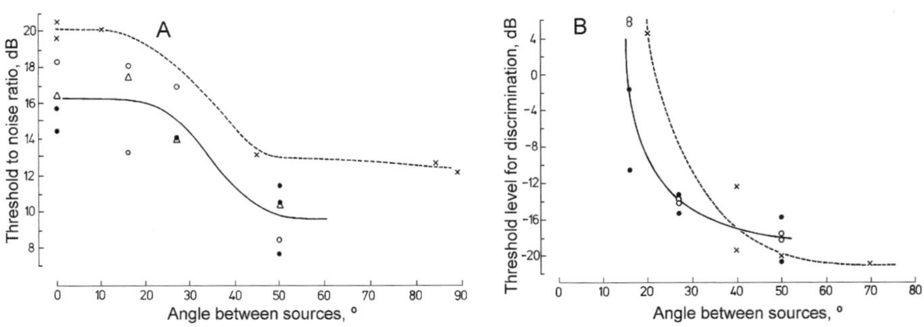

FIGURE 6.9. Angular discrimination by Atlantic cod in azimuth and elevation. A: threshold to noise ratio as a function of angle between two sound projectors transmitting a pure tone and masking noise, respectively. Symbols (○, •, Δ) represent data for 110 Hz from three fish with the projectors in the median vertical plane. Symbols (x) indicate comparable values with the projectors in the horizontal plane (data from Chapman and Johnstone 1974). The decrease in masking as the angular separation between tone and noise increases demonstrates that the nervous system is able to process differences in sound direction. (**B**) Sound pressure thresholds for detection of an angular change in the direction of a 110-Hz tone as a function of the angular change. Symbols (○, •) represent thresholds from two fish for changes in elevation. Symbols (x) indicate values for changes in azimuth (data from Chapman and Johnstone 1974). The steep increase in threshold toward the smaller angular separations indicates the limit of angular discrimination. (From Hawkins and Sand 1976.)

the swimbladder. At 200 Hz, the difference between detection threshold and the sound pressure required for directional discrimination was 23 dB, which corresponds to the reduction in microphonic sensitivity of about 20 dB at this frequency when the swimbladder in Atlantic cod is emptied (Sand and Enger 1973).

The behavioral studies on directional hearing performed during the first half of the 1970s were all designed to test discrimination ability in the horizontal plane. However, electrophysiological data (Sand 1974) indicated that the ability of fish to determine the elevation of a sound source should be at least as good as the resolution of azimuth, in contrast to humans, who are unable to discriminate between pure tones from sources at different elevations in the median vertical plane. Hawkins and Sand (1977) employed the cardiac conditioning technique to test this hypothesis, in experiments on Atlantic cod corresponding to those previously performed by Chapman and Johnstone (1974) in the horizontal plane. There was a significant decrease in auditory masking as the angular separation between tone and noise sources in the median vertical plane was increased (Fig. 6.9A), confirming the ability to perform an auditory discrimination based on directional cues. The power of angular resolution in the vertical plane, studied via directional change of a pulsed tone, was close to 16° (Fig. 6.9B), as compared to 20° previously reported for the horizontal plane. For fish living in a three-dimensional medium, in contrast to humans mainly confined to a surface, the ability to discriminate between source elevations is of course highly relevant.

The study by Hawkins and Sand (1977) prompted Buwalda et al. (1983) to test the validity of the phase model in Atlantic cod in three-dimensional space. They used the cardiac conditioning technique, combined with multiple, opposing pairs of sound projectors to generate both propagating waves and synthesized standing waves (Buwalda 1981). Switching a pulsed, pure tone from a reference source to an opposing source was detected under all conditions, whereas switching to a completely synthesized standing wave that simulated the phase relations of the reference source was not detected. However, switching to a standing wave simulating the phase relations of the opposite source was detected. It was concluded that the detection of sound propagation direction is based on the characteristic phase relationship between particle motion and sound pressure. Cues resulting from sound propagation itself are irrelevant, and apparently not perceived. The study demonstrated that Atlantic cod can resolve the 180° propagation ambiguity for all stimulus directions, which provides a basis for ambiguity-free directional detection in three-dimensional space. The authors acknowledged that the results did not provide irrefutable proof that fish are capable of determining the actual sound source position, and that there is a theoretical possibility that the observed phase discrimination was not related to directional detection and perception. However, the fact that the phase cue was so readily discriminated was accepted as evidence of its natural significance.

The study by Buwalda et al. (1983) also indicated a just noticeable phase difference between velocity and pressure of 20°–30°, which is much less the 180° associated with opposite source positions. This variable is a function of the distance from a source, and ranges from 0° to –90° (or from 180° to 90°) for far and close sources, respectively. Consequently, the investigators suggested that fish might utilize phase discrimination also for determining the distance to a sound source. This hypothesis was strengthened in a subsequent study by Schuijf and Hawkins (1983), who demonstrated that Atlantic cod can discriminate between pure tones emitted alternately from two aligned sound projectors at different distances from the fish. This kind of distance discrimination is lacking in terrestrial vertebrates, and it was suggested that the Atlantic cod possesses a real acoustical sense of space.

Although the behavioral studies referred to above have shown that fish can discriminate between different sound directions, and between sound sources at different distances at the same azimuth and elevation, it is still not settled if they are able to perceive the actual location of sound sources. Further, all these studies have employed monopole sound sources, which generate radial particle motions in both the near- and farfield, and the prevailing models for directional hearing in fish assume that the axis of particle motion points to the source. However, apart from the pulsating swimbladder in vocalizing species, most sources of biological significance are best approximated as dipoles or more complex sources. The reactions to such sources commonly occur within the nearfield, where the axis of particle motion is not necessarily radial to the source. Thus, it is doubtful if resolving the axis of particle motion enables fish to perceive the actual location for most biologically significant sound sources in the nearfield. Nearly 70 years ago, von Frisch and Dijkgraaf (1935) performed the very first scientific study that specifically addressed the problem of directional hearing in fish. The title of their paper is “Können Fische die Schallrichtung wahrnehmen?” It is amazing that this question (Can fish sense the sound direction?) is as relevant today as it was in 1935. It certainly is difficult for the terrestrial mammal human to envisage how fish perceive their environment.

4. Directional Fast-Start Escape Responses

Fish display different types of fast-start escape responses to close range predatory strikes, defined by the pattern of the initial body bending (Domenici and Blake 1993; Hale 2002). The C-start response is the most common, and the best studied regarding sensory motor control (see reviews by Faber et al. 1989, 1991; Korn and Faber 1996; Zottoli and Faber 2000; Eaton et al. 2001). C-starts are typically triggered by high-intensity acousticolateral, somatic, and visual stimuli activating either of the paired Mauthner cells (M-cells) in the brainstem. Each M-cell receives massive input from the eighth nerve from the ear on the same side. The Mauthner axon crosses the midline and projects into the spinal cord, where it connects to motoneurons that innervate trunk muscle on the side opposite the M-cell soma. A typical C-start (Fig. 6.10) is triggered by one of the M-cells firing a single action potential, which elicits a virtually instant contraction of the muscles on the opposite side along the entire length of the fish.

FIGURE 6.10. Movements during a C-start type of fast-start behavior. A startle stimulus (arrow) at the left side of a goldfish at rest triggers the startle response. The fish forms a C-shaped bend of its body before the first propulsive tail stroke, and accelerates away from the stimulus. Successive silhouettes viewed from above are displayed at 5-ms intervals and shifted a fixed distance to the right for clarity. (From Eaton et al. 2001.)

The M-cells belong to the reticulospinal system, which is a distributed network extending from the caudal midbrain to the spinal cord, and a C-start is probably initiated by parallel activity of the Mauthner neuron and commissural hind brain neurons (Kimmel et al. 1982; Metcalfe et al. 1986; Lee et al. 1993; Foreman and Eaton 1993). The propulsive phase of the C-response is probably controlled by more caudal medullar neurons with ipsilateral spinal projections (Forman and Eaton 1993; Eaton et al. 2001). An extensive and hierarchic brainstem escape network has recently been confirmed by using fluorescent calcium indicators to monitor the activity of reticulospinal neurons in the transparent larvae of zebrafish (Fetcho and O'Malley 1997; Liu and Fetcho 1999; Ritter et al. 2001; Gathan et al. 2002).

A striking feature of the C-starts in fish is the oriented response away from the aversive stimulus source. Directional responses to stimuli activating the touch, lateral line, and visual systems are easily explained by the topographic organization of the representation of cutaneous mechanoreceptors and the visual field in the central nervous system. The idea that the Mauthner system is directionally sensitive also to sound was originally suggested by Moulton and Dixon (1967), but remained controversial until directional C-starts away from controlled acoustic stimuli were convincingly demonstrated in Atlantic herring (*Clupea harrengus*) by Blaxter et al. (1981) and in goldfish by Eaton et al. (1981). Directional escape responses to acoustic stimuli have later been described in several other species (see reviews by Eaton 1995; Canfield and Rose 1996; Hale et al. 2002).

How is the fish able to utilize acoustic information to launch a directional escape away from an attacking predator? Eaton and Emberley (1991) addressed this problem by analyzing the relationship between the direction of the acoustic stimulus and the angular component of the initial escape movement. They suggested that the fish measures the angle to the sound source, which then determines the magnitude, or time span, of the initial, rotational phase of the C-response. At that time the phase model for directional hearing was well established, and Eaton (1995) and Guzik et al. (1999) adopted and developed this model to explain the directional escape responses in fish.

As noted in Section 2.2, the phase model is insufficient for the localization of dipoles and higher order sources within the acoustic nearfield, due to the directional ambiguity of the particle motions. The acoustics of a striking predator may be best approximated by a dipole, and a potential prey will certainly respond within the nearfield of the attacker. At first glance, these conditions may seem incompatible with the phase model for directional discrimination. Further, the requirement for a separate pressure channel may not be fulfilled at threshold levels for the predominantly low-frequency signals generated by a striking predator, because the auditory gain provided by a swimbladder declines toward lower frequencies. However, in spite of these shortcomings of the phase model, it may still be applicable for the escape responses. The M-cell system is far from a low-level signal detector, and is activated by large particle motions and pressure changes generated by an accelerating predator at close range. Although the swimbladder is not involved in detection of infrasound at auditory threshold levels (Sand and Karlsen 1986), this input channel may still provide the required pressure information at the high pressure levels associated with fast-start escape responses. Further, Eaton et al. (2002) have pointed out that the primary axis of motion points directly at the prey during a predator attack, and the direction of particle acceleration detected by the prey is therefore most likely approximately in line with the approach path of the predator. Observed escape trajectories commonly display a wide scatter relative to the stimulus direction (Domenici and Blake 1993; Fig. 6.11A, B), and the model is only required to perform a coarse estimation of direction. In this game, speed is much more essential than accurate directional discrimination. In fact, a wide scatter of potential escape trajectories in the general direction away from the attacker may reflect an adaptive advantage, as it makes it difficult for the predator to predict the flight path and thus reduces the probability of a successful strike. This may be compared to the zigzag flight path of a rabbit chased by a fox.

The current neural model for the directional discrimination by the Mauthner system in hearing specialists suggests how a transient acoustic stimulus originating on either the left or the right side of a fish results in an initial orientation of an escape response away from the side of the stimulus (Eaton et al. 1995; Guzik et al. 1999). The model predicts that the M-cell system receives afferents that convey compression and rarefaction of the pressure component, and acceleration afferents conveying both left-to-right and right-to-left information. Intracellular recordings from M-cells and other relevant brainstem neurons in goldfish have shown that these neurons receive both pressure and acceleration inputs, as predicted by the model (Casagrand et al. 1999). An important feature of the current model is that both initial compression and rarefaction may contribute to the activation of the M-cell. Thus, an attack from the right will produce an initial right to left acceleration combined with a pressure increase, while a suction type of predator at this position will cause left to right acceleration and a rarefaction. According to the model, both these combinations of initial sound pressure and acceleration will elicit the appropriate escape to the left by inhibition of the left and activation of the right M-cell.

Although the current neural model for the C-start escape responses seems reasonable, the postulated efficiency of initial rarefaction to initiate the response lacks behavioral support. Most of the behavioral studies performed to date have

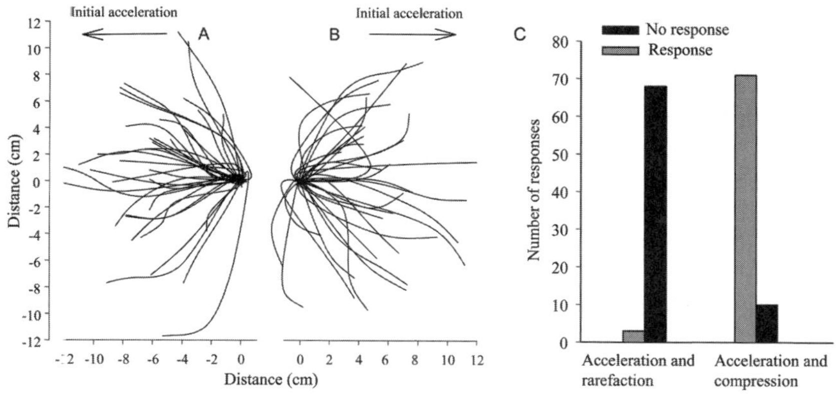

FIGURE 6.11. Startle trajectories in the horizontal plane displayed by juvenile roach in response to the initial half-cycle of an acceleration of 6.7 Hz. The fish were accelerated within a closed chamber at a stimulus level 15 dB above response threshold. The trajectories show movements of the head of the responding fish during 160 ms from the video frame before stimulus onset. (**A**) Trials with the initial acceleration to the left (push mode). (**B**) Tests with the initial acceleration to the right (pull mode). Startle responses in both stimulus situations were on average in the same direction as the initial acceleration. (**C**) Synergistic effects of acceleration and compression on triggering of the startle responses. The histograms present the number of responsive and nonresponsive fish in the leading (rarefaction) and trailing (compression) half of the test chamber, respectively. The fish mainly responded to the combination of linear acceleration and pressure increase. (From Karlsen et al. 2004.)

been hampered by insufficient control of the stimulus parameters. Generally, the frequency of the stimulus has been too high (100–2000 Hz) to reveal the relative effects of sound compression and rarefaction. Such frequencies may also be far above those associated with a real predator attack (Kalmijn 1989; Bleckmann et al. 1991). The otolith organs in fish are highly sensitive to the acceleration component of infrasound down to at least 0.1 Hz (Sand and Karlsen, 1986; Karlsen, 1992a,b), and typical behavioral threshold values are in the range of 10^{-5} m/s^2. At higher intensities around 10^{-2} m/s^2, infrasound may initiate strong avoidance responses in fish (see review by Sand et al., 2001).

For a prey fish, infrasonic acceleration may thus be a more realistic simulation of an approaching predator than the higher frequencies employed in previous studies, and Karlsen et al. (2004) have recently studied fast-start responses in the otophysan species roach (*Rutilus rutilus*) to infrasonic initial accelerations. The fish were accelerated in a controlled manner within a closed chamber suspended in a swing system (Karlsen 1992b). Typical C-start escape responses were indeed induced by accelerations within the infrasonic range, with a threshold of 0.023 m/s^2 for an initial acceleration at 6.7 Hz. The response trajectories displayed a wide scatter, but were on average in the same direction as the initial acceleration (Fig. 6.11A, B). Unexpectedly, startle responses occurred mainly in the trailing half of the test chamber, in which the fish were subjected to linear acceleration and compression (Fig. 6.11C). This combination characterizes the stimuli produced by an approaching predator. Very few responses were observed in the leading half of the test chamber, where the fish were subjected to acceleration and rarefaction. This type of stimulus is expected from a retracting predator. The lack of response to initial acceleration and rarefaction may also be an adaptive behavior, since an unnoticed pray fish may be easily spotted by a predator if an unnecessary escape response is initiated. It was concluded that particle acceleration is essential for the directionality of the startle response to infrasound, and that synergistic effects of acceleration and compression trigger the response. Since the current neuronal model for fast-start escape responses predicts that compression and rarefaction are equally efficient in triggering the response in combination with acceleration, it may need revision.

The sense of hearing in fish is an extremely sensitive sense that detects the faintest signals, including communication signals that are intended to be heard. The Mauthner system, on the other hand, may have evolved to do the opposite, namely to detect high intensity predatory signals that are intended to be concealed (Eaton and Popper 1995). A typical predator strike is characterized by a rapid acceleration of the head towards the prey, which generates low frequency compression and particle acceleration in the same direction as the strike, and Eaton and Popper (1995) have even suggested that aquatic predators might employ a "stealth strategy" to avoid acoustic detection. As predators accelerate towards the prey, various species open their oral cavities with a velocity equivalent to a 10–20 Hz signal. Rather than sucking the prey towards the mouth, this initial moth opening may reduce the acoustic and hydrodynamic noise associated with the acceleration of the predator. The final suction that pulls the prey into the oral cavity is not initiated before virtual contact with the prey.

5. The Lateral Line and Source Localization

The fish lateral line responds to midwater hydrodynamic events and to capillary surface waves (for reviews see Sand 1984; Bleckmann 1994; Coombs and Montgomery 1998). Since the physical properties of water surface waves and midwater hydrodynamic events are radically different, lateral line perception in midwater (and benthic) fish is treated separately from lateral line perception in surface feeding fish.

5.1 Surface Wave Perception

As first shown by Schwartz (1965, 1971), several teleost species of the families Cyprinodontidae, Hemirhamphidae, Gasteropelecidae and Pantodontidae can detect capillary surface waves with their cephalic lateral line (e.g., Fig. 6.12A). In their natural habitats such waves are usually caused by terrestrial (prey) insects fallen into the water. Capillary surface waves generated by a struggling insect often last for several seconds, are irregular in time course, have displacement amplitudes $< 100\,\mu m$, and contain frequencies in the range 10 to about 100 Hz (Lang 1980; Bleckmann 1985).

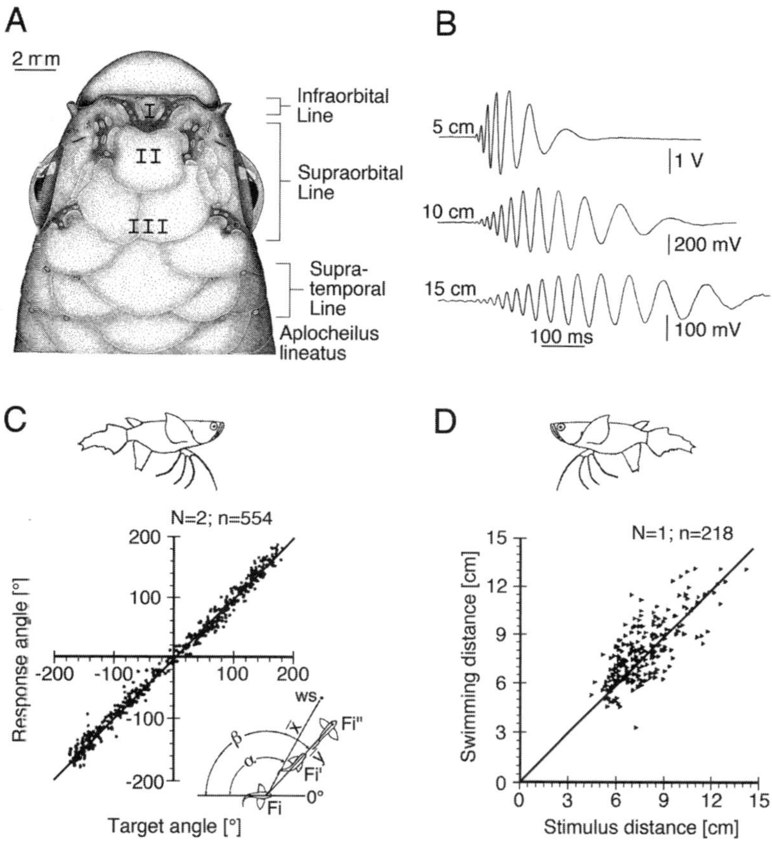

FIGURE 6.12. (**A**) Dorsal view of the head of the surface feeding fish *Aplocheilus lineatus*. The cephalic lateral line of *Aplocheilus* consists of three groups (labeled I, II, and III from rostral to caudal), each of which contains three single neuromasts. (The drawing was kindly provided by G. Tittel.) (**B**) A surface wave train (called a click) produced by dipping the tip of a small rod once into the water. The click stimulus was recorded at 5, 10, and 15 cm source distance. Note the differences in amplitude calibration. (Redrawn from Bleckmann and Schwartz 1982.) (**C, D**) Orientation behavior of mature, visually deprived *Pantodon buchholzi*. The accuracy of target angle determination (**C**) and distance determination (**D**) is shown. Each dot represents one response. Surface wave stimuli (clicks) were produced by dipping the tip of a small rod once into the water. Inset in **C**: Fi, animal at stimulus onset; Fi', animal during the phase of swimming; Fi", animal after stopping (identified by the spreading of pectoral fins); WS, wave source, X, source distance, Y:, swimming distance; α target angle, β, turning angle. (Redrawn from Bleckmann et al. 1989.)

5.1.1 Propagation of Surface Waves

Water surface waves radiate with dispersion, i.e., their propagation velocity is frequency dependent and has a minimum of 23 cm/s at a frequency f of 13 Hz (corresponding to a wave length λ of about 1.7 cm) (Lighthill 1980).

For frequencies higher than 13 Hz, this dispersion causes a frequency-downward modulation of the initial 7–9 wave cycles of any broadband wave stimulus (Bleckmann and Schwartz 1982). After traveling a certain distance, an initial dispersive wave group consists of a set of waves with different wavelengths and frequencies. Locally, λ can be defined by the crest-to-crest distance, but due to dispersion the distance between successive crests differs slightly (Fig. 6.12B). These differences reflect the distance-dependent frequency modulation of the wave group and the speed by which the wave packet spreads in space. The distance to a surface wave source can be determined unequivocally if the local frequency $\omega(= 2\pi f)$ and the frequency modulation around ω are known (Käse and Bleckmann 1987). Besides being dispersed, water surface waves are attenuated during propagation. Due to geometrical spreading, this attenuation is strongest in the vicinity of a wave source. In addition, attenuation increases with frequency (decreasing λ), i.e., the water surface behaves like a low pass filter (see Bleckmann et al. [1989] for a thorough treatment of surface wave physics).

5.1.2 The Cephalic Lateral Line of Surface Feeding Fish

Surface feeding fish detect capillary surface waves with their cephalic lateral line, which consists of superficial neuromasts (e.g., *Aplocheilus lineatus*, Fig. 6.12A), canal neuromasts (*Fundulus notatus*), or large neuromasts contained in cavities (*Pantodon buchholzi*) (Schwartz 1970). The sensory epithelium of lateral line neuromasts consists of hair cells that are separated and surrounded by numerous supporting cells. The apical end of lateral line hair cells contains 30 to 150 stereovilli and a single true kinocilium that project into a cupula extending several hundred micrometers into the surrounding water. Displacement of the stereovilli toward the kinocilium causes a depolarization, while displacement in the opposite direction hyperpolarizes the hair cell. Consequently, the responses of a single hair cell vary with the stimulus angle in a cosine fashion, as also described in Section 2.1. Within a neuromast, hair cells are usually oriented with their kinocilia pointing in the direction of the long (most sensitive) axis of the neuromast. In both superficial and canal neuromasts, the hair cells are oriented in two opposing directions, i.e. any displacement of the cupula will cause responses of opposite polarities from the two sets of cells, which work 180° out of phase. Consequently, lateral line neuromasts, just like individual hair cells, are directionally sensitive. A single lateral line afferent may innervate more than one hair cell, but is coupled only to hair cells aligned in the same direction. An afferent fiber therefore responds best (with a decrease or increase in spontaneous discharge rate) if the cupula moves in one of the two possible directions with respect to the most sensitive axis of the neuromast. Like auditory fibers, primary lateral line afferents respond to a sinusoidal wave stimulus with phase coupling. Stimulus intensity is encoded both by the degree of phase coupling and by the firing rate. Single lateral line neuromasts therefore encode the intensity (via phase coupling and spike rate) and frequency (via phase coupling) of a sinusoidal wave stimulus. Since different cephalic neuromasts of surface feeding fish are aligned in different directions (Schwartz 1970 and Fig. 6.12A), these fish are nearly equally sensitive to all wave directions.

5.1.3 Wave Source Localization

Even blinded surface feeding fish respond to a surface wave stimulus produced by dipping a rod once into the water with an orienting movement. At a distance up to about 15 cm, surface feeding fish can determine both the target angle and the distance to a wave source (Fig. 6.12C, D and Schwartz, 1971). They do so by exploiting the physical properties of the propagating wave stimulus.

5.1.3.1 Determination of Target Angle

According to the strong attenuation and low propagation velocity (23–50 cm/s in the relevant frequency range) of water surface waves, the direction to a surface wave source can be determined by:

- Measuring the intensity gradient of the stimulus
- Comparing the neuronal activity of primary lateral line afferents innervating cephalic neuromasts aligned in different directions
- Measuring arrival time differences between neuromasts

Localization via Intensity Gradients. Amplitude cues are well suited for target angle determination because amplitude gradients are steepest, and the distance to a wave source shortest, in the radial direction, i.e., perpendicular to the lines of equal stimulus intensity. Physiological experiments have shown that the steepness of amplitude curves obtained from the cephalic neuromasts of surface feeding fish increases with increasing stimulus frequency, i.e., high-frequency surface waves should be especially convenient for determining amplitude gradients (Mohr and Bleckmann 1998). Contrary to this consideration, the ability of surface feeding fish to determine the target angle does neither improve with increasing wave frequency nor with decreasing wave source distance (Bleckmann et al. 1989). It is therefore unlikely that amplitude cues are essential for target angle determination.

Localization via Angular Differences in Afferent Responses and Stimulus Arrival time. Up to about ±130° (0° is directly in front of the fish), target angle determination in an *Aplocheilus* with only one group of cephalic lateral line neuromasts left intact is not different from that of intact fish. In contrast, an *Aplocheilus* with only the supratemporal neuromasts and the neuromasts innervated by the dorsal branch of the trunk lateral line left intact, show a fairly accurate target angle determination only in the range 100°–160°. Regardless of target angle, unilaterally ablated fish always turn to the intact side (Schwartz 1965; Müller and Schwartz 1982).

An *Aplocheilus* with only one cephalic neuromast left intact remains sensitive to all wave directions, but it no longer determines the target angle (Fig. 6.13A). Instead, the degree of turning now correlates with the anterior–posterior position

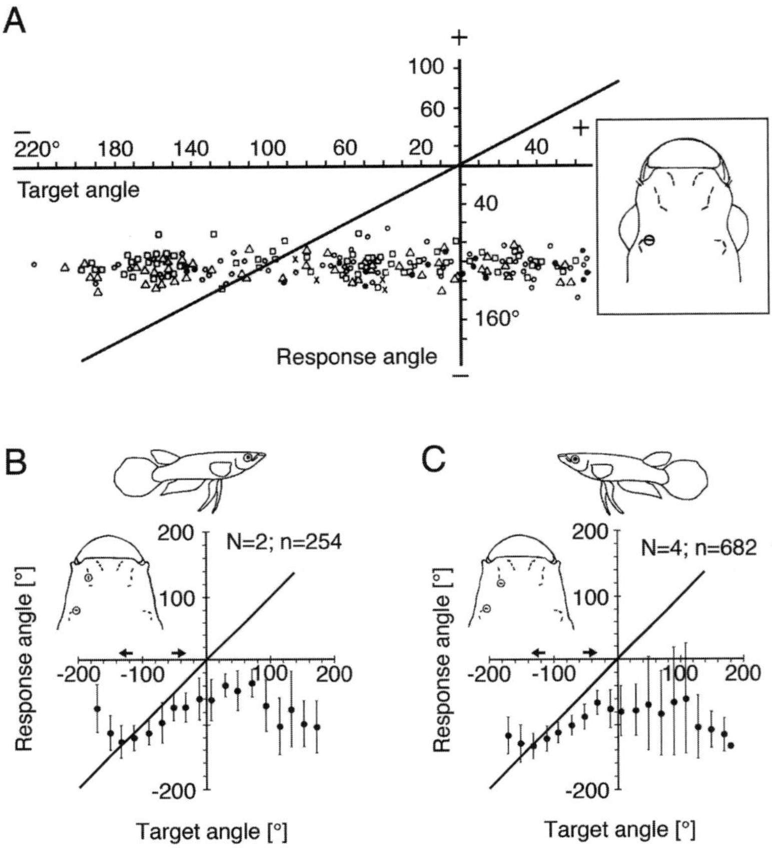

FIGURE 6.13. (**A**) Directional response of an *A. lineatus* with all but the encircled cephalic lateral line neuromasts destroyed (see inset; straight lines indicate the spatial arrangement of the neuromasts and neuromast orientations). Different symbols correspond to different test series performed with the same animal. 0° is in front, 180° is behind the fish. Positive angles are right, negative angles are left. (Redrawn from Müller and Schwartz 1982.) (**B, C**) Directional responses of an *A. lineatus* with all cephalic neuromasts destroyed except the two neuromasts encircled. Mean values and standard deviations are shown. The alignment of the long (most sensitive) axis of the two neuromasts differed by 83° in B and 19° in **C**. The distance of the two neuromasts from a fixed point at the fish's snout was 4.1 mm (**B**) and 2.8 mm (**C**). Arrowheads indicate the upper and lower limits of the dynamic range. (Redrawn from Bleckmann et al. 1989.)

of the particular neuromast in that a more caudal neuromast induces a larger response angle than a more rostral neuromast (Bleckmann et al. 1989). Thus, each neuromast appears to have a place value that determines the orienting response of the fish.

Within a certain dynamic range, an *Aplocheilus* with only two neuromasts left intact retains its ability to determine the target angle (Tittel et al. 1984).

Tittel (1991) performed carefully designed ablation experiments to test whether sensitivity differences due to neuromast directionality and/or arrival time differences between neuromast are used for target angle determination. The ablation combinations were chosen to give slight differences in neuromast position, but large differences in neuromast alignment (i.e., large differences in neuromast output), and vice versa (insets in Fig. 6.13B, C). The subsequent behavioral tests indicated that arrival time and/or phase differences between neuromasts – but not neuromast output differences - are the most important cues used for target angle determination. A second line of evidence also suggests that neuromast output differences based on neuromast directionality are less important: If single-frequency waves are presented, a 30-Hz stimulus leads to more accurate responses than a 80-Hz stimulus (Tittel 1985), despite the fact that input–output curves are steeper at higher wave frequencies (Mohr and Bleckmann 1998). It should be stressed that the correlation between arrival time differences at different neuromasts and the target angle is unequivocal only if the frequency-dependent wave propagation speed is taken into account (see preceding text). Obviously, surface-feeding fish do correct for this relationship.

5.1.3.2 Determination of Wave Source Distance

Close to the source, the distance to the center of a concentric surface wave stimulus can be determined by measuring the curvature of the stimulus. In addition, if the attenuation and distance-dependent frequency modulation of surface waves are known, the distance to a surface wave source can be determined by measuring:

- The relative amplitude decrease per unit of distance
- The mean frequency and the frequency modulation of the first 7–9 wave cycles
- The amplitude spectrum of a wave train

It should be stressed that in waves lacking higher frequencies the amplitude spectrum can give some information about the source distance only if compared with a commonly experienced standard.

Integration Time. The basic frequency of a wave train, its local frequency modulation, and its amplitude spectrum can be obtained only if the stimulus is integrated over at least a few wave cycles. The first stimulus-evoked muscle potentials can be recorded from a freely swimming *Aplocheilus* after the first 8.5 cycles of a wave train passing the head of the fish, independently of stimulus amplitude (Bleckmann and Schwartz 1981; Bleckmann 1982). Thus, the information sufficient for *Aplocheilus* to localize a wave source must be contained in these 8.5 wave cycles.

Responses to Artificial Wave Stimuli. Due to the low pass filter properties of the water surface (see Section 5.1.1), a rough estimation of source distance is possible by evaluating the amplitude and frequency content of a wave stimulus. For instance, the determination of source distance could be based on the assumption that high-amplitude, broad-bandwidth stimuli on average have traveled a shorter distance than low-amplitude, low-frequency and narrow-bandwidth stimuli. Hoin-Radkovski et al. (1984) tested this assumption by stimulating the surface feeding fish *P. buchholzi* with clicks of different upper frequency limits. They found that the relative localization errors (for definition, see Hoin-Radkovski et al. 1984) of *Pantodon* are independent of stimulus frequency content and source distance. Therefore, in this fish the determination of source distance cannot be based on the evaluation of the wave spectrum alone.

Responses to Single-Frequency Wave Trains. Decoding of local frequency and frequency modulation of a wave stimulus is a possible strategy to determine the distance a surface wave stimulus has traveled. To test this prediction, Bleckmann (1980) and Hoin-Radkovski et al. (1984) presented single-frequency wave stimuli to *Aplocheilus* and *Pantodon*. As expected, this led to an impaired distance determination in both species, with a tendency to underestimate the source distance if it exceeded 6–8 cm. However, there was still a weak but significant correlation between source distance and swimming distance. For *A. lineatus* this was valid only for wave frequencies below 50 Hz (Bleckmann 1980). In general, the relative localization error at a given source distance grew with frequency, while it increased with source distance at a given frequency (Bleckmann 1988). Assuming that the curvature of the wave front is also used as a cue for wave source distance and that it is determined through arrival time differences, this is exactly what is to be expected (Hoin-Radkovski et al. 1984). At a given source distance, surface-feeding fish swam progressively shorter in response to single-frequency wave stimuli of higher frequencies. This indicates that the amplitude spectrum is also evaluated, because, if no other cues are available, a high-frequency stimulus can be "expected" to have traveled a shorter distance than a low frequency stimulus. Considering the low pass filter properties of the water surface, this reflects a likely localization strategy. Application of frequency-upward modulated stimuli that contained high-frequency wave components also caused an underestimation of source distance, which supports the notion that surface-feeding fish do follow the strategy outlined in the preceding text.

Responses to Altered Wave Trains. When confronted with a computer controlled wave stimulus that was generated at a distance of 7 cm, but simulated the frequency modulation of a click that had traveled a distance of 15 cm, the fish swam on average 4–6 cm beyond the wave source (Bleckmann and Schwartz 1982; Hoin-Radkovski et al. 1984). This result was the final proof that local frequency modulation of a wave train is one of the cues used by surface feeding fish to determine the source distance. The use of frequency modulation for distance determination was further supported in experiments with *Aplocheilus*, in which all but one cephalic neuromast were destroyed. Although such a fish had no way to determine the curvature of the wave front or the amplitude decrease per unit of distance, it still showed an increase of swimming distance with wave source distance if clicks were presented (Müller and Schwartz 1982). Distance determination in ablated fish was, however, somewhat impaired. This indicates again that surface feeding fish use also other parameters than frequency modulation for distance determination.

Sound Waves. Any object that causes water motions also generates sound pressure waves. Since sound pressure propagates with a speed of about 1500 m/s in water and the propagation velocity of surface waves is in the cm per second range (see Section 5.1.1), fish could calculate the wave source distance by comparing the arrival times of pressure (acoustic) and surface waves. However, up to now the use of pressure waves for localization of surface wave sources has not been demonstrated in surface-feeding fish (for a detailed discussion of distance determination in surface feeding fish see Bleckmann 1988).

5.2 Perception of Subsurface Water Motions

Midwater and bottom dwelling fish also use the lateral line to detect and localize sound sources, provided the sound of these sources causes relative movements between the fish and the water surrounding the fish. At relevant stimulus frequencies, this is only the case if the fish is within the innermost part of the acoustic nearfield. At larger distances, the whole body motions of the fish deprive the lateral line of its stimulus (Sand 1981). If the light conditions are adequate, most fish primarily use visual cues for detection of a sound source (e.g., a prey or a predator). During daytime, piscivorous fish may initiate their pursuits from distances of several body lengths. However, in darkness a strike usually only occurs if the prey distance is less than 5–10 cm (Enger et al. 1989). In complete darkness, a fish in hunting mood typically glides slowly through the water, driven only by occasional tail flips. In the dark, an intact lateral line is essential for the initiation of a strike from some distance. If the lateral line is blocked with cobalt ions (Karlsen and Sand 1987), bluegills (*Lepomis macrochirus*) never make a successful attack at a preyfish, nor do they bite a simulated prey object, unless the simulated prey accidentally is touched (Enger et al. 1989).

5.2.1 Prey Detection in the Mottled Sculpin

The mottled sculpin (*Cottus bairdi*) is a benthic fish that exhibits a natural and unconditioned orienting response to both live prey and vibrating spheres (Coombs and Janssen 1990). In the absence of nonmechanosensory cues (such as vision), the initial orientation and approach behavior of mottled sculpins relies heavily, if not exclusively, on lateral line input (Coombs 1994). The peripheral lateral line of the mottled sculpin consists of superficial and canal neuromasts (Janssen et al. 1987). In blinded sculpin, the approach behavior to a dipole source depends largely on the initial orientation of the fish relative to the source (and the axis of source vibration). Indirect approaches in which the fish either keeps the source continuously to one side of the body (Fig. 6.14A), or alternates

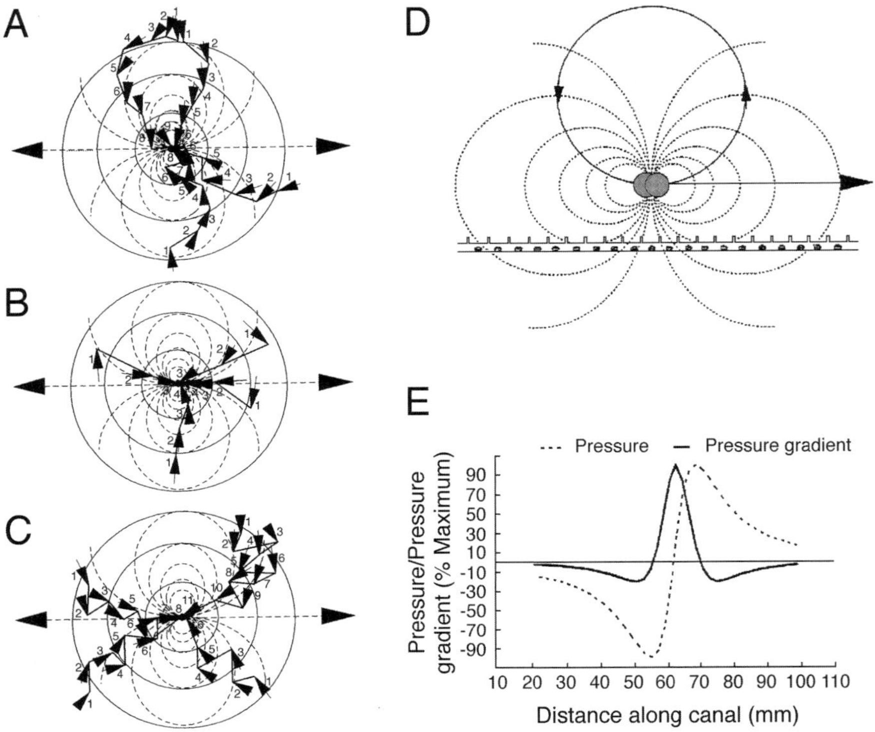

FIGURE 6.14. (**A–C**) Step-by-step approaches of a mottled sculpin to a dipole source. (**A**) Smoothly arching approaches in which the fish keeps its head to one side of the source. (**B**) Direct approach paths in which the vibrating sphere is kept mainly in front of the fish's head. (**C**) Zig-zag approaches in which the fish alternates between being to the left and to the right of the source. Dashed lines indicate flow lines about the source (center of graph), whereas thin-lined circles represent fixed radial distances of 3, 6, and 9 cm from the source center. (Redrawn from Coombs et al. 2000.) (**D**) Schematic representation of iso-pressure contours (dashed lines) and flow lines (solid lines with arrows) around a dipole source. Iso-pressure contours are depicted for a single plane that bisects the source along its axis of oscillation indicated by the large arrowhead to the right. A lateral line canal is modeled as a single tube with an array of pressure sampling points (canal pores) at 2-mm intervals (not to scale). In the example shown, the canal is confined to a single horizontal plane through the source center and its long axis is parallel to the axis of source oscillation. (**E**) Corresponding plots of pressure (dashed line) and pressure gradient (solid line) distribution across the modeled trunk lateral line canal. Note that the maximal pressure gradient is centered at the source, arbitrarily located at X distance = 61 mm along the modeled canal. (Redrawn from Coombs and Conley 1995.)

between keeping the source to the left and right side of the body, tend to occur when the fish is pointing toward the source at signal onset (Fig. 6.14C). When the source is to the side of the fish at the time of stimulus onset, mottled sculpin approach the source in a more direct path (Fig. 6.14B) (Coombs et al. 2000). Blinded sculpin not only determine the direction, but also the distance to a dipole source as long as the lateral line is intact on the side of the fish facing the dipole (Janssen and Corcoran 1993). The ability to estimate source distance is, however, restricted to short ranges (about one fish body length), as are most, but not all (e.g., hydrodynamic trail following, see later), lateral line sensing abilities.

As described by Webb, Montgomery, and Mogdans (Chapter 5), the fish lateral line encodes water movements and pressure gradients in spatially nonuniform flow fields. Figure 6.14D illustrates the isopressure contours and flow lines about a dipole source, and Figure 6.14E depicts the calculated pressure and pressure gradients across a hypothetical trunk lateral line canal. The pressure gradient pattern along this canal (solid line in Fig. 6.14E) was derived by computing the pressure difference across consecutive pairs of pores. The calculated pressure gradient distribution reveals a complex but predictable pattern consisting of a large, central, and positive peak surrounded by two smaller negative peaks on either side. At certain points the pressure gradient function passes through zero, i.e., at these points the sign of the pressure gradient passes from negative to positive, or vice versa. Theoretical calculations and pressure measurements with a miniature hydrophone show that the points of phase reversals are invariant as a function of source amplitude but change as a function of source distance. As source distance increases, the spatial separation between side peaks and phase reversal points increases. At the same time the peak amplitude decreases at a fall-off-rate of $1/\text{distance}^3$, i.e., at a rate predicted for dipole sources (Kalmijn 1988).

Recordings from primary lateral line afferents or of neuromast receptor potentials reveal that information about the position of a vibrating sphere relative to the fish is linearly coded in excitation patterns that reflects the spatial characteristics of the pressure gradients distributed along the lateral line canal (Sand 1981; Coombs et al. 1996; Curcic-Blake and van Netten 2006). This means that the excitation patterns of lateral line canal neuromasts can be predicted if one knows the course of lateral line canals on the animal's body, the interpore spacing, and the pressure distribution around the source (Coombs et al. 1996). An algorithm developed by Curcic-Blake and van Netten (2006) and Goulet et al. (2008) even shows that lateral line excitation patterns not only provide the information about source location but also about the direction of sphere vibration.

Calculations of lateral line excitation patterns for a linear array of canal neuromasts at different distances from the source indicate that the information about source azimuth is contained in the location of the maximum pressure-difference amplitude, whereas information about source distance is contained in the spread of excitation. This distance cue is robust and unambiguous; that is, if source vibration amplitude or size is increased at a given distance, the level, but not the spread, of lateral line excitation increases. Thus, although peak excitation levels may be identical for a distant, high-amplitude (or large) source and a near-by, low-amplitude (or small) source, there is sufficient information in the spread of excitation to distinguish between the two. Therefore, the spatial representation of source distance along a two-dimensional array of sensors may provide the mottled sculpin with a mechanism of depth perception like that already suggested for the lateral line of the blind cavefish *Anoptichthys jordani* (Hassan 1989). Unlike visual images, which get smaller as the source moves further away, hydrodynamic images get larger, as do electrosensory images (Emde et al. 1998).

The approach behavior (Fig. 6.14A–C), and the hypothesis that spatial excitation patterns along the lateral line system of mottled sculpin (Fig. 6.14D) play a major role in encoding both source direction and source distance, suggest the following strategies used by the mottled sculpin in finding a dipole source:

- Moving in a direction that increases the pressure difference along the head while keeping it consistently low across the head
- Narrowing the fish-to-source gap with each successive movement
- Keeping the source lateralized
- Avoiding approach positions that are perpendicular to the flow line or that place the fish in the pressure zero area of a dipole field

5.2.2 Source Localization by Wake Tracking

Whenever a fish moves, it involuntarily becomes a sound source. The low-frequency nearfields caused by a moving fish are, to a first approximation, dipolar. However, besides a dipolar flow field a swimming fish involuntarily generates a wake that may persist for up to several minutes (Fig. 6.15). Fishborne wakes consist of vortices and contain frequencies from below 10 Hz up to about 100 Hz and water velocities that reach several mm/s (Hanke et al. 2004). The sensitivity of the fish lateral line covers the amplitude and frequency range of the water motions found in the wakes of even a small fish (Bleckmann et al. 1991). In addition, the widespread spatial distribution of lateral line neuromasts on the head and body of fish should ease the analysis of complex three-dimensional water motions like those found in the wakes of fish.

Wake height and the lateral distance between the vortices in a fish wake correspond to the size of the tail fin of the fish that generated the wake. The specific structure of fish wakes also provides information about swimming style. Detection of rotation and traveling direction of the vortices and the direction of the dragged water give information on swimming direction. The mean velocity and maximum vorticity contain information about the time that has passed since the wake generator swam by (Hanke et al. 2000, Hanke and Bleckmann 2004). All these facts and the observation that many piscivorous fish hunt at night, or at depths where low light levels limit vision, led to the question whether the wakes left by swimming fish are used by some predators to track their piscine prey, analogous to the way by which dogs (Steen and Wilsson 1990) and snakes (Chiszar et al. 1990) follow the tracks left by their terrestrial prey. Recent behavioral experiments have shown that the nocturnal piscivorous European catfish (*Siluris glanis*) can track the wakes and thus the swim path of a prey fish (guppy, *Poecilia reticulata*) even in complete darkness (Fig. 6.16). Wakes up to 10 s old were followed over distances that covered up to 55 times the body length of the prey (Pohlmann et al. 2001). Blocking the lateral line with Co^{2+}

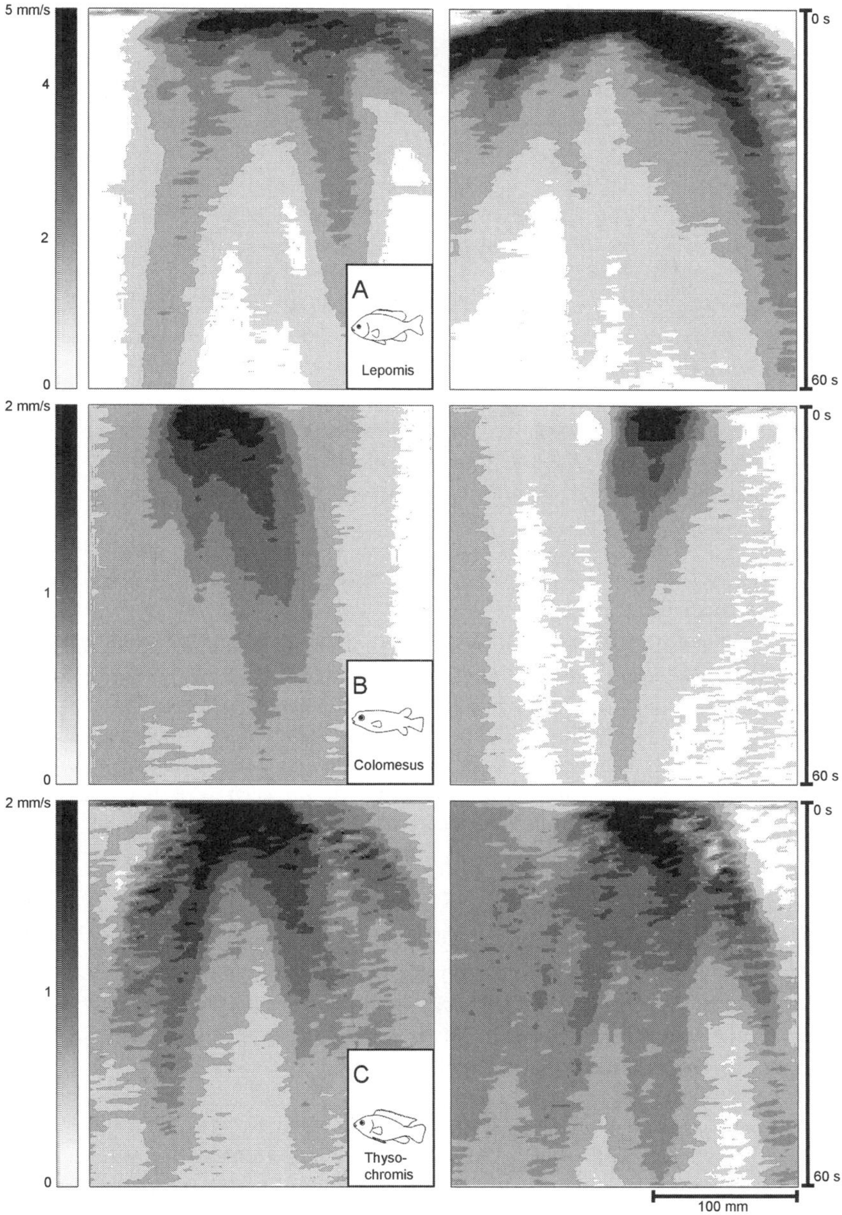

FIGURE 6.15. Spatial extent of wakes (x-axis) caused by swimming *Lepomis gibbosus* (**A**), *Colomesus psittacus* (**B**), and *Thysochromis ansorgii* (**C**) as function of time (y-axis). To resolve the low water velocities in aged trails, all velocities larger than 5 mm/s (**A**) or 2 mm/s (**B, C**) were coded in dark black. (Redrawn from Hanke and Bleckmann 2004.) (A color version of this figure can be found online at http://www.springer.com/978-0-387-73028-8.)

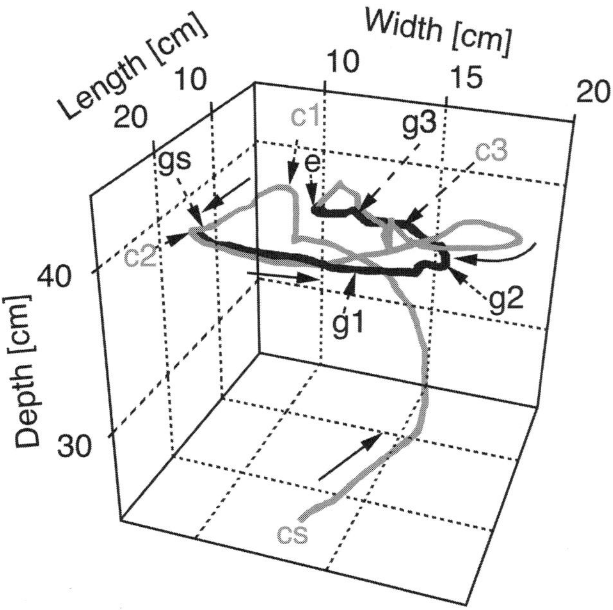

FIGURE 6.16. Three-dimensional plot with temporal information of an attack categorized as wake following. Gray, predator; black, prey. The numbers depict cm and correspond to calibrated positions in the test tank. Solid arrows indicate swimming direction. Three points in time (1 = 11.3 s, 2 = 8.6 s, 3 = 3.4 s before the attack) were chosen to indicate the locations of both fish. The positions of the catfish and the prey fish are indicated by c and g, respectively. The path-following appears to begin when the prey is at g2 and the predator is at c2. (Redrawn from Pohlmann et al. 2001.)

showed that lateral line input was indispensable for wake tracking (Pohlmann et al. 2004). It should be stressed that a predator tracking a wake usually does not perceive the instantaneous location of its prey. Therefore, it cannot approach the prey directly or in an arc, intercepting the prey's path. Nevertheless, wake tracking considerably extends the area in which prey is detectable and thus enhances the encounter probability (Hanke and Bleckmann 2004).

5.2.3 The Lateral Line and Spatial Orientation

Surgically blinded fish avoid aquarium walls without actual touch (Hofer 1908; Dijkgraaf and Kalmijn 1962). Ablation experiments have shown that blind fish use lateral line input for the avoidance of walls or other objects. The cavefish *Astyanax mexicanus* (formerly *Anoptichthys jordani*) is a champion in lateral line perception. Although this species lacks functioning eyes, it can pass through a barrier of rods without touching them (Hahn 1960). Hence, *Astyanax* can determine the position and shape of an object by using nonvisual cues. Studies by von Campenhausen and coworkers (von Campenhausen et al. 1981; Weissert and von Campenhausen 1981; Teyke 1985; Abdel-Latif et al. 1990;

Hassan 1992a, 1992b; Hassan et al. 1992; Hassan 1993) have shown that a gliding fish produces a flow field that is altered by nearby objects. Blind cavefish obviously use lateral line input to evaluate these alterations while gliding past an object. The ability of blind cavefish to detect, localize and discriminate between objects is remarkable. For instance, under favorable conditions *Astyanax* distinguishes between two "fences," each of which has six bars differing only slightly with respect to their relative positions.

Blind cavefish increase their swimming speed in a new environment (Teyke 1985). The increased water velocity across their body surface decreases the boundary layer thickness, which enhances lateral line perception (Teyke 1988, 1989). This behavior provides an excellent method for measuring the ability of cavefish to discriminate between objects and environments in space. By allowing the fish to learn a landmark array before changing the landmark positions within the configuration and recording swimming velocity, De Perera (2004) showed that blind cavefish use hydrodynamic information to develop an inner map of their environment. After learning the position of four landmarks, spaced equally apart, blind cavefish showed a significant increase in swimming velocity when exposed to landmark transformations. Therefore, the fish compare the environment they perceive with an internal representation of the environment they have learnt. The data also indicate that blind cavefish use lateral line input to encode the absolute distance between landmarks and possibly also shape within their spatial maps (De Perera 2004).

5.3 Central Processing of Lateral Line Input

5.3.1 Directional Coding

In surface feeding fish and aquatic amphibians (*Xenopus laevis*), individual lateral line neuromasts can be stimulated by water surface waves from many directions (Zittlau et al. 1986; Elepfandt and Wiedemer 1987). However, because the response of a single primary lateral line afferent depends on both stimulus amplitude and direction, it cannot signal the direction of surface wave propagation unequivocally. In contrast, some lateral line units recorded from the torus or tectum of aquatic amphibians (*Ambystoma*, *Xenopus*) show strong preference for a certain stimulus direction (e.g., Behrend et al. 2006). In the few cases tested, directional tuning was independent of stimulus intensity and/or stimulus frequency (Zittlau et al. 1986; Bartels et al. 1990). Hence, some tectal (midbrain) lateral line units unequivocally encode for stimulus direction. If the tectal recording sites were shifted from a caudomedial to a rostrolateral position in *Xenopus* and *Ambystoma*, the receptive fields shifted from caudal to rostral on the contralateral water surface (Zittlau et al. 1986; Bartels et al. 1990). Therefore, stimulus direction is a lateral line parameter mapped in the tectum of aquatic amphibians. Unfortunately, comparable studies in fish are lacking.

The physiologically identified tectal maps of wave direction in amphibians suggest involvement of the tectum in wave source localization, and small tectal lesions are sufficient to alter the turning responses of *Xenopus* (Claas et al. 1989).

For stimulus directions corresponding to the receptive fields of the destroyed neuronal populations, the response frequency was reduced and a precise localization of stimulus direction no longer occurred. Consequently, frogs with an ablated tectum completely failed to orient to a water surface wave source (Claas et al. 1989). Most likely, the tectum of *Xenopus* is part of the sensorimotor interface for orienting reactions.

5.3.2 Coding of Object Distance

Surface feeding fish determine the wave source distance up to about 15 cm. Hence, the question arises whether there are central lateral line units that encode wave source distance, but this has not been investigated in surface feeding fish. In *Xenopus*, however, some tectal units respond only at specific distances between the frog and a surface wave source (test range 6–16 cm) (Claas et al. 1989).

Little is known about the coding of wave source direction and wave source distance in midwater and bottom-dwelling fish. None of the medullary and midbrain lateral line units tested so far have shown signs of directional coding or distance coding. What has been found, however, are central units that encode the motion direction of an object passing the fish laterally (Bleckmann and Zelick 1993; Müller et al. 1993; Wojtenek et al. 1998; Engelmann and Bleckmann 2004) or the motion direction of a vibrating sphere (Meyer and Bleckmann, unpublished). Evidently, more data are needed to learn whether and how central lateral line neurons encode the position of an object in space.

6. Summary and Suggestions for Future Work

Although numerous studies have shown that fish can discriminate between different sound directions in three-dimensional space, it is still not settled if fish are able to perceive the actual location of a sound source at a distance. Current models for directional hearing in fish are based on neural calculation of the direction of particle movements of the incident sound by vectorial weighing of input from different regions of the sensory maculae. However, a simple particle motion detector is unable to discriminate between opposing sound sources (180° ambiguity). The directions of farfield particle motion and wave propagation coincide during acoustic compression. The phase model for directional hearing predicts that species with a swimbladder detect both the direction of the incident particle movements and the sound pressure, and the 180° ambiguity is resolved by decoding the phase difference between these components. Such phase analysis might also enable fish to detect the distance to a monopole source within the nearfield.

Most behavioral studies have focused on detection of changes in sound direction, rather than detection of the actual location of a sound source. In the acoustic farfield, the radial particle motions simplify possible computation of source location, whereas the situation is more complex in the nearfield of a dipole or higher order source. The unifying guidance model, which works for all types of sound sources, suggests that fish do not perceive the source location at a distance, but are instead guided to the source by turning in a manner that keeps a constant angle between the body axes and the incident particle acceleration detected by the inner ear.

Behavioral evidence supporting auditory source localization has mainly been obtained in studies of the fast C-start escape response to short-range acoustic stimuli of high intensity. Such stimuli activate either of the paired Mauthner cells in the brain stem, which then elicits instant contraction of muscles on the opposite side, followed by directed acceleration away from the source. Recent experiments indicate that simultaneous infrasonic acceleration and compression, i.e., characteristics of stimuli produced by an approaching predator, is more efficient in triggering the response than combined acceleration and rarefaction.

Surface feeding fish use their cephalic lateral line to detect the capillary surface waves caused by terrestrial insects fallen into the water. Target angle determination is based on arrival time and/or phase differences between neuromasts. This is feasible due to the low propagation speed of surface waves. The distance to the wave source (determined up to about 15 cm) is determined by evaluating the distance-dependent frequency modulation of the signal.

Midwater (and benthic) species may use their lateral line to localize moving objects at close range (within about one body length). Information about source azimuth is contained in the location of the maximum pressure-difference amplitude, whereas information about source distance is contained in the spread of excitation along linear arrays of neuromasts.

The lateral line may also detect the vortices in a fish wake, which provide information about fish size, swimming direction, and the time since the wake generator passed by.

Finally, a gliding fish produces a flow field that is sensed by the lateral line. This flow field is altered by nearby objects, and fish may thus use lateral line input to evaluate these alterations while gliding past an object. This ability is particularly well developed in blind cavefish, which perceive their physical environment, and construct an internal representation of it, based on lateral line input.

The field of directional hearing and sound source localization in fish is stilled riddled by numerous unsolved problems, although it has been an active research field for 70 years. Also regarding the reception of hydrodynamic stimuli by the lateral line, many basic questions remain unanswered. Among the most important questions that should be addressed in future experiments, are the following:

1. To which extent are fish able to perceive the actual location of a sound source at a distance? This problem relates to determination of both direction and distance, and must be addressed for farfield and nearfield detection of both monopole and dipole sources.
2. Does the brain use information from all otolith organs in its computation of sound direction? Although afferents from all otolith organs are sufficiently

sensitive to respond to particle motions associated with sounds of normal intensity, this question is still not settled.

3. Is the elevation of a sound source determined by a monaural mechanism? Some studies of the optimal axes of primary afferents indicate that computation of azimuth requires information from both ears, whereas one ear might suffice for computation of sound source elevation. The first assumption is supported by behavioral studies (ablation experiments), whereas the second assumption still needs experimental clarification.
4. How, and to which degree, is the directional information in the incident particle acceleration protected against masking by the amplified secondary particle motions radiating from the swimbladder? Both peripheral (appropriate alignment of hair cell populations) and central mechanisms (common mode rejection) may be involved, but the possible existence of such mechanisms lacks experimental verification.
5. Are only fish utilizing the swimbladder as an accessory hearing organ able to resolve the 180° ambiguity, or to discriminate between sources at different distances? So far, these questions, which relate to discrimination rather than to absolute localization, have only been addressed for species in which the swimbladder has an auditory function.
6. Is the advantage of a swimbladder in lowering auditory thresholds at the expense of acute directional hearing? As opposed to hearing nonspecialists, single units in the torus semicircularis in goldfish lack the diversity in the axes of optimal vibration direction observed among primary afferents. Does this finding reflect poor directional hearing in goldfish (and other hearing specialists), or is it due to inadequate sampling of higher order auditory neurons? So far, no behavioral studies have determined the limits for angular discrimination in otophysans.
7. How are directional information processed in the central nervous system? Some studies have addressed this question, but central processing of directional information is still incompletely understood. This also applies to possible multimodal processing that may integrate directional information from otolith organs and the lateral line.
8. If fish are unable to localize a sound source at a distance, are they still able to approach a source by the mechanism postulated by the guidance model? This question may be addressed by tracking of the approach paths toward concealed monopole and dipole sources.
9. Do predators striking prey at close range employ a "stealth strategy" to boost their success rate? It has been suggested that the predator may delay triggering of fast start escape responses by gradually opening the mouth during the strike prior to the final suction, thereby reducing its own bow wave that alerts the prey. This interesting hypothesis lacks experimental verification.
10. What is the relationship between acceleration and pressure in stimuli that triggers fast start escape responses in different species? Recent observations indicate that synergistic effects of initial acceleration and compression trigger

such responses in otophysans, while the current model for computation of the escape direction does not discriminate between compression and rarefaction. Comparison of species with different swimbladder anatomy, combined with experiments allowing independent control of acceleration and pressure at low frequencies, may clarify this issue.

11. What potential information about the sender (e.g., size, species, direction and velocity of movement) do hydrodynamic stimuli sensed by the lateral line contain? Studies of the physical properties (time course, frequency, and amplitude content), three-dimensional extension, and aging of biologically relevant hydrodynamic stimuli (like the wakes left by swimming fish) are only in their beginning.
12. What kinds of hydrodynamic noise do fish encounter in their natural habitats, and how does the lateral line cope with different noise levels? Discrimination between signal and noise is a general problem in sensory systems. In addition, at the behavioral level little is known about the kind of information fish can extract with their lateral line from the ambient subsurface and surface water motions. This question is especially intriguing in view of the many peripheral specializations seen in the lateral line, i.e., what kind of ethoecological adaptations are there?
13. How is lateral line input processed in the central nervous system? Some of the specific questions are: How are simple and complex hydrodynamic stimuli coded in single channels, and across channels? Are there both parallel and distributed processing of hydrodynamic input at successive levels of the brain? Are there differences between species in central wave source localization algorithms, e.g., in midwater and surface feeding fish? What kinds of simple or multiple mappings—computed as well as topographic—are there in the central lateral line pathway? Are there subsystems for wave source recognition and wave source localization? How are lateral line and auditory information integrated in the CNS?

References

Abdel-Latif H, Hassan ES, von Campenhausen C (1990) Sensory performance of blind Mexican cavefish after destruction of the canal neuromasts. Naturwissenschaften 77:237–239.

Bartels M, Münz H, Claas B (1990) Representation of lateral line and electrosensory systems in the midbrain of the axolotl, *Ambystoma mexicanum*. J Comp Physiol A 167:347–356.

Behrend O, Branoner F, Zhivkov Z, Ziem U (2006) Neural responses to water surface waves in the midbrain of the aquatic predator *Xenopus laevis*. Eur J Neurosci 23: 729–744

Blaxter JHS, Gray JAB, Denton EJ (1981) Sound and startle responses in herring shoals. J Mar Biol Assoc UK 61:851–869.

Bleckmann H (1980) Reaction time and stimulus frequency in prey localization in the surface-feeding fish *Aplocheilus lineatus*. J Comp Physiol A 140:163–172.

Bleckmann H (1982) Reaction time, threshold values and localization of prey in stationary and swimming surface feeding fish *Aplocheilus lineatus* (Cyprinodontidae). Zool Jahrb Abt Physiol 86:71–81.

Bleckmann H (1985) Discrimination between prey and non-prey wave signals in the fishing spider *Dolomedes triton* (Pisauridae). In: Kalmring K, Elsner N (eds) Acoustic and Vibrational Communication in Insects. Berlin: Paul Parey, pp. 215–222.

Bleckmann H (1988) Prey identification and prey localization in surface-feeding fish and fishing spiders. In: Atema J, Fay RR, Popper AN, Tavolga WN (eds) Sensory Biology of Aquatic Animals. New York: Springer-Verlag, pp. 619–641.

Bleckmann H (1994) Reception of hydrodynamic stimuli in aquatic and semiaquatic animals. In: Rathmayer W (ed) Progress in Zoology, Vol 41. Stuttgart: Gustav Fischer, pp. 1–115.

Bleckmann H, Schwartz E (1981) Reaction time of the topminnow *Aplocheilus lineatus* to surface waves determined by video- and electromyogram recordings. Experientia 37:362–363.

Bleckmann H, Schwartz E (1982) The functional significance of frequency modulation within a wave train for prey localization in the surface-feeding fish *Aplocheilus lineatus*. J Comp Physiol A 145:331–339.

Bleckmann H, Zelick R (1993) The responses of peripheral and central mechanosensory lateral line units of weakly electric fish to moving objects. J Comp Physiol A 172:115–128.

Bleckmann H, Tittel G, Blübaum-Gronau E (1989) The lateral line system of surface-feeding fish: anatomy, physiology, and behavior. In: Coombs S, Görner P, Münz H (eds) The Mechanosensory Lateral Line. Neurobiology and Evolution. New York: Springer-Verlag, pp. 501–526.

Bleckmann H, Breihaupt T, Blickman R, Tautz J (1991) The time course and frequency content of local flow fields caused by moving fish, frog and crayfish. J Comp Physiol A 168:749–757.

Buwalda RJA (1981) Segregation of directional and nondirectional acoustic information in the cod. In: Tavolga WN, Popper AN, Fay RR (eds) Hearing and Sound Communication in Fishes. New York: Springer-Verlag, pp. 139–172.

Buwalda RJA, van der Steen J (1979) The sensitivity of the cod sacculus to directional and non-directional stimuli. Comp Biochem Physiol A 64:467–471.

Buwalda RJA, Schuijf A, Hawkins AD (1983) Discrimination by the cod of sound from opposing directions. J Comp Physiol A 150:175–184.

Canfield JG, Rose GJ (1996) Hierarchical sensory guidance of Mauthner-mediated escape responses in goldfish (*Carassuis auratus*) and cichlids (*Haplochromis burtoni*). Brain Behav Evol 48:137–156.

Casagrand JL, Guzik AL, Eaton RC (1999) Mauthner and reticulospinal responses to onset of acoustic pressure and acceleration stimuli. J Neurophysiol 82:1422–1437.

Chapman CJ (1973) Field studies of hearing in teleost fish. Helgoländer wiss Meeresunters 24:371–390.

Chapman CJ, Hawkins AD (1973) A field study of hearing in the cod, *Gadus morhua* L. J Comp Physiol 85:147–167.

Chapman CJ, Johnstone ADF (1974) Some auditory discrimination experiments on marine fish. J Exp Biol 61:521–528.

Chapman CJ, Sand O (1974) Field studies of hearing in two species of flatfish *Pleuronectes platessa* (L.) and *Limanda limanda* (L.) (Family Pleuronectidae). Comp Biochem Physiol A 47:371–385.

Chiszar, David, Ted Melcer, Robert Lee and David Duvall (1990) Chemical cues used by prairie rattlesnakes (*Crotalus viridis*) to follow trails of rodent prey. J Chem Ecol 16: 79–86.

Claas B, Münz H, Zittlau KE (1989) Direction coding in central parts of the lateral line system. In: Coombs S, Görner P, Münz H (eds) The Mechanosensory Lateral Line. Neurobiology and Evolution. New York: Springer-Verlag, pp. 409–419.

Coombs S (1994) Nearfield detection of dipole sources by the goldfish (*Carassius auratus*) and the mottled sculpin (*Cottus bairdi*). J Exp Biol 190:109–129.

Coombs S, Conley RA (1995) Source distance determination by the mottled sculpin lateral line. In: Burrows M, Mattheson T, Newland PL, Schuppe H (eds) Nervous Systems and Behaviour. Stuttgart: Thieme Verlag, pp. 349.

Coombs S, Janssen J (1990) Behavioral and neurophysiological assessment of lateral line sensitivity in the mottled sculpin, *Cottus bairdi*. J Comp Physiol A 167:557–567.

Coombs S, Montgomery JC (1998) The enigmatic lateral line system. In: Fay RR, Popper AN (eds) Comparative Hearing: Fish and Amphibians. New York: Springer-Verlag, pp. 319–362.

Coombs S, Hastings M, Finneran J (1996) Modeling and measuring lateral line excitation patterns to changing dipole source locations. J Comp Physiol A 178:359–371.

Coombs S, Finneran JJ, Conley RA (2000) Hydrodynamic imaging formation by the lateral line system of the Lake Michigan mottled sculpin, *Cottus bairdi*. Philos Trans Roy Soc B 355:1111–1114.

Curcic-Blake B, van Netten SM (2006) Source location encoding in the fish lateral line canal. J Exp Biol 209: 1548–1559.

de Munck JC, Schellart NAM (1987) A model for the nearfield acoustics of the fish swimbladder and its relevance for directional hearing. J Acoust Soc Am 81:556–560.

De Perera T (2004) Spatial parameters encoded in the spatial map of the blind Mexican cavefish, *Astyanax fasciatus*. Anim Behav 68:291–295.

de Vries H (1950) The mechanics of the labyrinth otoliths. Acta Oto-Laryngol 38:262–273.

Dijkgraaf S, Kalmijn AJ (1962) Verhaltensversuche zur Funktion der Lorenzinischen Ampullen. Naturwissenschaften 49:400.

Domenici P, Blake RW (1993) The kinematics and performance of fish fast-start swimming. J Exp Biol 200:1165–1178.

Eaton RC, Emberley DS (1991) How stimulus direction determines the trajectory of the Mauthner-initiated escape response in a teleost fish. J Exp Biol 161:469–487.

Eaton RC, Popper AN (1995) The octavolateralis system and Mauthner cell: interactions and questions. Brain Behav Evol 46:124–130.

Eaton RC, Lavender WA, Wieland CM (1981) Identification of Mauthner initiated response patterns in goldfish: evidence from simultaneous cinematography and electrophysiology. J Comp Physiol A 144:521–531.

Eaton RC, Canfield JG, Guzik AL (1995) Left-right discrimination of sound onset by the Mauthner system. Brain Behav Evol 46:165–179.

Eaton RC, Lee RKK, Foreman MB (2001) The Mauthner cell and other identified neurons of the brainstem escape network of fish. Prog Neurobiol 63:467–485.

Eaton RC, Casagrand JL, Cummins GI (2002) Neural implementation of the phase model for localizing impulse sounds by the Mathner system. Bioacoustics 12:209–212.

Edds-Walton PL (1998) Anatomical evidence for binaural processing in the descending octaval nucleus of the toadfish (*Opsanus tau*). Hear Res 123:41–54.

Edds-Walton PL, Fay RR (2003) Directional selectivity and frequency tuning of midbrain cells in the oyster toadfish, *Opsanus tau*. J Comp Physiol A 189:527–543.

Edds-Walton PL, Fay RR (2005a) Sharpening of directional responses along the auditory pathway of the oyster toadfish, *Opsanus tau*. J Comp Physiol A 191:1079–1086.

Edds-Walton PL, Fay RR (2005b) Projections to bimodal sites in the torus semicircularis of the toadfish, *Opsanus tau*. Brain Behav Evol 66:73–87.

Edds-Walton PL, Fay RR, Highstein S (1999) Dendritic arbors and central projections of physiologically characterized auditory fibers from the saccule of the toadfish, *Opsanus tau*. J Comp Neurol 411:212–238.

Elepfandt A, Wiedemer L (1987) Lateral-line responses to water surface waves in the clawed frog, *Xenopus laevis*. J Comp Physiol A 160:667–682.

Emde G von der, Schwarz S, Gomez L, Budelli R, Grant K (1998) Electric fish measure distance in the dark. Nature 395:890–894.

Engelmann J, Bleckmann H (2004) Coding of lateral line stimuli in the goldfish midbrain in still and running water. Zoology 107: 135–151.

Enger PS (1976) On the orientation of hair cells in the labyrinth of perch (*Perca fluviatilis*). In: Schuijf A, Hawkins AD (eds) Sound Reception in Fish. Amsterdam: Elsevier, pp. 49–62.

Enger PS, Hawkins AD, Sand O, Chapman C J (1973) Directional sensitivity of saccular microphonic potentials in the haddock. J Exp Biol 59:425–434.

Enger PS, Kalmijn AJ, Sand O (1989) Behavioral investigations on the functions of the lateral line and inner ear in predation. In: Coombs S, Görner P, Münz H (eds) The Mechanosensory Lateral lLine—Neurobiology and Evolution. New York: Springer-Verlag, pp. 575–587.

Faber DS, Fetcho JR, Korn H (1989) Neural networks underlying the escape response in goldfish. Ann NY Acad Sci 563:11–33.

Faber DS, Korn H, Lin JW (1991) Role of medullary networks and postsynaptic membrane properties in regulating Mauthner cell responsiveness to sensory excitation. Brain Behav Evol 37:286–297.

Fay, RR (1984) The goldfish ear codes the axis of acoustic particle motion in three dimensions. Science 225:951–954.

Fay RR, Edds-Walton PL (1997) Directional response properties of saccular afferents of the toadfish, *Opsanus tau*. Hear Res 111:1–21.

Fay RR, Edds-Walton, PL (2000) Directional encoding by fish auditory systems. Philos Trans Roy Soc Lond B 355:1281–1284.

Fay RR, Olsho LW (1979) Discharge patterns of lagenar and saccular neurons of the goldfish eighth nerve: displacement sensitivity and directional characteristics. Comp Biochem Physiol A 62:377–386.

Fetcho JR., O'Malley DM (1997) Imaging neuronal networks in behaving animals. Curr Opin Neurosci 7:832–838.

Flock Å (1965) Electromicroscopical and electrophysiological studies on the lateral line canal organ. Acta Otolaryngol Suppl 199:1–90.

Forman MB, Eaton RC (1993) The direction change concept for reticulospinal control of goldfish escape. J Neurosci 13:4101–4113.

Gathan E, Sankrithi N, Campos, JB, O'Malley DM (2002) Evidence for a widespread brain stem escape network in larval zebrafish. J Neurophysiol 87:608–614.

Goulet J, Engelmann J, Chagnaud B, Franosch J-MP, Suttner MD, van Hemmen JL (2008) Object localization through the lateral line system of fish: theory and experiment. J Comp physical A 194:1–17.

Guzik AL, Eaton RC, Mathis DW (1999). A connectionist model of left-right sound discrimination by the Mauthner system. J Comput Neurosci 6:121–144.
Hahn G (1960) Ferntastsinn und Strömungssinn beim augenlosen Höhlenfisch *Anoptichthys jordani* Hubbs und Innes im Vergleich zu anderen Teleosteern. Naturwissenschaften 47:611.
Hale M E (2002) S- and C-start escape responses of the muskellunge (*Esox masquinongy*) require alternative neuromotor mechanisms. J Exp Biol 205:2005–2016.
Hale ME, Long Jr JH, McHenry M J, Westneat MW (2002) Evolution of behaviour and neural control of the fast-start escape response. Evolution 56:993–1007.
Hanke W, Bleckmann H (2004) The hydrodynamic trails of *Lepomis gibbosus* (Centrarchidae), *Colomesus psittacus* (Tetraodontidae) and *Thysochromis ansorgii* (Cichlidae) investigated with scanning particle image velocimetry. J Exp Biol 207:1585–1596.
Hanke W, Brücker C, Bleckmann H (2000) The ageing of the low frequency water disturbances caused by swimming goldfish and its possible relevance to prey detection. J Exp Biol 203:1193–1200.
Harris GG (1964) Considerations on the physics of sound produced by fishes. In: Tavolga WN (ed) Marine Bio-acoustics. Oxford: Pergamon Press, pp. 233–247.
Hassan ES (1989) Hydrodynamic imaging of the surroundings by the lateral line of the blind cavefish *Anoptichthys jordani*. In: Coombs S, Görner P, Münz H (eds) The Mechanosensory Lateral Line. Neurobiology and Evolution. New York: Springer-Verlag, pp. 217–228.
Hassan ES (1992a) Mathematical description of the stimuli to the lateral line system of fish, derived from a three-dimensional flow field analysis. I. The case of moving in open water and of gliding towards a plane surface. Biol Cyber 66:443–452.
Hassan ES (1992b) Mathematical description of the stimuli to the lateral line system of fish, derived from a three-dimensional flow field analysis. II. The case of gliding alongside or above a plane surface. Biol Cyber 66:453–461.
Hassan E-S (1993) Mathematical description of the stimuli to the lateral line system of fish, derived from a three-dimensional flow field analysis. III. The case of an oscillating sphere near the fish. Biol Cyber 69:535–538.
Hassan ES, Abdel-Latif H, Biebricher R (1992) Studies on the effects of Ca^{++} and Co^{++} on the swimming behavior of the blind Mexican cavefish. J Comp Physiol A 171:413–419.
Hawkins AD, Horner K (1981) Directional characteristics of primary auditory neurons from the cod ear. In: Tavolga WN, Popper AN, Fay RR (eds) Hearing and Sound Communication in Fishes. New York: Springer-Verlag, pp. 311–328.
Hawkins AD, Sand O (1977) Directional hearing in the median vertical plane by the cod. J Comp Physiol A 122:1–8.
Hofer B (1908) Studien über die Hautsinnesorgane der Fische I. Die Funktion der Seitenorgane bei den Fischen. Ber königl Bayer biol Versuchsstation München 1:115–168.
Hoin-Radkovski I, Bleckmann H, Schwartz E (1984) Determination of source distance in the surface-feeding fish *Pantodon buchholzi* (Pantodontidae). Anim Behav 32:840–851.
Horner K, Sand O, Enger PS (1980) Binaural interaction in the cod. J Exp Biol 85:323–331.
Hudspeth AJ, Corey DP (1977) Sensitivity, polarity and conductance change in the response of vertebrate hair cells to controlled mechanical stimuli. Proc Natl Acad Sci USA 74:2407–2411.

Janssen J, Corcoran J (1993) Lateral line stimuli can override vision to determine sunfish strike trajectory. J Exp Biol 176:299–305.

Janssen J, Coombs S, Hoekstra D, Platt C (1987) Anatomy and differential growth of the lateral line system of the mottled sculpin, *Cottus bairdi* (Scorpaeniformes: Cottidae). Brain Behav Evol 30:210–229.

Kalmijn AJ (1982) Electric and magnetic field detection in elasmobranch fishes. Science 218:916–918.

Kalmijn AJ (1988) Hydrodynamic and acoustic field detection. In: Atema J, Fay RR, Popper AN, Tavolga WN (eds) Sensory Biology of Aquatic Animals. New York: Springer-Verlag, pp. 83–130.

Kalmijn AJ (1989) Functional evolution of lateral line and inner ear sensory systems. In: Coombs S, Görner P, Münz H (eds) The Mechanosensory Lateral Line—Neurobiology and Evolution. New York: Springer-Verlag, pp. 187–215.

Kalmijn AJ (1997) Electric and near-field acoustic detection, a comparative study. Acta Physiol Scand 161(Suppl) 638: 25–38.

Karlsen HE (1992a) The inner ear is responsible for detection of infrasound in the perch (*Perca fluviatilis*). J Exp Biol 171:163–172.

Karlsen HE (1992b) Infrasound sensitivity in the plaice (*Pleuronectes platessa*). J Exp Biol 171:173–187.

Karlsen HE, Sand O (1987) Selective and reversible blocking of the lateral line in freshwater fish. J Exp Biol. 133:249–267.

Karlsen HE, Piddington RW, Enger PS, Sand O (2004) Infrasound initiates directional fast-start escape responses in juvenile roach (*Rutilus rutilus*). J Exp Biol. 207:4185–4193.

Käse R, Bleckmann H (1987) Prey localization by surface wave-ray tracing—fish track bugs like oceanographers track storm. Experientia 43:290–293.

Kimmel CB, Powell SL, Metcalfe, WK (1982) Brain neurons which project to the spinal cord in young larvae of the zebrafish. J Comp Neurol 205:112–127.

Kleerekoper H, Chagnon EC (1954) Hearing in fish with special reference to *Semotilus atromaculatus*. J Fish Res Bd Can 11:130–152.

Korn H, Faber DS (1996) Escape behaviour—brainstem and spinal cord circuitry and function. Curr Opin Neurobiol 6:826–832.

Lang HH (1980) Surface wave discrimination between prey and nonprey by the back swimmer *Notonecta glauca* L. (Hemiptera, Heteroptera). Behav Ecol Sociobiol 6:233–246.

Lee RKK, Eaton RC, Zottoli SJ (1993) Segmental arrangement of reticulospinal neurons in the goldfish hindbrain. J Comp Neurol 327:1–18.

Lighthill J (1980) Waves in Fluids. Cambridge, UK: Cambridge University Press.

Liu KS, Fetcho JR (1999) Laser ablations reveal functional relationships of segmental hindbrain neurons in zebrafish. Neuron 23:325–335.

Lu Z, Popper AN (1998) Morphological polarizations of sensory hair cells in the three otolithic organs of a teleost fish: fluorescent imaging of ciliary bundles. Hear Res 126:47–57.

Lu Z, Popper AN (2001) Neural response directionality correlates of hair cell orientation in a teleost fish. J Comp Physiol A 187:453–465.

Lu Z, Song J, Popper AN (1998) Encoding of acoustic directional information by saccular afferents of the sleeper goby, *Dormitator latifrons*. J Comp Physiol A 182:805–815.

Lu Z, Xu Z, Buchser WJ (2003) Acoustic response properties of lagenar nerve fibers in the sleeper goby, *Dormitator latifrons*. J Comp Physiol A 189:889–905.

Lu Z, Xu Z, Buchser WJ (2004) Coding of acoustic particle motion by utricular fibers in the sleeper goby, *Dormitator latifrons*. J Comp Physiol A 190:923–938.

Ma WLD, Fay RR (2002) Neural representations of the axis of acoustic particle motion in nucleus centralis of the torus semicircularis of the goldfish, *Carassius auratus*. J Comp Physiol A 188:301–313.

McCormick CA, Hernandez DV (1996) Organization of inner ear endorgan projections in the goldfish, *Carassius auratus*. Brain Behav Evol 43:189–205.

Metcalfe WK, Mendelson B, Kimmel CB (1986) Segmental homologies among reticulospinal neurons in the hindbrain of the zebrafish larva. J Comp Neurol 251:147–159.

Mohr C, Bleckmann H (1998) Electrophysiology of the cephalic lateral line of the surface-feeding fish *Aplocheilus lineatus*. Comp Biochem Physiol A 119:807–815.

Moulton JM, Dixon RH (1967) Directional hearing in fishes. In: Tacvolga WN (ed) Marine Bioacoustics, Vol 2. Oxford: Pergamon Press, pp. 187–232.

Müller U, Schwartz E (1982) Influence of single neuromasts on prey-localizing behavior of surface-feeding fish *Aplocheilus lineatus*. J Comp Physiol A 149:399–408.

Müller HM, Fleck A, Bleckmann H (1993) The responses of central octavolateralis cells to moving sources. J Comp Physiol A 179: 455–471.

Myrberg AA Jr, Banner A, Richaard JD (1969) Shark attraction using a video-acoustic system. Mar Biol 2:264–276.

Myrberg AA Jr, Gordon CR, Klimley AP (1976) Attraction of free ranging sharks by low frequency sound, with comments on its biological significance. In: Schuijf A, Hawkins AD (eds) Sound Reception in Fish. Amsterdam: Elsevier, pp. 205–228.

Nelson DR, Gruber SH (1963) Sharks: attraction by low-frequency sounds. Science 142:975–977.

Parvulescu A (1967) The acoustics of small tanks. In: Tavolga WN (ed) Marine Bioacoustics II. Oxford: Pergamon Press, pp. 7–14.

Piddington RW (1972) Auditory discrimination between compressions and rarefactions by goldfish. J Exp Biol 56:403–419.

Pohlmann K, Grasso FW, Breithaupt T (2001) Tracking wakes: the nocturnal predatory strategy of piscivorous catfish. Proc Natl Acad Sci USA 98:7371–7374.

Pohlmann K, Atema J, Breithaupt T (2004) The importance of the lateral line in nocturnal predation of piscivorous catfish. J Exp Biol 207: 2971–2978.

Popper AN, Coombs S (1982) The morphology and evolution of the ear in Actinopterygian fishes. Am Zool 22:311–328.

Popper AN, Salmon M, Parvulescu A (1973) Sound localization by two species of Hawaiian squirrelfish, *Myripristis berndti* and *M. argyromus*. Anim Behav 21:86–97.

Reinhardt F (1935) Über Richtungswahrnehmung bei Fischen, besonders bei der Elritze (*Phoxinus laevis* L.) und beim Zwergwels (*Amiurus nebulosus* Raf.). Z Vergl Physiol 22:570–603.

Ritter DA, Bhatt DH, Fetcho JR (2001) In vivo imaging of zebrafish reveals differences in the spinal networks for escape and swimming movements. J Neurosci 21:8956–8965.

Rogers PH, Popper AN, Cox M, Saidel WM (1988) Prosessing of acoustic signals in the auditory system of bony fish. J Acoust Soc Am 83:338–349.

Sand O (1974) Directional sensitivity of microphonic potentials from the perch ear. J Exp Biol 60:881–899.

Sand O (1981) The lateral-line and sound reception. In: Tavolga WN, Popper AN, Fay RR (eds) Hearing and Sound Communication in Fishes. New York: Springer-Verlag, pp. 459–480.

Sand O (1984) Lateral line systems. In: Bolis L, Keynes RD, Maddrell SHP (eds) Comparative Physiology of Sensory Systems. Cambridge, UK: Cambridge University Press, pp. 3–32.

Sand O, Enger PS (1973) Evidence for an auditory function of the swim bladder in the cod. J Exp Biol 59:405–414.

Sand O, Hawkins AD (1973) Acoustic properties of the cod swim bladder. J Exp Biol 58:797–820.

Sand O, Karlsen HE (1986) Detection of infrasound by the Atlantic cod. J Exp Biol 125:197–204.

Sand O, Karlsen HE (2000) Detection of infrasound and linear acceleration in fish. Philos Trans Roy Soc Lond B 355:1295–1298.

Sand O, Enger PS, Karlsen HE, Knudsen FR (2001) Detection of infrasound in fish and behavioural responses to intense infrasound in juvenile salmonids and European silver eels: a minireview. Am Fish Soc Symp 26:183–193.

Schellart NAM, Buwalda RJA (1990) Directional variant and invariant hearing thresholds in the rainbow trout (*Salmo gairdneri*). J Exp Biol 149:113–131.

Schellart NAM, de Munck JC (1987) A model for directional and distance hearing in swim bladder-bearing fish based on the displacement orbits of the hair cells. J Acoust Soc Am 82:822–829.

Schuijf A (1975) Directional hearing of cod (*Gadus morhua*) under approximate free field conditions. J Comp Physiol 98:307–332.

Schuijf A (1976) The phase model of directional hearing in fish. In: Schuijf A, Hawkins AD (eds) Sound Reception in Fish. Amsterdam: Elsevier, pp. 63–86.

Schuijf A (1981) Models of acoustic localization. In: Tavolga WN, Popper AN, Fay RR (eds) Hearing and Sound Communication in Fishes. New York: Springer-Verlag, pp. 267–310.

Schuijf A, Buwalda RJA (1975) On the mechanism of directional hearing in cod (*Gadus morhua*). J Comp Physiol 98:333–344.

Schuijf A, Hawkins AD (1983) Acoustic distance discrimination by the cod. Nature 302:143–144.

Schuijf A, Siemelink M (1974) The ability of cod (*Gadus morhua*) to orient towards a sound source. Experientia 30:773–774.

Schuijf A, Baretta JW, Wildschut JT (1972). A field investigation on the discrimination of sound direction in *Labrus berggylta* (Pisces: Perci-formes). Netherlands J Zool 22:81–104.

Schuijf A, Visser C, Willers AFM, Buwalda RJA (1977) Acoustic localization in an ostariophysian fish. Experientia 33:1062–1063.

Schwartz E (1965) Bau und Funktion der Seitenlinie des Streifenhechtlings (*Aplocheilus lineatus* Cuv. u. Val.). Z vergl Physiol 50:55–87.

Schwartz E (1970) Ferntastsinnesorgane von Oberflächenfischen. Z Morphol Tiere 67:40–57.

Schwartz E (1971) Die Ortung von Wasserwellen durch Oberflächenfische. Z vergl Physiol 74:64–80.

Steen JB, Wilsson E (1990) How do dogs determine the direction of tracks? Acta Physiol Scand 139:531–534.

Teyke T (1985) Collision with and avoidance of obstacles by blind cavefish *Anoptichthys jordani* (Characidae). J Comp Physiol A 157:837–843

Teyke T (1988) Flow field, swimming velocity and boundery layer: parameters which affect the stimulus for the lateral line organ in blind fish. J Comp Physiol A 163:53–61.

Teyke T (1989) Learning and remembering the environment in the blind cavefish *Anoptichthys jordani*. J Comp Physiol A 164:655–662.

Tittel G (1985) Determination of stimulus direction by the topminnow, *Aplocheilus lineatus*. A model of two dimensional orientation with the lateral line system. In: Barth FG (ed) Verhandlungen der Deutschen Zoologischen Gesellschaft 78. Stuttgart: Gustav Fischer, p. 242.

Tittel G (1991) Verhaltensphysiologische, ultrastrukturelle und ontogenetische Studien am Seitenliniensystem von *Aplocheilus lineatus*. Ein vorläufiges Modell zur Richtungsdetermination von Beuteobjekten. PhD Thesis, University of Giessen, 445 pp.

Tittel G, Müller U, Schwartz E (1984) Determination of stimulus direction by the topminnow *Aplocheilus lineatus*. In: Varju D, Schnitzler HU (eds) Localization and Orientation in Biology and Engineering. New York: Springer-Verlag, pp. 69–72.

van Bergeijk WA (1964) Directional and nondirctional hearing in fish. In: Tavolga WN (ed) Marine Bio-acoustics. Oxford: Pergamon Press, pp. 281–299.

van Bergeijk WA (1967) The evolution of vertebrate hearing. In: Neff WD (ed) Contributions to Sensory Physiology, Vol. 2. New York: Academic Press, pp. 1–49.

von Campenhausen C, Riess I, Weissert R (1981) Detection of stationary objects in the blind cavefish *Anoptichthys jordani* (Characidae). J Comp Physiol A 143:369–374.

von Frisch K, Dijkgraaf S (1935) Können Fische die Schallrichtung Wahrnehmen? Z Vergl Physiol 2:641–655.

Weeg MS, Fay RR, Bass AH (2002) Directionality and frequency tuning of primary saccular afferents of a vocal fish, the plainfin midshipman (*Porichthys notatus*). J Comp Physiol A 188:631–641.

Weissert R, von Campenhausen C (1981) Discrimination between stationary objects by the blind cavefish *Anoptichthys jordani*. J Comp Physiol A 143:375–382.

Wojtenek W, Mogdans J, Bleckmann H (1998) The responses of midbrain lateral line units of the goldfish *Carassius auratus* to moving objects. Zoology 101:69–82.

Wubbels RJ, Schellart NAM (1997) Neuronal encoding of sound direction in the auditory midbrain of the rainbow trout. J Neurophysiol 77:3060–3074.

Wubbels RJ, Schellart NAM (1998) An analysis of the relationship between the response characteristics and topography of directional- and non-directional auditory neurons in the torus semicircularis of the rainbow trout. J Exp Biol 201:1947–1958.

Wubbels RJ, Schellart NAM, Goossens JHHLM (1995) Mapping of sound direction in the trout lower midbrain. Neurosci Lett 199:179–182.

Yan H, Fine ML, Horn NS, Colón WE (2000) Variability in the role of the gasbladder in fish audition. J Comp Physiol A 186:435–445.

Zittlau KE, Claas B, Münz H (1986) Directional sensitivity of lateral line units in the clawed toad *Xenopus laevis* Daudin. J Comp Physiol A 158:469–477.

Zottoli SJ, Faber DS (2000) The Mauthner cell: what has it taught us? Neuroscientist 6:26–38.

7
Multipole Mechanisms for Directional Hearing in Fish

Peter H. Rogers and David G. Zeddies

1. Introduction

If fish are to behave appropriately with respect to objects and events in their environment they must process an acoustic scene that is often complex (Fay and Popper 2000). A presumptively important part of such behavior is the ability to determine properly the direction from which a sound emanates. Although the question regarding mechanisms for sound-source localization in fishes has been of interest since Karl von Frisch and Sven Dijkgraaf (1935) performed behavioral studies in European minnows (*Phoxinus laevis*), the mechanisms remain poorly understood, with relatively few biologically plausible models. A localization mechanism that exploits the amplitude, time, or phase difference between the ears as employed by terrestrial vertebrates is not available to fish because the ears are very close together, the speed of sound in water is more than three times faster than in air, and the close impedance match between the fish's body and water precludes usable diffracted paths (van Bergeijk 1964, 1967; and see Sand and Bleckmann, Chapter 6). Another major difficulty that any model must address is a resolution of the so called "180° ambiguity" that arises because the axis of particle motion associated with a passing sound points both toward and away from the sound source (for review of sound localization by fish see Fay 2005; Sand and Bleckmann, Chapter 6). Current models of directional hearing in fish with mechanisms to resolve the 180° ambiguity include the "phase model" proposed by Schuijf and colleagues (e.g., Chapman and Hawkins 1973; Schuijf 1975; Schuijf and Buwalda 1975) that compares the phase of the pressure and particle motion components of sound or the phase of the direct-path particle motion and the particle motion of sound reflected from surfaces or objects; an "orbital" model by Schellart and de Munck (1987; de Munck and Schellart 1987) in which sound pressure and particle motion together cause the otolith orbits to rotate either clockwise or counterclockwise depending on whether the source is to the left or right; a computational model by Rogers et al. (1988) that also uses both pressure and particle motion; and, a more algorithmic approach pointed out by Kalmijn (1997) by which a fish could make its way to a sound source by maintaining a constant angle with respect to the axis of vibration even if the axis of vibration does not point toward the source.

Each of these models has drawbacks and limitations. The "phase model" and its variants require either pressure sensitivity from a gas-filled chamber or that the fish remain near surfaces and objects. The "orbital" model also requires pressure detection and only works for sinusoidal signals, and Kalmijn's guided approach is limited to locating sources that broadcast throughout the duration of approach. This chapter examines mechanisms by which fishes with and without pressure detection could determine the direction to a sound source and resolve the 180° ambiguity.

Acoustic sensors that are small compared to the wavelengths of the signals they sense can be described in terms of "multipoles," that is, monopoles, dipoles, quadrupoles, and combinations of monopoles, dipoles, and quadrupoles. Section 2 of this chapter is a self-contained presentation of the theory of multipole sensors. As fish ears are small compared to the wavelengths of the sounds detected, Sections 3 and 4 explain how the formalism of multipole sensors may be applied to the fish ear, with particular attention to the concept of quadrupole hearing (Section 4). It is shown in Section 4 that a quadrupole mechanism could be used by hair cells lacking an overlying otolith to resolve the 180° ambiguity without the need for a gas-filled chamber to sense pressure, and in some circumstances provide an improved signal to noise ratio (S/N) resulting in a lower detection threshold.

This chapter assumes a basic knowledge of underwater acoustics. A reader who is unfamiliar with the physics of sound is encouraged to consult any of a number of textbooks on acoustics (e.g., Beranek 1954; Kinsler and Frey 1962; and, especially, Pierce 1981), as well as chapters that review underwater acoustics in the context of fish hearing (Kalmijn 1988; Rogers and Cox 1988; Clark and Bass 2003).

2. Theory of Multipole Sensors

A "disturbance" is any perturbation from the equilibrium state of a medium, and a "wave" is defined as a propagating disturbance. Sound is a "compressional wave" because the disturbance involves a localized change in the ambient density of the medium. (Shear waves in solids and gravitational waves in fluids are examples of mechanical waves that do not involve changes in density and hence are not considered to be sound.) In addition to compression of the medium, an acoustic (sound) disturbance involves perturbations from the equilibrium values of pressure, displacement, and velocity of the medium as well as strain, strain rate, viscous stress, and temperature. In water, acoustic disturbances propagate at around 1500 m/s.

A sensor may detect sound by responding to any of the acoustic perturbations, with the directional dependence of the response determined by the perturbation sensed. For an acoustically small sensor, the response to each of the perturbations can be expressed in terms of spatial/temporal derivatives of the incident pressure field. For example, the acoustic density perturbation, ρ_A, is directly proportional to the acoustic pressure perturbation,

$$\rho_A(x, t) = p_A(x, t)/c^2, \tag{7.1}$$

while, from Newton's law, the particle acceleration is proportional to the gradient of the pressure,

$$\rho_o \frac{\partial \vec{u}}{\partial t} = -\vec{\nabla} p, \tag{7.2}$$

and the time derivative of shear strain rate is proportional to a second spatial derivative of the pressure,

$$\rho_o \frac{\partial \vec{\nabla} \vec{u}}{\partial t} = -\vec{\nabla}\vec{\nabla} p. \tag{7.3}$$

Note that Eq. (7.1) is a scalar equation, Eq. (7.2) is a vector equation, and Eq. (7.3) is a tensor or dyadic equation. Once a coordinate system [e.g., Cartesian; $\vec{r} = (x, y, z)$ or spherical $\vec{r} = (r, \theta, \phi)$] has been defined, the various derivatives are specified and the various directional responses are determined.

It is customary to characterize an acoustic sensor by its response to an incident plane wave. However, not all sound waves are well approximated by plane waves, and thus it is important to be able to determine the response of a sensor to any incident acoustic field, e.g., a spherical wave, an evanescent wave, a shallow water normal mode, or the sound field in a small tank. The response for each type of sensor to a plane wave is given. In addition, the response in terms of the underlying incident field derivatives is given so that the results can be used to determine the response to any incident field.

A plane acoustic wave propagating in the $\vec{n}$ direction can be expressed by

$$p(\vec{r}, t) = p_o f\left(t - \frac{\vec{r} \cdot \vec{n}}{c}\right), \tag{7.4}$$

where f is any function. The vector $\vec{n}$ is a unit vector parallel to the direction of propagation of the wave.

A sensor is referred to as a monopole, dipole, or quadrupole depending on whether it responds to the pressure, the pressure gradient, or a second derivative of the pressure. The monopole, dipole, quadrupole nomenclature usually appears in the literature in the context of acoustic sources rather than receivers (see, e.g., Chapter 4 in Pierce 1981), but by reciprocity it is applicable to both. And, in the recent literature, combinations of monopole and dipole sensors have been referred to as "vector sensors" and combinations of monopole dipole and quadrupole sensors as "dyadic sensors" (Cray et al. 2003; McEachern 2003).

The response of a monopole sensor is simply proportional to the incident pressure. The proportionality constant does not depend on the nature or direction of the incident wave and thus the response is independent of the direction of incidence. The response of a dipole sensor to an arbitrary incident field is proportional to the pressure gradient. A dipole sensor is characterized by a vector $\vec{d}$, called the dipole moment, and the directivity factor is given by

$$D_{\mathrm{d}}(p, t) = (\vec{d} \cdot \vec{\nabla}) p_{\mathrm{inc}}. \tag{7.5}$$

For the special case of an incident plane wave:

$$\begin{aligned} D_{\mathrm{d}}(\vec{n}, t) &= p_{\mathrm{o}}(\vec{d} \cdot \vec{\nabla}) f\left(t - \frac{\vec{r} \cdot \vec{n}}{c}\right) = p_{\mathrm{o}}(\vec{d} \cdot \vec{n}) f'\left(t - \frac{\vec{r} \cdot \vec{n}}{c}\right) \\ &= p_{\mathrm{o}}(d_x n_x + d_y n_y + d_z n_z) f'\left(t - \frac{\vec{r} \cdot \vec{n}}{c}\right) \end{aligned}. \tag{7.6}$$

In general, the dipole response is proportional to the cosine of the angle between $\vec{d}$ and $\vec{n}$. If $\vec{d}$ is in the z direction the response is simply proportional to $\cos\theta$.

The response of a quadrupole is determined by its quadrupole moment tensor, $\mathbf{Q}$, defined in general by its nine components:

$$\mathbf{Q} = \begin{pmatrix} Q_{xx} & Q_{xy} & Q_{xz} \\ Q_{yx} & Q_{yy} & Q_{yz} \\ Q_{zx} & Q_{zy} & Q_{zz} \end{pmatrix}. \tag{7.7}$$

Each term in this tensor describes the proportionality of the response to a certain second derivative of the pressure,

$$\begin{aligned} D_Q(\vec{n}, t) = Q_{xx}\frac{\partial^2 p}{\partial x \partial x} + Q_{xy}\frac{\partial^2 p}{\partial x \partial y} + Q_{xz}\frac{\partial^2 p}{\partial x \partial z} \\ + Q_{yx}\frac{\partial^2 p}{\partial y \partial x} + Q_{yy}\frac{\partial^2 p}{\partial y \partial y} + Q_{yz}\frac{\partial^2 p}{\partial y \partial z}, \\ + Q_{zx}\frac{\partial^2 p}{\partial z \partial x} + Q_{zy}\frac{\partial^2 p}{\partial z \partial y} + Q_{zz}\frac{\partial^2 p}{\partial z \partial z} \end{aligned} \tag{7.8}$$

or, for incident plane waves [combining Eq. (7.4) with Eq. (7.8)] this results in directivity factors given by

$$\begin{aligned} D_Q(\vec{n}, t) = (&Q_{xx}\, n_x n_x + Q_{xy} n_x n_y + Q_{xz} n_x n_z \\ &+ Q_{yx} n_y n_x + Q_{yy} n_y n_y + Q_{yz} n_y n_z \\ &+ Q_{zx} n_z n_x + Q_{zy} n_z n_y + Q_{zz} n_z n_z)\, p_o f''\left(t - \frac{\vec{r} \cdot \vec{n}}{c}\right) \end{aligned}. \tag{7.9}$$

In spherical coordinates $(n_x, n_y, n_z) = (\sin\theta\cos\phi, \sin\theta\sin\phi, \cos\theta)$ so that, for example, a quadrupole receiver which only had a Q_{xx} term would have a

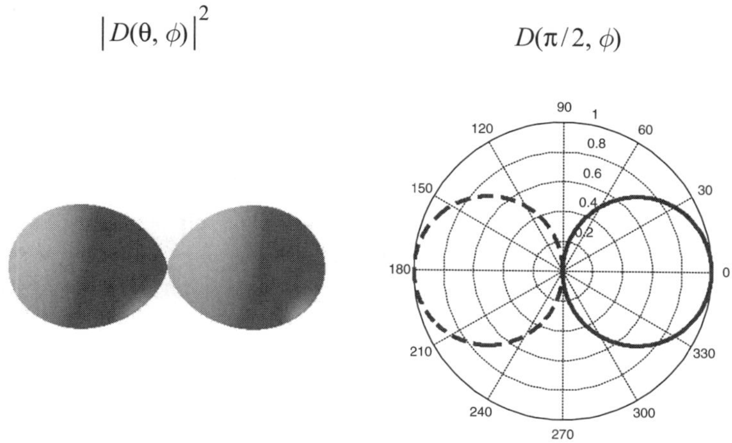

FIGURE 7.1. Directivity factor (right) and its square-magnitude (left) for a dipole receiver oriented in the x direction. Dashed and solid lines have opposite polarity.

response proportional to $\sin^2\theta\cos^2\phi$, a quadrupole receiver that had only a Q_{xy} term would have a response proportional to $\sin^2\theta\sin\phi\cos\phi$ and a quadrupole receiver that only had a Q_{zz} term would have a response proportional to $\cos^2\theta$. A Q_{xx} type quadrupole is called a linear quadrupole and a Q_{xy} quadrupole is called a lateral quadrupole. The directivity factor and its magnitude squared for a dipole pointed in the x direction and a lateral quadrupole (Q_{xy}) are shown in Figures 7.1 and 7.2, respectively. The lateral quadrupole has four lobes whereas the dipole has two, and the quadrupole is seen to be more directional than the

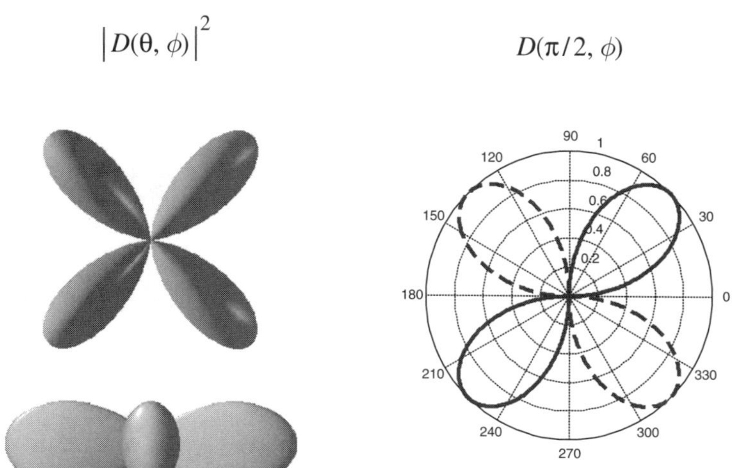

FIGURE 7.2. Directivity factor in x–y plane (right) and two views of its square-magnitude (left) for a lateral quadrupole (Q_{xy}) receiver. Dashed and solid lines have opposite polarity.

dipole. Note the changes in polarity in the figures (solid lines are positive and dashed lines are negative). Also note that cross sections of the quadrupole lobes are elliptical, not circular.

3. Monopole and Dipole Mechanisms in Fish Hearing

The acoustic mechanosensor in fish is the sensory hair cell (see Popper and Schilt, Chapter 2). The principal mechanism for detecting sound is the otolithic mechanism illustrated schematically in Figure 7.3, wherein the overlying otolith

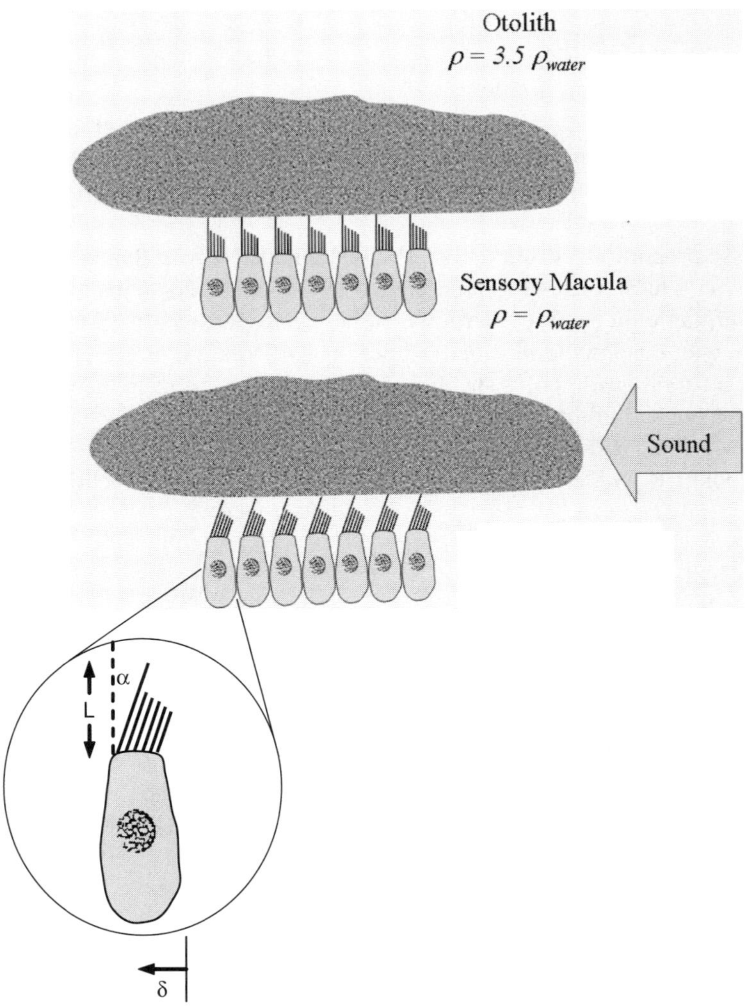

FIGURE 7.3. The otolithic hearing mechanism. The dense otolith lags behind the macula, which tends to move with the fluid. The cell output is proportional to the pivot angle $\alpha = \delta/L$. The sensor is a dipole.

functions as the proof mass of a displacement sensor or accelerometer (e.g., de Vries 1950; van Bergeijk 1964). The stereociliary bundle of the hair cell is constrained to pivot about a fixed axis with the cell responding proportionally to the pivot angle α (Corey and Hudspeth 1977). The cell body moves with the fluid, while the otolith, due to its greater inertia, lags behind. The force that drives the relative motion between the otolith and the cell body is the pressure gradient. The sensor response is thus dipole in nature, and is proportional to the cosine of the angle between the acoustic propagation direction and the hair cell orientation direction.

Acoustic particle displacements involved in underwater sound are usually very small. A typical SPL threshold for a fish species with good pressure sensitivity at 100 Hz is 80 dB re: 1 μPa (Fay 1988). The acoustic particle displacement associated with the passage of such a wave is just 0.01 nm. Even a very loud sound, say the signal produced 100 m away from a large Navy sonar with a source level of 208 dB re: 1 μPa at 1 m, the particle displacement is only 0.25 μm. As mentioned in the preceding text, the stimulus for the hair cell is the angle $\alpha = \delta/L$, where δ is the displacement of the tip of the kinocilium and L is its length as shown in Figure 7.3. The tip displacement δ is, at most, equal to the relative displacement of the sensory macula with respect to the otolith. As shown later, this must always be less than the fluid particle displacement for a hearing generalist (nonspecialist).

The hair cell transduction mechanism does not permit it to respond directly to acoustic pressure. To get a pressure response a so-called "indirect" hearing mechanism is required. In this indirect mechanism, fish make use of gas-filled chambers, such as the swimbladder. Gas-filled chambers are much more compliant than water and hence expand and contract appreciably in response to oscillatory pressure. Assume for simplicity that the gas-filled chamber is spherical and has a radius a that is small in comparison with a wavelength in water. Below the resonant frequency of the chamber, which should be well above the fish's hearing range, such a chamber behaves like a spring. The compression of the chamber due to acoustic overpressure, p_{inc}, is governed by the bulk modulus of the gas within the chamber, $\rho_{\text{air}} c_{\text{air}}^2$, which is defined by

$$\frac{\delta V}{V} = \frac{p_{\text{inc}}}{\rho_{\text{air}} c_{\text{air}}^2}, \tag{7.10}$$

where V is the volume of the chamber. For a spherical chamber where $V = (4\pi/3)\, a^3$, $\delta V/V = 3\delta a/a$, and the displacement of the surface is given by

$$\delta a = \frac{a p_{\text{inc}}}{3\rho_{\text{air}} c_{\text{air}}^2}, \tag{7.11}$$

and its velocity is

$$u_a = \frac{\omega a p_o}{3\rho_{\text{air}} c_{\text{air}}^2} = \frac{2\pi a c_{\text{water}} p_{\text{inc}}}{3\lambda_{\text{water}} \rho_{\text{air}} c_{\text{air}}^2}, \tag{7.12}$$

where λ_{water} is the characteristic wavelength of the incident sound field. The particle velocity at the surface of the gas chamber is thus related to the incident particle velocity $[u_{inc} = p_{inc}/\rho_{water}c_{water}]$ by

$$u_a = \left(\frac{2\pi}{3}\right)\left(\frac{a}{\lambda_{water}}\right)\left(\frac{\rho_{water}c^2_{water}}{\rho_{air}c^2_{air}}\right)u_{inc}. \tag{7.13}$$

For example, at 100 Hz, λ_{water} is 15 m, and a typical value of a is 1 cm; making u_a over 20 times larger than u_{inc}. This result comes about primarily because the bulk modulus of water is 16,000 times lager than that of air. At higher frequencies, the amplification of particle velocity is proportionately even larger. If the air chamber is close to the ear, or if it is directly coupled to the ear, as it is in otophysan fishes (e.g., goldfish, catfish), the indirect path not only provides a signal that is monopole in nature, it also provides increased sensitivity to sound, especially at higher frequencies. Indeed, otophysans, as well as other species that have air bubbles close to the ear (e.g., mormyrids, anabantids), generally have more sensitive hearing that extends to higher frequencies when compared to other fishes (Fay 1988; Popper et al. 2003; Ladich and Popper 2004).

As discussed in the preceding text, a monopole sensor is incapable of ascertaining directional information from an incident sound because its directivity factor is independent of angle. The $\cos\theta$ angular response of a dipole sensor, however, does provide a basis for determining the direction from which a sound originates. There are at least four different ways in which this discrimination can occur:

1. The sensory epithelia (maculae) of otolithic end organs contain groups of hair cells with the same or similar morphological orientation (see Popper et al. 2003; Popper and Schilt, Chapter 2). These "orientation groups" are typically oppositely oriented across a dividing line. The utricle and lagena of most fishes follow a conserved pattern (e.g., Fig. 7.4, left), but the saccule is more varied (see Popper and Fay 1999; Popper et al. 2003; Popper and Schilt, Chapter 2). If, as is typical of the lagena and utricle, the hair bundles are oriented over a wide range of directions so that the moment vectors of

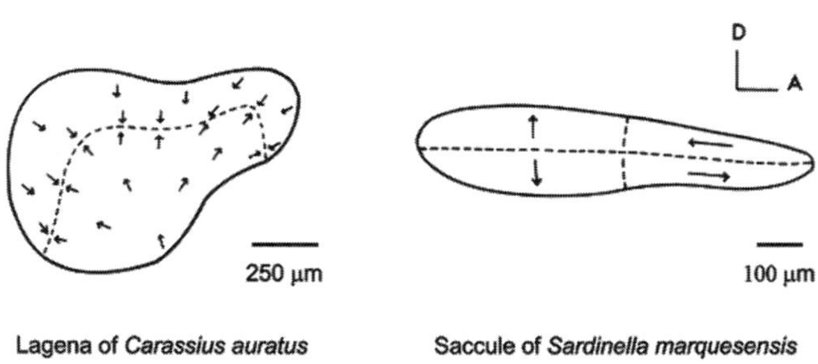

FIGURE 7.4. Hair cell orientation patterns of sensory epithelia (maculae). Hair cell groups point in a variety of directions on the lagena of goldfish (*Carassius auratus*, left), while hair cell groups have simple orthogonal groupings on the saccule of the sardine (*Sardinella marquesensis*, right). Dotted lines represent striola. (Goldfish adapted from Popper 1983, sardine adapted from Platt and Popper 1981.)

the dipoles span all directions of interest, then the direction of the moment vector of the hair cell with the largest response points toward (or away from) the direction of incidence. The accuracy of this method is highly dependent on the signal-to-noise ratio.

2. If, as is typical of the saccule, the hair cell groups are orthogonally oriented (e.g., Fig. 7.4, right), the ratio of the signals from the two directions will be proportional to the tangent of the angle of incidence and the arctangent of the ratio will yield the bearing angle of the sound.
3. If, as in case (2) there are orthogonal orientation groups, the signals from the two groups can be combined to produce an equivalent dipole cosine pattern that points in any direction since the cosine pattern can be rotated by an arbitrary angle β using the trigonometric identity: $\cos(\theta - \alpha) = \cos\theta\cos\alpha - \sin\theta\sin\alpha$, as shown in Figure 7.5. The direction with the largest signal corresponds to the incident direction.
4. With any hair cell bundle orientation configuration the interaction of the acoustic particle velocity with the shape and structure of the otolith can induce complex steady state and oscillatory flow patterns whose structure depends on the direction of the incident field. These flow patterns are reflected in the pattern of hair cell excitation. The problem of directionalization becomes a matter of pattern recognition with the sensory macula functioning like an acoustical analog of the retina (Kotas et al. 2007).

However, all of these suffer from an inability to distinguish sources separated by 180°. The consequence of an inability to resolve the 180° ambiguity could be immense; a fish may turn directly toward a predator that it is trying to avoid, or it may turn away from a food source or potential mate. For plane wave (or farfield) incident sound the 180° ambiguity can be resolved by combining a monopole receiver with a dipole receiver that has equal sensitivity along its main response axis as shown in Figure 7.6. The sum of the output of the omnidirectional monopole sensor plus the cosine response of a dipole yields a response with a cardioid beam pattern given by

$$D_c(\theta) = 1 + \cos\theta, \tag{7.14}$$

which has a maximum response at $\theta = 0$ (i.e., in the direction of $\vec{d}$) and zero response for $\theta = \pi$. The cardioid directivity of Eq. (7.14) is shown in Figure 7.6.

It was shown earlier that orthogonal dipoles can be combined to produce a dipole "steered" to point in any direction. By combining a monopole with a

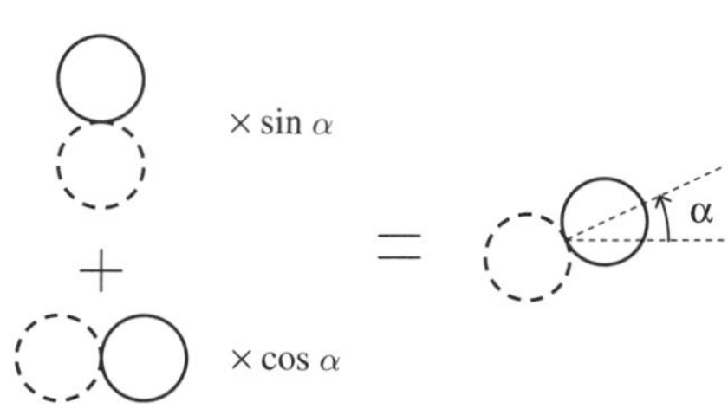

FIGURE 7.5. Combining two orthogonal dipoles to form a rotated dipole.

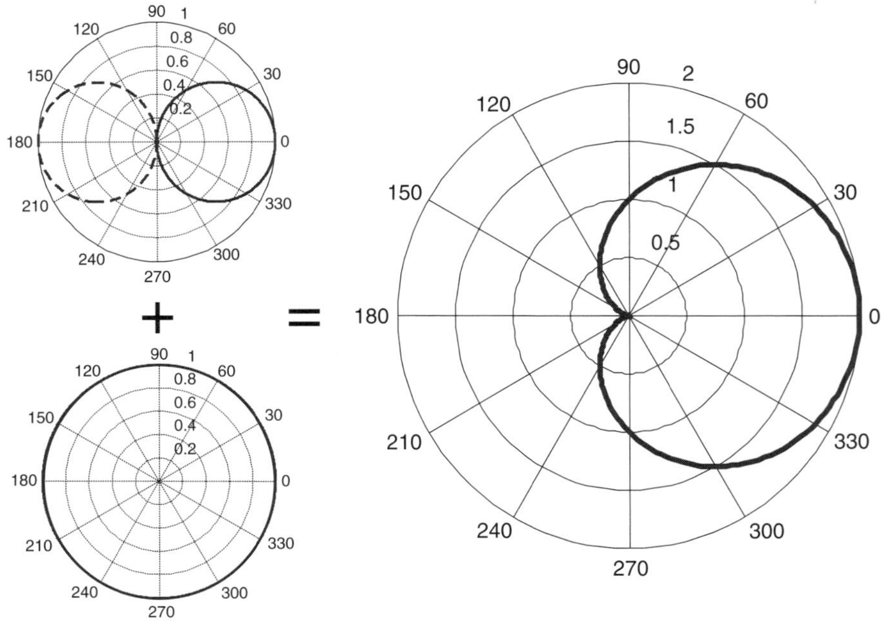

FIGURE 7.6. A unidirectional sensor can be created by combining the output of a monopole and a dipole sensor of equal sensitivity (left). The resulting combined sensor has the "cardioid" response shown on the right.

steered dipole, it is also possible to produce a cardioid that has its maximum pointed in any direction.

Any of the four methods discussed in the preceding text for using dipoles to determine the bearing of a sound source can be modified to incorporate data from a monopole channel to resolve the 180° ambiguity. In the case of the "auditory retina," (described in Section 4), the ability to resolve the ambiguity would rely on the flow pattern associated with the pressure signal being distinctly different from the pattern associated with the direct signal. If both patterns can be recognized simultaneously, the direction of incidence can be ascertained. It should be noted that the flow pattern associated with pressure signals is consistent because the driving forces resulting from an incident pressure have an invariant direction and spatial distribution. For example, in a fish with a swimbladder but without special coupling to the ear, the driving forces would always be directed along the axis toward the swimbladder.

4. Quadrupole Mechanisms in Fish Hearing

It was shown that the 180° ambiguity can be resolved if the fish is able to detect both particle velocity and pressure. It was also shown that only fishes that possess a gas-filled chamber can detect acoustic pressure. Many fish, such as sharks, some tuna, and many bottom fish do not have gas-filled chambers (Popper et al. 2003; Ladich and Popper 2004; Song et al. 2006). Can fish without gas-filled chambers resolve the 180° ambiguity, and, if so, how?

In the absence of an overlying otolith a hair cell is hypothesized to respond to sound as a lateral quadrupole (Rogers and Cox 1988). The underlying rationale is illustrated in Figure 7.7. Consider a hair cell oriented in the y direction whose stereocilia are oriented in the x direction at equilibrium. The stereocilia are assumed to be constrained to pivot in the x–y plane. For sound incident in the x, y, or z directions, it is evident that the pivot angle of the stereocilia, α, would not change and the cell would not respond. However, for sound incident in the x–y plane at 45° to either axis α would change and the cell would respond. If the acoustic particle displacement vector is given by $\vec{\xi}(x, t)$, it is not difficult to show that for acoustic disturbances, the pivot angle α is given by

$$\alpha = \frac{\partial \xi_x}{\partial y}. \tag{7.15}$$

Using Eq. (7.3), Eq. (7.15) becomes

$$\ddot{\alpha} = -\frac{1}{\rho_0} \frac{\partial^2 p}{\partial y \partial x}, \tag{7.16}$$

so, for sinusoidal signals

$$\alpha = \frac{1}{\rho_0 \omega^2} \frac{\partial^2 p}{\partial y \partial x}, \tag{7.17}$$

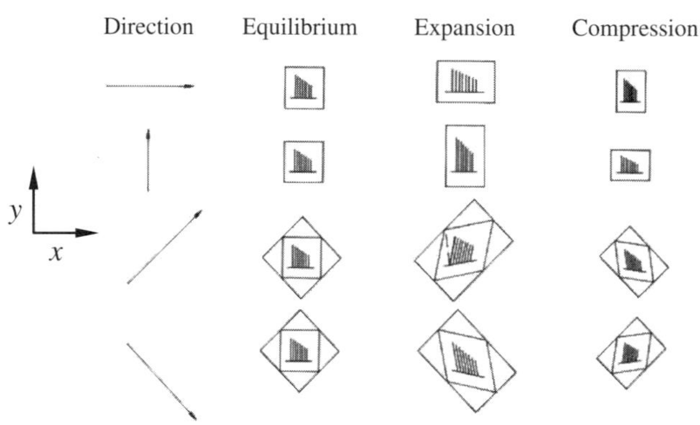

FIGURE 7.7. An uncovered hair cell is a quadrupole sensor. Sound incident in the x or y directions do not result in pivoting of the kinocilium and hence produce no response. Sound incident at 45° to either of these directions produce pivoting and, hence, a response. (Adapted from Rogers and Cox 1988.)

and, for an incident plane wave

$$\begin{aligned}\alpha &= \frac{k_x k_y}{\rho_0 \omega^2} p_o = \frac{n_x n_y}{\rho_0 c^2} p_o \\ &= \sin^2\theta \sin\phi \cos\phi \frac{p_o}{\rho_0 c^2}\end{aligned} \quad (7.18)$$

Equation (7.18) exhibits the lateral quadrupole nature of the response of the uncovered hair cell (see Fig. 7.2). Interestingly, α is simply proportional to the pressure and is completely independent of frequency. The response is maximal for sound incident in the x–y plane at an angle of 45° with respect to either axis. For sound incident in these directions,

$$\alpha_{\max} = \frac{p_o}{2\rho_0 c^2}. \quad (7.19)$$

This angle is always extremely small since $2\rho_0 c^2$ is of the order of 5 GPa.

It is evident that an uncovered hair cell would be rather insensitive. Why might such sensor be useful to a fish?

1. A quadrupole and a dipole sensor can resolve the 180° ambiguity in much the same way as a monopole and a dipole. Uncovered hair cells could thus provide a mechanism for fish without a gas-filled chamber to resolve the 180° ambiguity. The underlying reason why a monopole and a dipole can resolve the ambiguity is that the beam pattern of a dipole is antisymmetric and the beam pattern of a monopole is symmetric. The beam pattern of a lateral quadrupole is symmetric when rotated 45° in the x–y plane. That is, if the main response axis of the lateral quadrupole is normalized and aligned with the axis of a dipole the sum of the two patterns will be a unidirectional, cardioid-like pattern (see Fig. 7.8; the magnitude-squared directivity dipole–quadrupole cardioid is shown in Fig. 7.9). Moreover, just as the responses of two orthogonal dipoles can be combined to produce a dipole response pointing in any azimuthal direction, two lateral quadrupoles in the same plane, rotated 45° with respect to one another, can be combined to produce a lateral quadrupole pointing in any azimuthal direction (Fig. 7.10). Thus two quadrupoles and two dipoles, properly oriented and weighted, can be summed to produce a unidirectional response directed in any azimuthal direction (Fig. 7.11). Also note from Figure 7.12 that a quadrupole–dipole cardioid is far more directive than a monopole–dipole cardioid and thus is more effective in localization.
2. A quadrupole has higher directivity than a monopole. This is not only useful for determining the direction of a sound source, as mentioned in the preceding text, but perhaps more importantly it is also useful for improving the signal-to-noise ratio (S/N). Signals usually originate from a single direction whereas ambient noise is omnidirectional. The noise against which a signal must be detected is the sum of noise signals coming from essentially all directions. A directional sensor such as a dipole, quadrupole, or cardioid that is oriented

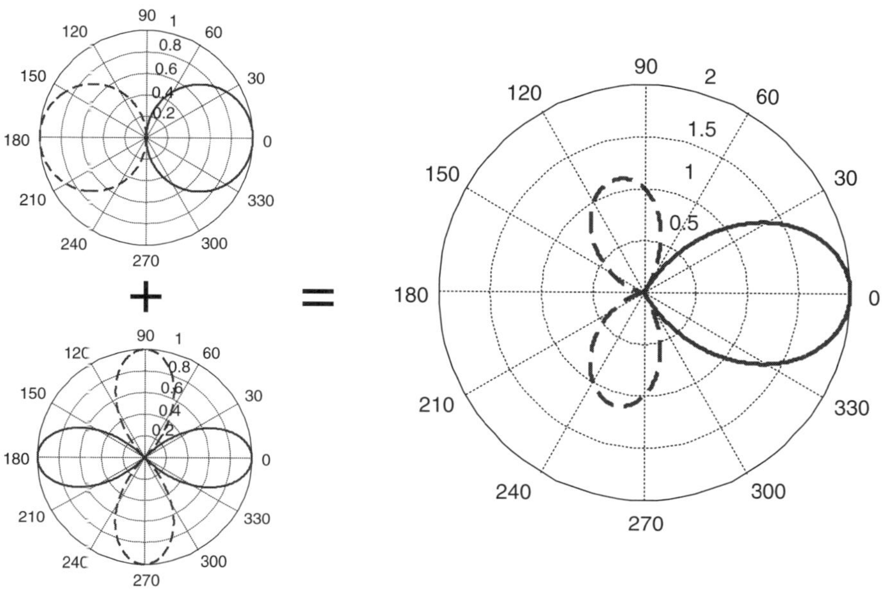

FIGURE 7.8. A dipole summed with a normalized lateral quadrupole that has been rotated by 45° (left) forms a unidirectional cardioid-like directivity pattern (right).

so that its main response axis is in the direction of the signal will reduce or eliminate noise signals originating from other directions. A directional sensor will therefore increase the S/N ratio in the presence of omnidirectional ambient noise. The directivity index (DI) of a receiver is the integral of the square of its normalized directivity pattern. The DI turns out be the improvement in S/N relative to that of an omnidirectional sensor for the receiver in omnidirectional noise. The DI for a dipole sensor or a dipole–monopole cardioid is equal to 4.8 dB. That is, the S/N for such a sensor is 4.8 dB higher than it would be for a monopole receiver. The DI for a dipole–quadrupole cardioid it is 8.2 dB. It is evident that dipole–quadrupole cardioids offer a S/N advantage in situations where ambient noise dominates. Somewhat higher DIs can be achieved for optimized non-cardioid monopole–dipole combinations and monopole–dipole–quadrupole combinations (Cray et al. 2003).

3. One of the problems with the otolith dipole detection mechanism is that the detector is essentially an accelerometer. The detector is thus highly sensitive to whole-body acceleration and vibration of any kind, whether induced by sound or not. A quadrupole sensor responds to sound-induced shear strain but not to acceleration or pressure. It thus responds to sound but not to whole-body acceleration or vibration. Lack of response to acceleration can result in a higher S/N in situations where whole-body motion is a significant noise source.

It is important to understand that sensitivity per se is rarely a dominating consideration in sensors. The important quantity is the S/N ratio. Sensitivity is

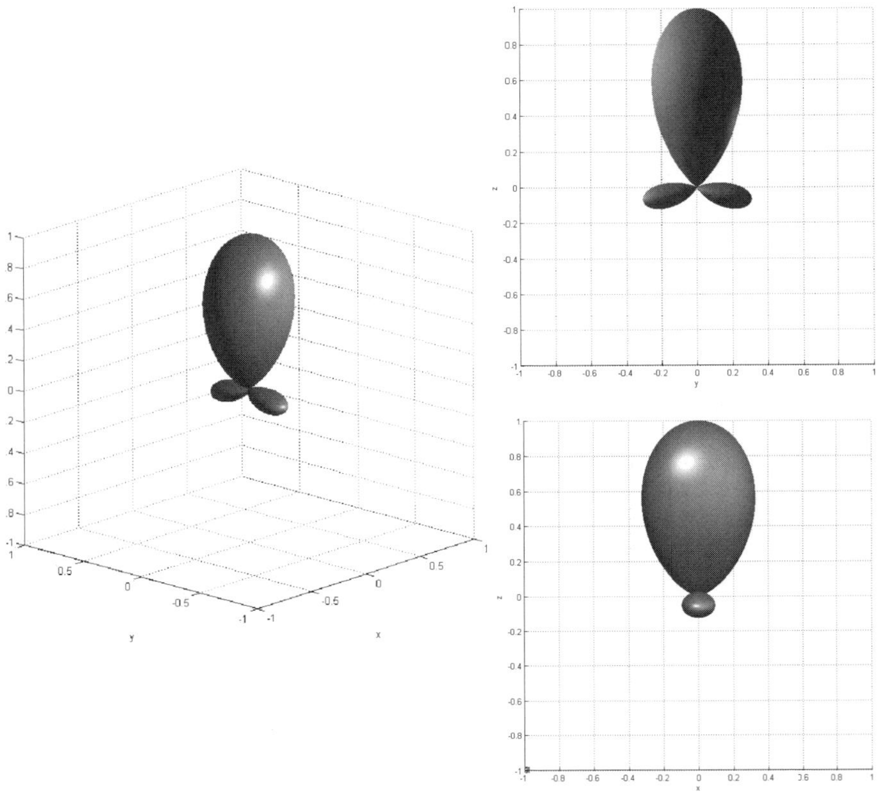

FIGURE 7.9. Magnitude squared directivity of the dipole–quadrupole cardioid of Figure 7.8 as seen from three different aspects.

important only when a sensor is self-noise limited because in this case sensitivity and S/N are related. Yet, when self noise is the limiting factor, low sensitivity can be overcome by large numbers of sensors because uncorrelated noise will be "averaged out." The single sensor S/N in this case is increased by a factor

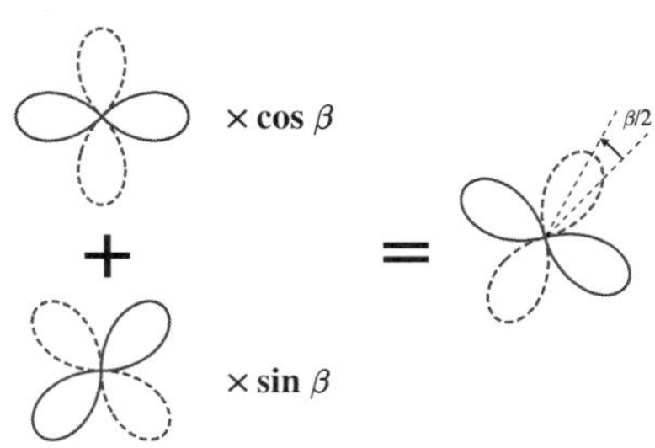

FIGURE 7.10. Two quadrupoles with a relative orientation of 45° can be combined to produce a lateral quadrupole oriented in any direction.

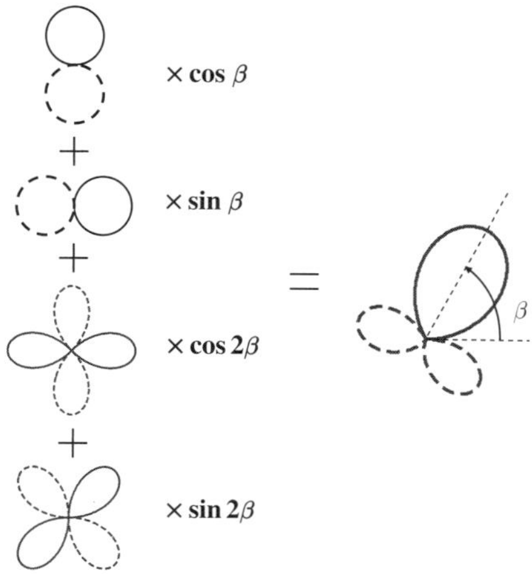

FIGURE 7.11. Two lateral quadrupoles and two dipoles, properly weighted and oriented, can be combined to produce a dipole–quadrupole cardioid oriented in any direction.

equal to the number of sensors. Thus, in the fish ear, the overall S/N would be greatly increased over the single hair cell S/N by having a large number of hair cells.

Sharks orient toward and are attracted to certain sounds (e.g., Nelson 1967). They do not have swimbladders or other gas-filled chambers and are not believed to be pressure sensitive; rather it is thought that they rely on the particle motion component of the sound (Fay et al. 1974; Popper and Fay 1977; Corwin 1981; Casper and Mann 2007; although see van den Berg and Schuijf 1983 for

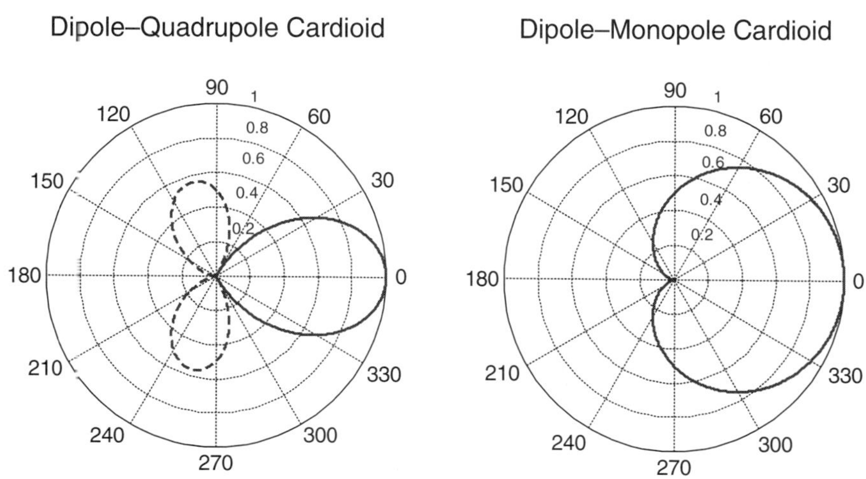

FIGURE 7.12. Comparison of the directivity of a dipole–quadrupole cardioid and a dipole–monopole cardioid.

evidence of pressure detection). The macula neglecta, which in most fish contains few hair cells and whose function is largely unknown, is large in sharks, containing millions of hair cells (Corwin 1977, 1981), and is very likely to be involved in hearing (Fay et al. 1974; Popper and Fay 1977; Corwin 1981). The macula neglecta does not have an overlying otolith (or otoconial mass) and the mechanism by which it may detect sounds remains speculative. The quadrupole hearing mechanism thus offers an attractive hypothesis for shark hearing, especially considering the number of hair cells on the macula neglecta.

In many fishes the otolith overlays the entire macula, but in a number of species a portion of the sensory macula is not covered by the otolith. The macula not covered by the otolith is covered by the otolithic membrane (AN Popper, personal observations), but the role, if any, that the otolithic membrane plays in stimulating hair cells is unknown. And, what little is known about the physical properties of the otolithic membrane makes it unlikely that hair cells away from the otolith could be stimulated via the membrane. In several deep-sea species, a large fraction of the saccular or lagenar macula is not covered by the otolith, as shown in Figure 7.13 (Popper 1980; Deng et al. 2003). These uncovered hair cells are of interest because if they are not useful in detecting sounds then they likely add noise and would have a detrimental effect on hearing. For deep-sea fishes it should also be pointed out that, even when present, the swimbladder may not confer much pressure sensitivity because it would stiffen at great depths. Stiffening would occur because the pressure of the gas within the swimbladder must be equal to the ambient pressure, which is very high at great depth. For an ideal gas (at constant temperature) the density is proportional to the pressure and thus the density of the gas within the swimbladder must also be very high at great depths. It was shown in Eq. (7.10) that the volume fluctuations, $\delta V(t)$, of the

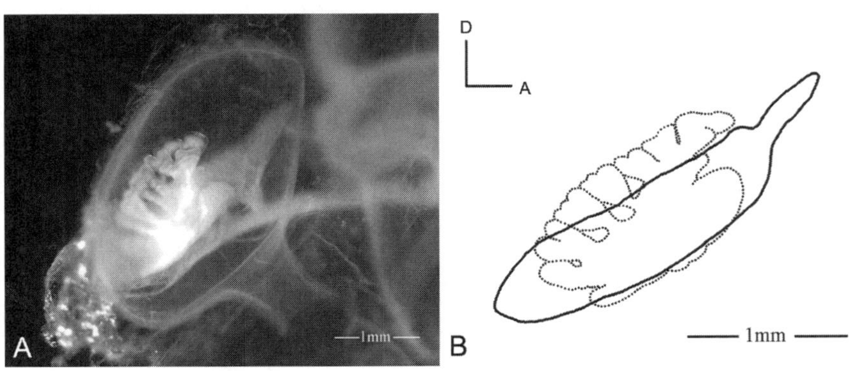

FIGURE 7.13. (**A**) Lagena of a deep-sea fish (*Coryphaenoides rupestris*). (**B**) Line drawing of the sensory epithelium (solid line) and the overlying otolith (dotted line) shown in the same orientation as in **A**. The sensory epithelium extends well beyond the otolith, resulting in a large number of hair cells not covered by the otolith. (Image and drawing provided by Xiaohong Deng, University of Maryland.)

swimbladder induced by an incident acoustic pressure are inversely proportional to the density and so would be much smaller at depth.

Uncovered hair cells acting as quadrupole sensors would contribute to hearing without the need for pressure input. Hence, it may not be surprising to find uncovered hair cells when pressure detection is absent or reduced; such as occurs at great depths or in fish without gas-filled chambers.

Although sensitivity may not be the most important consideration, one must nonetheless address the question of whether a quadrupole sensor is hopelessly insensitive. Since the cellular transduction mechanism is identical, and the number of sensors is comparable for both the dipole and quadrupole mechanisms, one could conclude that the quadrupole would be hopelessly insensitive if a single "quadrupole hair cell" were significantly less sensitive to sound than a single "dipole hair cell."

At first glance, it would appear that this is the case. If the otolith is stationary and the hair cell body moves with the fluid then the maximum rotation angle α_d induced by a sound wave of amplitude p_o and frequency ω is given by

$$\alpha_{\mathrm{d}} = \frac{\delta}{L} = \frac{\delta_{\mathrm{A}}}{L} = \frac{p}{\rho c \omega L}, \tag{7.20}$$

where δ is the displacement of the tip of the kinocilium and is here assumed to be equal to δ_{A}, the acoustic particle displacement, and L is the length of the kinocilium. The maximum rotation angle for the quadrupole mechanism, α_{q}, is given by Eq. (7.19) for any frequency. Hence,

$$\alpha_{\mathrm{q}} = \frac{kL\alpha_{\mathrm{d}}}{2}, \tag{7.21}$$

where $k = \omega/c$ is the acoustic wave number. For a frequency of 100 Hz and $L = 6\,\mu\mathrm{m}$ this yields $\alpha_{\mathrm{q}} \approx \alpha_{\mathrm{d}}/800,000$, which is significantly smaller than α_{d}. However, in Eq. (7.21) the true value of α_{d} has been overestimated because: (1) the density of the otolith is not infinite hence it is not stationary, (2) the fluid boundary layer is larger than the length of the kinocilium so the cell body does not move at the acoustic particle velocity, and (3) the system is actually an accelerometer not a displacement sensor since its natural frequency is most likely above the range of hearing. The true value for α_{d} can be estimated as follows:

$$\alpha_{\mathrm{dtrue}} = \left(\frac{g-1}{g+1/2}\right)\left(\frac{L}{L_{\mathrm{BL}}}\right)\left(\frac{f}{f_{\mathrm{res}}}\right)^2 \alpha_{\mathrm{d}}, \tag{7.22}$$

where g is the specific gravity of the otolith (~ 3.5), L_{BL} is the thickness of the boundary layer, and f_{res} is the resonance frequency of the otolith system. The first term in parentheses on the right-hand side of Eq. (7.22), which accounts for the finite density of the otolith, is $\sim$0.6; the second term, which accounts for the finite length of the boundary layer, is $\sim$0.01. If f_{res} is 2 kHz, then at

100 Hz, $\alpha_{\mathrm{dtrue}} \approx 0.000015\alpha_{\mathrm{d}}$, or $\alpha_{\mathrm{d}} \approx 67,000\alpha_{\mathrm{dtrue}}$. Substituting to find α_{q} yields that $\alpha_{\mathrm{q}} \approx 67,000\alpha_{\mathrm{dtrue}}/800,000$, or $\alpha_{\mathrm{q}} \approx 0.1\alpha_{\mathrm{dtrue}}$. Further, if f_{res} is 7 kHz then $\alpha_{\mathrm{d}} \approx \alpha_{\mathrm{q}}$. Quadrupoles may be insensitive, but not hopelessly so.

5. Summary

The theory of multipole detection of acoustic signals by acoustically small sensors was presented and applied to directional hearing mechanisms in fish. While monopole and dipole mechanisms have been extensively discussed in the literature (often referred to as "indirect" and "direct" hearing mechanisms, respectively), quadrupole mechanisms are lesser known. A dipole sensor alone can be used to obtain directivity in several plausible ways, but always with a 180° ambiguity. It was shown, however, that a dipole sensor can be combined with a monopole sensor to form a cardioid sensor that resolves the ambiguity. Hair cells that do not have an overlying otolith likely function as lateral quadrupole sensors. And, when combined with a dipole, the quadrupole can be used to resolve the 180° ambiguity in much the same way as the monopole–dipole sensor. Since monopole sensing requires that a fish possess a gas-filled chamber, the quadrupole provides a mechanism by which a fish without a swimbladder such as a shark or flatfish can resolve the 180° ambiguity. The dipole–quadrupole cardioid is much more directive than the monopole–dipole cardioid, and it was shown that the quadrupole mechanism can improve S/N when omnidirectional ambient noise or acceleration noise dominates. Although the sensitivity of a quadrupole sensor may be low, it was shown that it is possible that the quadrupole sensitivity may begin to approach that of the dipole.

References

Beranek LL (1954) Acoustics. New York: McGraw-Hill.

Casper BM, Mann DA (2007) Dipole hearing measurements in elasmobranch fishes. J Exp Biol 210:75–81.

Chapman CJ, Hawkins AD (1973) A field study of hearing in the cod, *Gadus morhua*. J Comp Physiol A 85:147–167.

Clark CW, Bass AH (2003) The physical acoustics of underwater sound communication. In: Simmons AM, Fay RR, Popper AN (eds) Acoustic Communication. New York: Springer-Verlag, pp 15–64.

Corey D, Hudspeth AJ (1977) Sensitivity, polarity, and conductance change in the response of vertebrate hair cells to controlled mechanical stimuli. Proc Natl Acad Sci USA 74:2407–2411.

Corwin JT (1977) Morphology of the macula neglecta in sharks of the genus *Carcharhinus*. J Morphol 152:341–362.

Corwin JT (1981) Peripheral auditory physiology in the lemon shark: evidence of parallel otolithic and non-otolithic sound detection. J Comp Physiol 142A:379–390.

Cray BA, Evora VM, Nuttall AH (2003) Highly directional acoustic receivers. J Acoust Soc Am 113:1526–1532.

de Munck JC, Schellart NAM (1987) A model for the nearfield acoustics of the fish swim bladder and its relevance for directional hearing. J Acoust Soc Am 81:556–560.

Deng X, Wagner HJ, Popper AN (2003) Variation in hair cell bundle characteristics in the saccules and lagenae of Macrouridae deep-sea fishes. Assoc Res Otolaryngol Absract 1015.

de Vries HL (1950) The mechanics of the labyrinth otoliths. Acta Oto-Laryngol 38:262–273.

Fay RR (1988) Hearing in Vertebrates: A Psychophysics Databook. Winnetka, IL: Hill-Fay Associates.

Fay RR (2005) Sound source localization by fishes. In: Popper AN, Fay RR (eds) Sound Source Localization. New York: Springer-Verlag, pp. 36–66.

Fay RR, Popper AN (2000) Evolution of hearing in vertebrates: the inner ears and processing. Hear Res 149:1–10.

Fay RR, Kendall JL, Popper AN, Tester AL (1974) Vibration detection by the macula neglecta of sharks. Comp Biochem Physiol A. 47:1235–1240.

Kalmijn AJ (1988) Hydrodynamic and acoustic field detection. In: Atema J, Fay RR, Popper AN, Tavolga WN (eds) Sensory Biology of Aquatic Animals. New York: Springer-Verlag, pp. 83–130.

Kalmijn AJ (1997) Electric and near-field acoustic detection, a comparative study. Acta Physiol Scand 161 (Suppl 638):25–38.

Kinsler LE, Frey AR (1962) Fundamentals of Acoustics. New York: John Wiley & Sons.

Kotas CW, Yoda M, Rogers PH (2007) Visualization of steady streaming near oscillating spheroids. Exp Fluids 42:111–121.

Ladich F, Popper AN (2004) Parallel evolution in fish hearing organs. In: Manley GA, Popper AN, Fay RR (eds) Evolution of the Vertebrate Auditory System. New York: Springer-Verlag, pp. 95–127.

McEachern JF (2003) Overview of vector sensors for undersea applications. J Acoust Soc Am 114:2450 (A).

Nelson DR (1967) Hearing thresholds, frequency discrimination and acoustic orientation in the lemon shark, *Negaprion brevirostris*. Bull Mar Sci 17:741–768.

Pierce (1981) Acoustics – An Introduction to Its Physical Principles and Applications. New York: McGraw-Hill.

Platt C, Popper AN (1981) Otolith organ receptor morphology in herringlike fishes. In: Gualterotti T (ed) Vestibular System: Function and Morphology. New York: Springer-Verlag, pp. 64–74.

Popper AN (1980) Scanning electron microscopic studies of the sacculus and lagena in several deepsea fishes. Am J Anat 157:115–136.

Popper AN (1983) Organization of the inner ear and processing of acoustic information. In: Northcutt RG, Davis RE (eds) Fish Neurobiology and Behavior. University of Michigan, Press, Ann Arbor, MI, pp. 125–178.

Popper AN, Fay RR (1977) Structure and function of the elasmobranch auditory system. Am Zool 17:443–452.

Popper AN, Fay RR (1999) The Auditory Periphery in Fishes. In: Fay RR, Popper AN (eds) Comparative Hearing: Fish and Amphibians. New York: Springer-Verlag, pp 43–100.

Popper AN, Fay RR, Platt C, Sand O (2003) Sound detection mechanisms and capabilities of teleost fishes. In: Collin SP, Marshall NJ (eds) Sensory Processing in Aquatic Environments. New York: Springer-Verlag, pp. 3–38.

Rogers PH, Cox M (1988) Underwater sound as a biological stimulus. In: Atema J, Fay RR, Popper AN, Tavolga WN (eds) Sensory Biology of Aquatic Animals. New York: Springer-Verlag, pp. 83–130.

Rogers PH, Popper AN, Cox M, Saidel WM (1988) Processing of acoustic signals in the auditory system of bony fish. J Acoust Soc Am 83:338–349.

Schellart NAM, de Munck JC (1987) A model for directional and distance hearing in swimbladder-bearing fish based on the displacement orbits of the hair cells. J Acoust Soc Am 82:822–829.

Schuijf A (1975) Directional hearing of cod (*Gadus morhua*) under approximate free field conditions. J Comp Physiol A 98:307–332.

Schuijf A, Buwalda RJA (1975) On the mechanism of directional hearing in cod (*Gadus morhua*). J Comp Physiol A 98:333–344.

Song J, Matieu A, Soper RF, Popper AN (2006) Structure of the inner ear of bluefin tuna (*Thunnus thynnus*). J Fish Biol 68:1767–1781.

van Bergeijk WA (1964) Directional and nondirectional hearing in fish. In: Tavolga WA (ed) Marine Bioacoustics. London: Pergamon Press, pp. 269–301.

van Bergeijk WA (1967) The evolution of vertebrate hearing. In: Neff WD (ed) Contributions to Sensory Physiology. New York: Academic Press, pp. 1–49.

Van den Berg AV, Schuijf A (1983) Discrimination of sounds based on the phase difference between particle motion and acoustic pressure in the shark *Chiloscyllium griseum*. Proc Roy Soc Lond B 218:127–134.

von Frisch K, Dijkgraaf S (1935) Can fish perceive sound direction? Z Vergl Physiol 22:641–655.

8 Vocal–Acoustic Communication: From Neurons to Behavior

Andrew H. Bass and Friedrich Ladich

1. Introduction

There is a long history of many elegant studies of acoustic communication in fish, and more specifically teleosts. Since Fine et al. (1977) and Myrberg (1981) reviewed the behavioral biology of sonic/vocal fish, an increasing number of studies have investigated the neural, muscular, and hormonal mechanisms that underlie acoustic communication. Several recent reviews have provided overviews of this work, which included investigations of auditory (Bass and McKibben 2003; Bass et al. 2005b; Ladich and Bass 2003a,b; Lu 2004; Bass and Lu 2007), vocal motor (Bass and McKibben 2003; Ladich and Bass 2003b; Ladich and Fine 2006), neuroendocrinological (Fine 1997; Forlano et al. 2006; Remage-Healey and Bass 2006a), and behavioral (Bass and McKibben 2003; Ladich and Myrberg 2006; Myrberg and Lugli 2006) mechanisms. Bass and Clark (2003) and Mann (2006) also reviewed the physical principles shaping acoustic communication in shallow and deep-water aquatic habitats and the general influence of the abiotic environment on vocal communication among fishes. Thus, we have decided to highlight some of their most salient points while integrating some of the more recent developments emerging from studies of the temporal encoding of auditory signals, species diversity of vocal control systems, and neuroendocrine influences on both auditory and vocal mechanisms that contribute to successful social communication.

2. Audition in Vocal Fishes

The vocal communication signals of teleost fishes, like those of other vertebrates (Bradbury and Vehrencamp 1998), are diverse in their spectral and temporal properties (Fish and Mowbray 1970; Ladich 1997b; Amorim 2006). Most fish sounds tend to be pulsatile, broad-band signals, but there are a few instances of long-duration, multiharmonic calls such as the "hum" of the plainfin midshipman fish *Porichthys notatus* and the "moan" of the mormyrid fish *Pollimyrus adspersus* (Fig. 8.1; see also Bass et al. 1999). The ascending

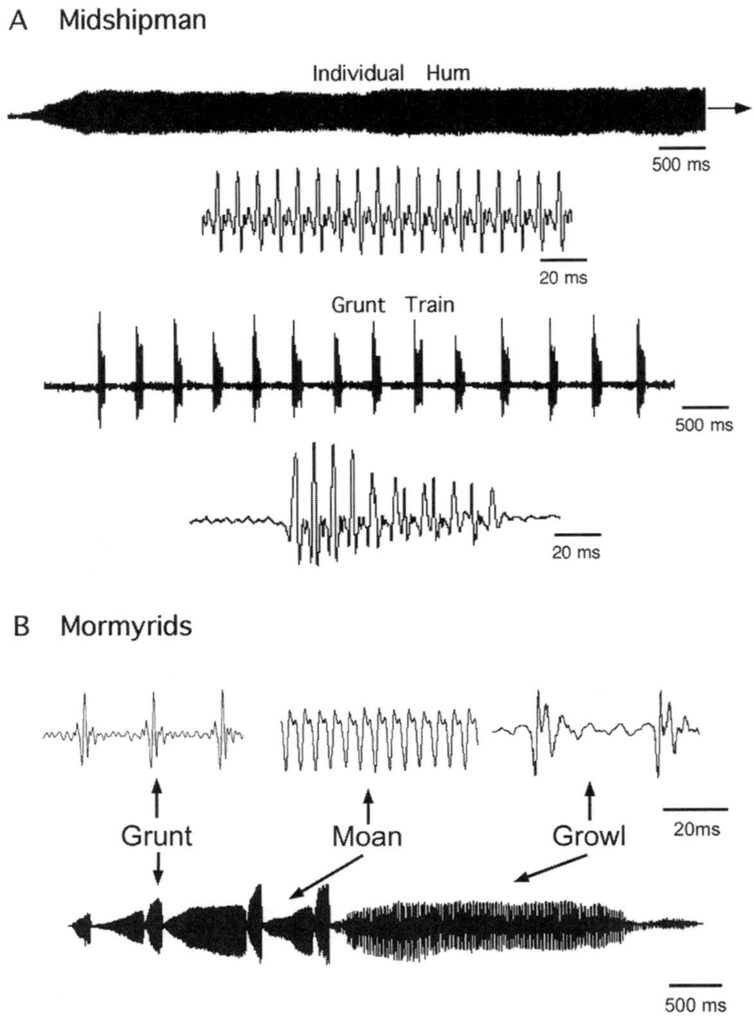

FIGURE 8.1. Teleost fish vocalizations. Field recordings on two time scales of midshipman hum and grunt train and mormyrid grunt, moan and growl. (From Bass et al. [1999, 2005b], in part, with permission of Springer Science + Business Media.)

auditory system of vocal teleosts largely resembles that of nonvocal teleosts and of vertebrates in general (for review of nonvocal species, see McCormick 1999; for review of vocal teleosts see Bass and McKibben 2003; Bass et al. 2005b; Bass and Lu 2007; see also Fay and Edds-Walton, Chapter 3). The neuroanatomical and neurophysiological studies of vocal midshipman fish (*P. notatus*) and the Gulf toadfish (*Opsanus beta)* have also identified sites of integration between the auditory and vocal motor systems at forebrain, midbrain, and hindbrain levels (Fig. 8.2; Bass et al. 1994, 2000; Goodson and Bass 2002). Together, this pattern

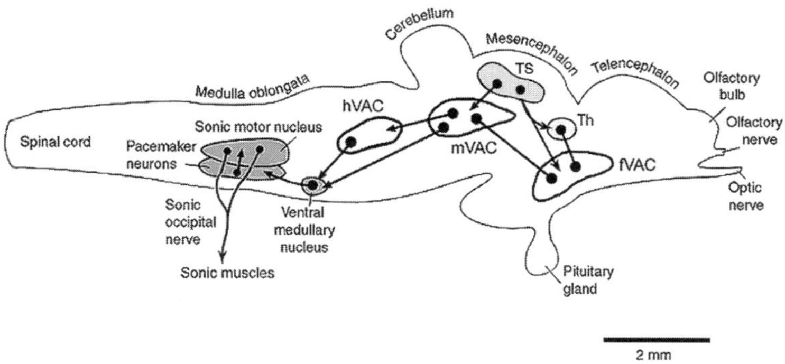

FIGURE 8.2. Organization of vocal–acoustic system in vocal fish. Sagittal view of the brain of a batrachoidid fish (midshipman and toadfish), showing a general outline of forebrain (fVAC), midbrain (mVAC), and hindbrain (hVAC) vocal–acoustic integration centers; each center includes several nuclei (see Goodson and Bass 2002 for details). The midbrain's torus semicircularis (TS) and the diencephalon's dorsal thalamus (Th) (lightly shaded) provide auditory input to the fVAC and mVAC. The eighth cranial nerve (not shown) is the major source of auditory input to the hVAC from the sacculus, the main organ of hearing in this group of teleosts (Cohen and Winn 1967). The vocal pattern generator is also indicated (dark shading) and includes a ventral medullary nucleus, pacemaker neurons, and a sonic motor nucleus that innervates the sonic swimbladder muscle via paired occipital nerve roots that are considered homologous to the hypoglossal nerve of tetrapods (Bass and Baker 1997). (From Bass and McKibben [2003] with permission from Elsevier.)

of connectivity forms a distributed network of vocal–acoustic centers in the brain. As we discuss in the text that follows, recent studies have focused on the auditory encoding of the temporal attributes of these signals and the influences of steroid hormones on that encoding.

2.1 Temporal Encoding of Vocal Signals

One of the major advantages that teleost fishes offer to studies of the neural mechanisms of acoustic communication is the opportunity to investigate the coevolution of the temporal mechanisms for both the encoding and production of vocal signals. As Capranica (1992, p. 402) stated: "It is the periodic function of time that is generated by the source and it is the time-varying waveform that has evolved through natural selection for species-specific intelligent communication.... The production and physical reality of animal signals lie in the time domain under active ongoing neural control.... The generation of that temporal waveshape is the signal that the nervous system controls.... The two systems [vocal and auditory] co-evolved and we should expect them to share the same underlying code for signal generation and recognition."

Myrberg and colleagues studied natural populations of damselfishes and showed the importance of pulse number and interval to the recognition of species differences in their multipulse, broadband "chirps" (e.g., see Myrberg and Spires 1972 for the bicolor damselfish *Eupomacentrus partitus*). Temporal features of an entire call, such as duration, may also be important as shown in playback studies of signals that mimic the multiharmonic "hum" of the plainfin midshipman (McKibben and Bass 1998) and the multiharmonic "boatwhistle" of toadfishes (*O. tau* and *O. beta*: Fish 1972; Winn 1972; Remage-Healey and Bass 2005). The simplicity of teleost vocalizations has been particularly advantageous because they can be mimicked by synthetic signals. Thus, pure tones with varying durations, frequency, and harmonic structure as well as short transients (clicks) with varying pulse periods, have been used to identify the relevant temporal parameters of calls in both behavioral studies and in neurophysiological investigations of the encoding properties of peripheral and central auditory neurons (see Bass and McKibben 2003; Wysocki 2006).

Recent studies in vocal mormyrids, the plainfin midshipman, and toadfishes (*O. tau* and *O. beta*) show that auditory neurons encode multiple temporal parameters of vocalizations including pulse repetition rate, duration, depth, and rate of amplitude modulations and waveform envelope shape (Fay and Edds-Walton 1997a,b; McKibben and Bass 1999, 2001a; Bass et al. 2001; Suzuki et al. 2002; Weeg et al. 2002; Edds-Walton and Fay 2003). For example, the medulla and midbrain of the weakly electric mormyrid fish *Pollimyrus adspersus* have interval-coding neurons that provide a central representation, as reflected in their spike rate, of the repetition rates that characterize their pulsatile vocalizations (Crawford 1997; Kozloski and Crawford 2000) (see Fig. 8.1). Behavioral studies are consistent with this neural sensitivity. The discrimination of interclick intervals by *P. adspersus* has been assessed using modulations of the fish's electric organ discharge rate as an indicator of discrimination capability (Marvit and Crawford 2000). These fishes are sensitive to small differences in click train stimuli that mimic the interclick intervals corresponding to their grunts. Discriminable differences are as small as 0.3 ms at interclick intervals of 15 ms and increase to about 3 ms for 40 ms interclick interval trains. The mean interclick interval of *P. adspersus* grunts is 17.9 ms compared to 22.7 ms in the sympatric, and closely related species *P. isidori* (Crawford et al. 1997). The authors conclude that this temporal discrimination ability is sufficient for species recognition, and in combination with frequency information mediated by multiharmonic moans (a sound emitted regularly in alternation to grunts during courtship), might provide the basis for individual recognition. The temporal discrimination threshold of *Pollimyrus* (see preceding text) is quite similar to the temporal resolution ability of the mormyrid *Gnathonemus petersii* for which physiologically defined, minimum resolvable click periods of 0.5 ms are found at 20 dB above hearing threshold (Wysocki and Ladich 2002).

The plainfin midshipman fish has midbrain neurons that encode the depth and rate of amplitude modulation of long-duration acoustic beats that are generated in their native habitat when the long-duration (min $\geq$ 1 h), multiharmonic advertisement hums of neighboring males overlap in time (Bodnar and Bass 1997, 1999, 2001a). A temporal code is exhibited by the degree of phase locking of the spike train of individual neurons to the beat modulation rate that is established by the difference frequency (dF) between the fundamental frequencies of the concurrent hums. Midbrain neurons are most sensitive to dFs that overlap the range of dFs observed in wild populations of plainfin midshipman during the breeding season (< 10 Hz). A neuron's selectivity cannot be explained on the basis of its frequency tuning properties. Eighth nerve afferents do not exhibit dF selectivity (McKibben and Bass 2001a), suggesting that it arises at central levels. Underwater playback studies are consistent with this neural sensitivity and show that individual fish distinguish dFs as small as 0.5 Hz (McKibben and Bass 1998, 2001b).

One temporal parameter that is highly divergent among the vocal signals of the plainfin midshipman is duration, with hums that typically last for minutes and agonistic grunts that have a duration of only 50–100 ms (see Fig. 8.1). Behavioral playback studies show the importance of duration and interpulse interval for signal discrimination (McKibben and Bass 1998, 2001b). Neurophysiological studies show that both peripheral and midbrain neurons encode signals of widely varying duration. Most neurons show no loss, and even an improvement, in the degree of temporal encoding of signals of increasing duration that mimic either hums or beats (McKibben and Bass 1999; Bodnar and Bass 2001b).

Multiple sound parameters are also encoded by the auditory brainstem response (ABR), which represents the summed activity of the auditory periphery (inner ear maculae, auditory nerve afferents) and brainstem auditory pathways (Kenyon et al. 1998). Most of these studies have been conducted in species known as "hearing specialists" (e.g., mormyrids), which have peripheral adaptations (e.g., either bony elements or swimbladder diverticula) that enhance the detection of the pressure component of sound. In contrast, "hearing generalists" (e.g., midshipman and toadfishes) lack these adaptations (see reviews in Popper and Fay 1999; Ladich and Bass 2003a; Ladich and Popper 2004; see also Braun and Grande, Chapter 4). Playbacks of double-click stimuli with varying click periods show that the physiologically defined, minimum resolvable click period is below 1.5 ms in five species of hearing specialists (the goldfish *Carassius auratus*, the catfish *Platydoras costatus*, the blue gourami *Trichogaster trichopterus*, the croaking gourami *Trichopsis vittata*, the mormyrid *Gnathonemus petersii*) and is independent of the ability to vocalize (Wysocki and Ladich 2002). A subsequent study (Wysocki and Ladich 2003) used playbacks of complex, species-specific sounds to see if sounds consisting of several repeated pulses varying in pulse period and amplitude could be reliably represented within the auditory system. Five vocal species were investigated, including four hearing specialists (two catfishes—*P. costatus* and *Pimelodus pictus*, the loach *Botia modesta*, the croaking gourami *T. vittata*) and one hearing generalist (the pumpkinseed sunfish *Lepomis gibbosus*). The experiments revealed that for all of these species, except the hearing generalist, each pulse of a sound produced by a conspecific elicited a separate response with the onset of each auditory evoked potential corresponding closely to the onset of each sound pulse (Fig. 8.3). The absence of a correlated response in the sunfish should not be taken as an absence of temporal encoding

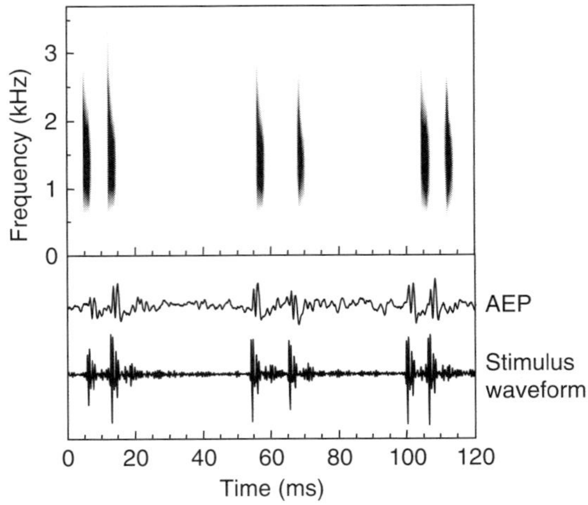

FIGURE 8.3. Sonogram (above) and oscillogram (stimulus waveform) of a *Trichopsis vittata* sound and the auditory evoked potential (AEP) of an animal to this sound. Sampling frequency 25 kHz; filter bandwidth 450 Hz, 50% overlap. (Adopted from Wysocki and Ladich [2003] and Wysocki [2006].)

among hearing generalists; batrachoidids are also hearing generalists but, as we discussed earlier for midshipman fish, single-unit recording studies show a precise mechanism for encoding the temporal parameters of sound.

2.2 Hormonal Influences on Audition

Recent studies of the plainfin midshipman show how steroid hormones can influence the hearing mechanisms of vocal teleosts. Each spring, males migrate from deep offshore sites (at least 200 m; see Sisneros et al. 2004b) into the shallow intertidal zone, where they build nests under rocky shelters and acoustically court females with their hums. Motivated by the discovery that only gravid females that are ready to spawn show positive phonotaxis to underwater playbacks of natural and synthetic hums (McKibben and Bass 1998), Sisneros and Bass (2003) investigated whether such a dramatic change in reproductive state and attention to an auditory stimulus might have a neurophysiological correlate at the level of auditory neurons. Sisneros and Bass (2003) reported seasonal differences in the degree of temporal encoding by eighth nerve, primary afferents that innervate the sensory epithelium of the sacculus, the main end organ of hearing in the midshipman's inner ear (Cohen and Winn, 1967). While afferents encode frequency in both spike rate and the extent of phase locking to pure tone stimuli, phase locking is the most reliable encoding mechanism among teleosts (see Fay 1978 for goldfish and Bass and McKibben 2003 for recent review of teleosts in general). Sisneros and Bass (2003) also showed that saccular afferents in nonreproductive females (those without mature eggs and found in sites distant from nesting grounds during the nonbreeding seasons of the fall and winter) are sensitive mainly to frequencies close to or below 100 Hz, the fundamental frequency of their calls. In contrast, recordings from saccular afferents in reproductive females (those carrying mature eggs and prepared to mate with a nesting male) showed a dramatic increase in phase locking values for frequencies > 100 Hz, which included the second and third harmonics of male advertisement calls that often contain as much or even more spectral energy than the fundamental frequency (Bass et al. 1999). These results suggest that the female's inner ear becomes sensitized to the upper harmonics of the mate call during the mating season. Sisneros and Bass (2003) propose that this enhanced sensitivity would improve a female's ability to detect a male calling from his nest in the shallow waters of the intertidal zone because higher harmonics will have a greater transmission distance in shallow water (reviewed in Bass and Clark 2003; see Fine and Lenhardt 1983 for field studies of the closely related toadfish *O. tau*).

The discovery of seasonal plasticity in frequency encoding prompted the question as to whether or not steroid hormones might influence this seasonally dependent trait. Like other teleost species, the plainfin midshipman exhibits seasonal changes in plasma levels of steroid hormones (Sisneros et al. 2004b). During the spring months just before the spawning season, females show peak levels of both 17β-estradiol and testosterone at the time when their eggs are becoming mature. Sisneros et al. (2004a) have demonstrated that the saccular afferents of nonreproductive females treated with either 17β-estradiol or testosterone showed robust encoding over a frequency range that included the upper harmonics of a male's advertisement call so that their auditory phenotype completely resembled that of reproductive females (Fig. 8.4). The steroid-induced effect manifests itself gradually over a period of about 1 month (Sisneros et al. 2004a). We expect that a similar time course of events occurs in natural populations during the spring, prenesting period when plasma steroid levels begin to rise. Steroids are essentially priming the inner ear of females for detecting and localizing the nests of calling males in the shallow waters of the intertidal zone. Males may also show similar seasonal shifts in the temporal encoding properties of the inner ear (males maintained in captivity through the fall and winter have an auditory phenotype like that of nonreproductive females; McKibben and Bass 1999). Males would benefit from knowing where suitable nest sites are located as well as the detection of competing males.

Other studies have also shown the presence of estrogen receptor α in the saccular epithelium following the cloning and sequencing of a partial cDNA for this receptor in midshipman fish (Sisneros et al. 2004a; Forlano et al. 2005). While studies in mammals, including humans, show estrogen receptor α in the hair cell epithelium of the cochlea (Stenberg et al. 1999, 2001), its functional significance has remained unknown, although some have suggested a role for steroids in affecting the sensitivity of female audition during the menstrual cycle (see Haggard and Gaston 1978; McFadden 1998). Studies of the midshipman fish now provide some potential insights into the neurophysiological correlates of this phenotype. The observed effect of steroids on frequency encoding in midshipman may yet be entirely due to the effects of 17β-estradiol, which circulates at higher levels during the spring and summer (Sisneros et al. 2004b). Like other teleosts,

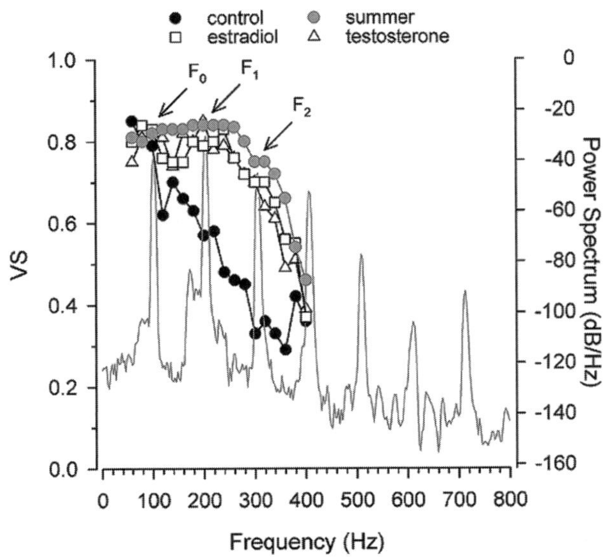

FIGURE 8.4. Coupling between vocal parameters and the degree of temporal encoding to frequency in eighth nerve, saccular afferents in midshipman fish. The phase-locking precision of saccular afferents is measured by the vector strength of synchronization (VS) as indicated on the *y*-axis to the left. The amplitude spectrum of a hum advertisement call from a type I male is indicated in relative dB values on the *y*-axis to the right. Frequency is plotted along the *x*-axis for both sets of measures. Shown here are median VS values obtained from recordings from nonreproductive females that were treated with either testosterone (triangles) or 17β-estradiol (squares). Also shown for comparison are VS values from wild caught reproductive females collected during the summer breeding period (light-shaded circles) and nonreproductive females (dark-shaded circles). Nonreproductive females show a high degree of temporal encoding only for frequencies close to the fundamental frequency (F_0). By contrast, reproductive females and steroid-treated females show robust encoding of F_0 as well as the second and third harmonics (F_1, F_2) of the hum advertisement call. (Adapted from Sisneros et al. [2004a].)

midshipman fish have unusually high brain levels of aromatase, the enzyme that converts testosterone to estrogen (Schlinger et al. 1999; Forlano et al. 2001). Aromatase is also localized to auditory ganglion cells, suggesting the potential for a local site of estrogen synthesis in the inner ear close to the hair cell epithelium (Forlano et al. 2005).

3. Vocal Control Systems

We use the terms vocal and vocalization to make comparisons to the vocal behaviors and neural mechanisms of tetrapods. This comparison rests on the evidence showing that the vocal muscles of teleosts with a swimbladder mechanism for vocalization (see later) are derived embryonically from the same mesodermal components as the muscles of the syrinx and larynx, while vocal neurons in the caudal hindbrain and rostral spinal cord may also share embryonic origins with those of tetrapods (Bass and Baker 1997; Bass et al. 2005a).

3.1 A Diversity of Vocal Organs

Fishes possess the largest diversity of sound generating (sonic) mechanisms among vertebrates. Despite a wealth of knowledge accumulated since Aristotle, new mechanisms and an increasing diversity among known mechanisms continue to be reported. Ladich and Fine (2006) recently proposed a classification scheme for sonic mechanisms in fishes that is based on morphological structures adapted exclusively for acoustic signaling and thus intraspecific communication (excluding unintentionally produced sounds such as those emitted during swimming/ hydrodynamic movements or feeding). The main group includes sonic swimbladder mechanisms with their numerous morphological variations. Swimbladders are vibrated by striated ("drumming" or sonic) muscles that have morphological and physiological traits adapted for rapid contraction (e.g., Bass and Marchaterre 1989; Walsh et al. 1995; Lewis et al. 2003; Nahirney et al. 2006; Rome 2006). Based on the origin and insertion of these muscles, various muscle types are distinguished. Intrinsic drumming muscles attach solely to the swimbladder walls (e.g., in Batrachoididae: toadfishes and midshipman fish; e.g., Fig. 8.5A), whereas extrinsic types originate on other structures such as the skull, ribs, vertebrae, and reportedly muscles. Extrinsic muscles that attach to the swimbladder and other structures provide a directly vibrating mechanism and include those in pimelodid catfishes (Ladich and Bass 1998; Ladich 2001) and holocentrid squirrelfishes (Carlson and Bass 2000) (e.g., Fig. 8.5B). If sonic muscles lack a direct attachment site on the swimbladder they are called indirectly vibrating mechanisms and include those in doradid catfishes (Ladich and Bass 1998) and piranhas (Ladich and Bass 2005) (e.g., Fig. 8.5C). Some species with extrinsic muscles may have both directly and indirectly vibrating mechanisms (e.g., Parmentier et al. 2003).

The second major group of adaptations for sound production involve the pectoral girdle, pectoral fin rays or fin tendons. For example, sculpins (family Cottidae), which lack a swimbladder, vibrate the pectoral girdle by rapidly contracting a muscle extending between the skull and the cleithrum (musculus cephaloclavicularis; see Bass and Baker 1991). Representatives of the majority of catfish families possess an enhanced first pectoral fin ray, the spine (e.g., Fig. 8.5D). The base of this spine possesses a dorsal process that bears series of ridges. Pressing these ridges against a groove in the pectoral girdle during a fin sweep results in the emission of stridulatory sounds (Fine and Ladich 2003). Fine et al. (1997) propose that each sound pulse is generated by the collision of a single ridge against the rough surface. Different species utilize different combinations of muscle movements to produce sound (Ladich 1997a; Heyd and Pfeiffer 2000). Croaking gouramis *Trichopsis* spp., representatives of the perciform family Osphronemidae (labyrinth fishes or gouramis), generate

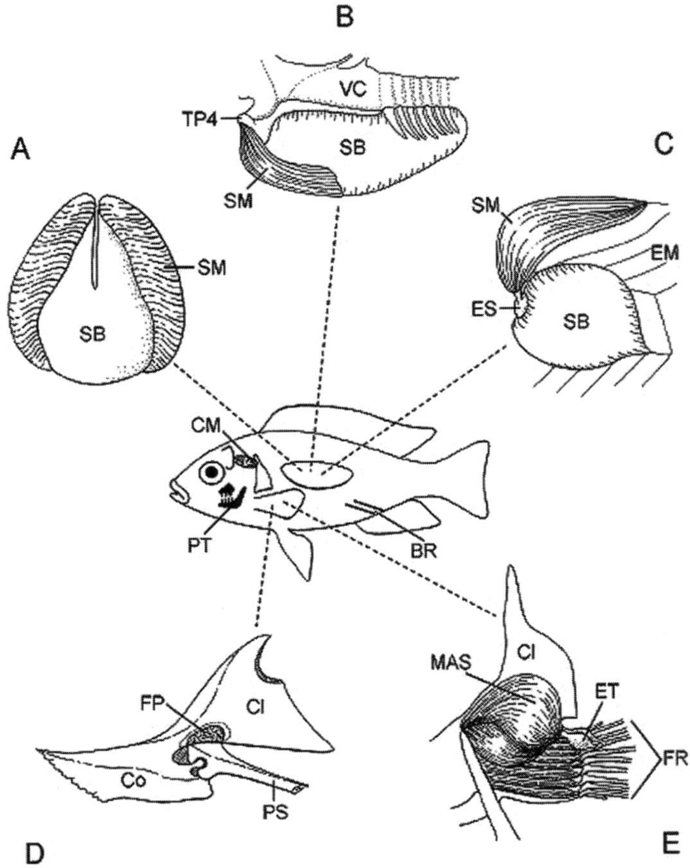

FIGURE 8.5. Overview of the main groups of sound generating (sonic) mechanisms in teleost fishes based on the classification proposed by Ladich and Fine (2006). The upper row shows swimbladder vibrating mechanisms. (**A**) Intrinsic sonic muscles (SM) attached to the walls of the swimbladder (SB) as found in the midshipman *Porichthys notatus*. (**B**) Extrinsic directly vibrating mechanism in the catfish *Pimelodus* sp. (**C**) Extrinsic indirectly vibrating mechanism in the catfish *Synodontis* sp.: the swimbladder is vibrated by a thin bony plate, the elastic spring (ES). The lower row shows pectoral mechanisms. (**D**) Pectoral spine (PS) stridulating mechanisms in many tropical catfishes. (**E**) Pectoral fin tendon plucking in croaking gouramis genus *Trichopsis*. Pectoral mechanisms also include vibrating the pectoral girdles by the cephaloclavicular muscle (CM) in the sculpins (family Cottidae). Sounds might also be generated by rubbing of pharyngeal teeth (PT). The release of air bubbles (BR) through the anus in herrings is proposed to serve in communication. Cl, cleithrum; Co, coracoid; EM, epaxial muscles; ET, enhanced tendons; FP, friction process; FR, fin rays; MAS, superficial adductor muscle; TP4, transverse process of the fourth vertebrum; VC, vertebral column. (Adapted from Fine and Ladich [2003] and Ladich [2004].)

pulsatile sounds by two enhanced pad-like tendons of the fourth and fifth pectoral fin rays (Kratochvil 1978) (e.g., Fig. 8.5E). Two sound pulses are emitted during abduction when the enhanced superficial adductor muscle stretches both tendons of one fin and snaps them over bony elevations of the base of the second and third fin rays.

Several investigations have also proposed that the grating of pharyngeal teeth results in the production of communication sounds in perciforms such as carangids, haemulids, labrids, pomacentrids, cichlids, and centrarchids. Rice and Lobel (2002) have found sexually dimorphic muscle morphology and enzyme activity associated with the pharyngeal jaws in a cichlid, which is suggestive of a pharyngeal mechanism of sound production. Other studies describe potentially new mechanisms including low-frequency sound production by water expulsion through the gill cover in a goby (Stadler 2002). In addition, bubbles emitted from the anus in herring *Clupea harengus* produce a stereotyped series of high-frequency pulses, which might have some communicative as well as hydrostatic value, but this still needs to be demonstrated (Wahlberg and Westerberg 2003; Wilson et al. 2004). The cod *Gadus morhua* produce high-frequency clicks of 7 kHz (Vester et al. 2004); the mechanism used here and among other species that make clicks (e.g., the skunk loaches *Botia horae*; Valinsky and Rigley 1981) awaits definition.

3.2 Central Vocal Pathways

Neuroanatomical studies have mapped a descending vocal motor system in batrachoidids (midshipman and toadfishes) that extends from the forebrain to motor neurons positioned in the caudal hindbrain and rostral spinal cord (Fig. 8.2). The discovery that small molecular weight biotin compounds (neurobiotin and biocytin) could cross multiple synapses in the vocal motor system led to the initial demonstration of a descending vocal pathway that overlapped central auditory neurons (Bass et al. 1994). Neurophysiological studies show that electrical stimulation in the anatomically-identified vocal-acoustic centers (VACs, Fig. 8.2) evoke a rhythmic motor output from a hindbrain-spinal, pacemaker–motor neuron circuit (see Goodson and Bass 2002; Kittelberger et al. 2006 and references therein).

The fundamental "morphophysiological" unit of the vocal system is a pacemaker–motor neuron circuit that extends from the caudal hindbrain into the rostral spinal cord (Fig. 8.2). Intracellular recording and staining studies in batrachoidids identified a midline pair of sonic (vocal) motor nuclei (SMN) that are innervated by nearby premotor neurons that generate pacemaker-like action potentials (Bass and Baker 1990). Each SMN innervates the ipsilateral sonic muscle that is attached to the lateral wall of the swimbladder (see Fig. 8.5A). Individual pacemaker neurons densely innervate both motor nuclei, thereby providing for extensive coupling of both motor neuron populations that, in turn, can lead to the simultaneous contraction of both sonic muscles (Pappas and Bennett 1966; Bass and Baker 1990). The vocal pacemaker neurons establish the firing rate of the motor neurons. The synchronous activity of motor neurons produces a rhythmic vocal motor volley that reflects the oscillatory-like activity of the pacemaker–motor neuron circuit. The motor volley can be readily evoked by electrical stimulation in vocal-acoustic centers (VACs, Fig. 8.2) and recorded intracranially by placing silver ball electrodes on the exiting occipital roots that give rise to the sonic nerve (Bass and Baker 1990, 1991). The frequency and duration of the motor volley is predictive of, respectively, the rate and duration of contraction of the sonic muscles (Bass and Baker 1990, 1991). Sonic muscle contraction rate and duration establish, in turn, the fundamental frequency and duration of natural vocalizations (e.g., Cohen and Winn 1967); hence, the vocal motor volley is referred to as a "fictive vocalization".

While an expansive vocal motor network may only be present among batrachoidid fishes, comparable motor circuits likely exist in other sonic fishes. A series of comparative studies has revealed diversity in the pattern of organization of the sonic motor nucleus among teleosts. The organization of the sonic motor nucleus has been identified in 12 taxonomically diverse species of teleosts. The teleosts studied so far include Osteoglossomorphs (Osteoglossiformes [mormyrids]), Ostariophysans (Siluriformes [catfishes] and Characiformes [e.g., piranhas]), Paracanthopterygii (Batrachoidiformes [midshipman fish and toadfishes]), and Acanthopterygii (Scorpaeniformes [scorpaenids, cottids and triglids], Beryciformes [holocentrids], and Perciformes [osphronemids]) (Bass and Baker 1991; Ladich and Fine 1992, 1994; Ladich and Bass 1996, 1998, 2005; Yoshimoto et al. 1999; Carlson and Bass 2000).

Two major patterns of organization have so far been identified for the sonic motor nucleus (Fig. 8.6). The first is shared by batrachoidids (midshipman and toadfishes), most of the catfish species so far studied (ariids, mochokids, doradids) and characids (piranhas)—all have a motor nucleus located on or close to the midline that innervates swimbladder-associated muscles. The second pattern includes motor neurons that are positioned within a ventral motor column that innervates non-swimbladder–associated musculature (holocentrids, scorpaenids, cottids, triglids, osphronemids). Pimelodid catfishes (*P. blochi, P. pictus*) show both patterns—a midline nucleus associated with a swimbladder mechanism and a ventrolateral nucleus associated with movement of a pectoral spine.

FIGURE 8.6. Comparative organization of sonic (vocal) motor nucleus. Cladistic relationships among teleostean families with identified patterns of organization for the sonic motor nucleus (SMN). (**A**) Ostariophysine families. (**B**) Non-ostariophysine families. There are four patterns recognized: central lateral (SMNlc), ventrolateral motor column (SMNv), lateral (SMNl), and medial (SMNm). *The only mormyrid studied so far is not known to be sonic, and the SMNm is therefore putative. Systematics follow Nelson (2006). (Adapted from Ladich and Bass [1998, 2005] and Carlson and Bass [2000], with permission from S. Karger AG, Basel.)

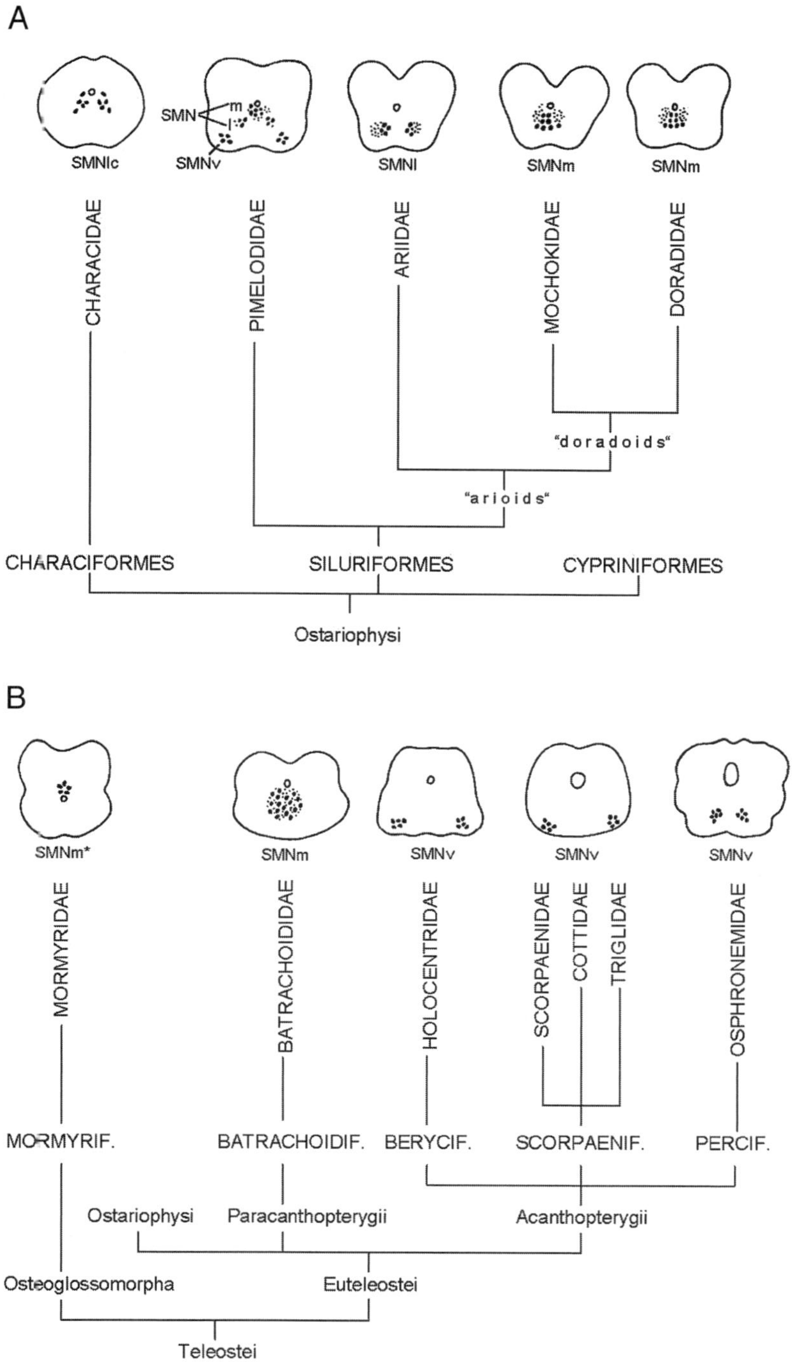
A
SMNlc
SMN
m
l
SMNv
SMNl
SMNm
SMNm
CHARACIDAE
PIMELODIDAE
ARIIDAE
MOCHOKIDAE
DORADIDAE
"doradoids"
"arioids"
CHARACIFORMES
SILURIFORMES
CYPRINIFORMES
Ostariophysi
B
SMNm*
SMNm
SMNv
SMNv
SMNv
MORMYRIDAE
BATRACHOIDIDAE
HOLOCENTRIDAE
SCORPAENIDAE
COTTIDAE
TRIGLIDAE
OSPHRONEMIDAE
MORMYRIF.
BATRACHOIDIF.
BERYCIF.
SCORPAENIF.
PERCIF.
Ostariophysi
Paracanthopterygii
Acanthopterygii
Osteoglossomorpha
Euteleostei
Teleostei

It is important to recognize that neuroanatomical patterns are not predictive of neurophysiological performance. Although batrachoidids (midshipman and toadfishes) and sculpin (the longhorn sculpin *Myoxocephalus scorpius* and the Pacific staghorn sculpin *Leptocottus armatus*) diverge neuroanatomically, they share a neurophysiological pattern of having synchronous firing of both sonic motor nuclei (Bass and Baker 1991) (Fig. 8.7). Although the northern searobin *Prionotus carolinus* (Scorpaeniformes, family Triglidae) shares a neuroanatomical pattern with the longhorn and Pacific staghorn sculpin (Scorpaeniformes, family Cottidae), they diverge neurophysiologically. The paired sonic motor nuclei fire out of phase with one another in searobins but in phase in sculpins (Bass and Baker 1991) (Fig. 8.7). The searobin's neurophysiological phenotype is correlated with an asymmetry in the morphology of the swimbladder (Tower 1908; Bass and Baker 1991) and is predictive of its alternate pattern of sonic muscle contraction during sound production (Connaughton 2004). This physiological pattern might yet be shared with another toadfish (the Lusitanian toadfish, *Halobatrachus didactylus*), which has an asymmetric swimbladder like that of searobins (Dos Santos et al. 2000). Assuming this toadfish species has a midline sonic motor nucleus like that of other batrachoidids, this would be yet another demonstration of how the gross pattern of organization of the SMN is not predictive of physiological performance in the vocal system of teleosts. This work also highlights the potential diversity, and thus the need for additional studies of the mechanisms of sound production among teleosts, especially with reference to an understanding of the premotor circuitry that leads to patterns of motor neuron activity (Bass and Baker 1990).

3.3 Hormonal Influences on Vocal Motor Patterning

A major advantage of the fictive vocalization preparation (see Section 3.2) is that it allows one to directly predict the behavioral relevance of central manipulations of the vocal motor system. This has been especially important in recent studies that have shown how neuropeptides and steroid hormones influence the operation of the vocal motor system in the plainfin midshipman fish that has two male reproductive morphs (Bass 1996). Type I males build nests in the intertidal zone and court females with their hum advertisement calls; type II males are about 50% smaller and attempt to steal fertilizations from type I males by either satellite-spawning form the periphery of a nest or sneak-spawning by entering a nest (Brantley and Bass 1994; Lee and Bass 2006; see also Lee and Bass (2004) for satellite- and sneak-spawning by small type I males). Type I males also produce agonistic grunts that are broadband, pulsatile signals 50–100 ms in duration (Fig. 8.1A) and growls that are multiharmonic, show amplitude and frequency modulations, and have durations on the order of hundreds of milliseconds (see Bass et al. 1999). Type II males and females are only known to make low-amplitude grunts in nonreproductive contexts (Brantley and Bass 1994). Motor neurons and pacemaker neurons have larger somata, dendrites, axon diameters, and neuromuscular junctions in type I males compared to type II males and

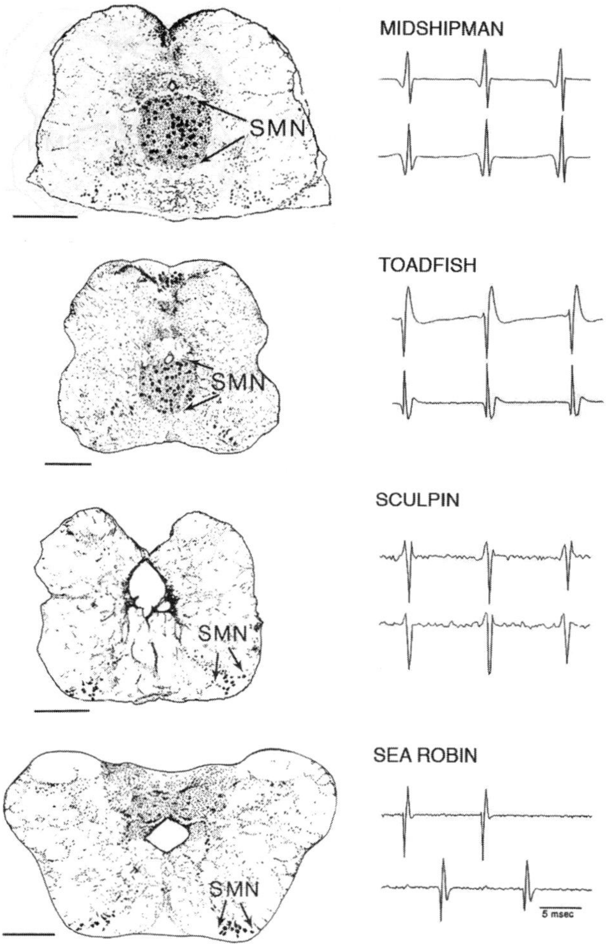

FIGURE 8.7. Divergent patterns of sonic motor nuclei (SMN) and neurophysiological activity. Shown to the left are line drawings of the SMN in several species of vocal fish. The bilaterally paired SMN are contiguous along the midline in batrachoidid fish (midshipman and toadfish) but are positioned ventrolaterally in scorpaenids (sculpins and searobins). Shown to the right are recordings of the motor volley (fictive vocalization) from a ventral occipital nerve root on the left (top trace) and right (bottom trace) sides of the hindbrain that were evoked by midbrain stimulation. Midshipman and toadfish exhibit a bilaterally synchronous motor volley. Sculpins also show a bilaterally synchronous pattern of activity, whereas their close relative the searobin shows asynchronous activity. The bar scales for the line drawings represent 500 μm; the time scale for the neurophysiological traces is indicated in the bottom trace. (Adapted from Bass and Baker [1991], with permission from S. Karger AG, Basel.)

females, which resemble each other (Bass 1996). Thus, the divergence in vocal behavior among reproductive morphs (type I males versus type II males and females) is paralleled by a divergence in vocal neuron traits.

Goodson and Bass (2000) investigated the influences of the closely related nine amino acid peptides arginine vasotocin (AVT) and isotocin, the teleost homologs of mammalian arginine vasopressin and oytocin, on vocal patterning in midshipman fish. These neuropeptides were chosen for study, in part, because forebrain and midbrain vocally active sites receive a dense input from AVT and isotocin-like immunoreactive neurons (Goodson et al. 2003). Injections of AVT, but not isotocin, into a vocally active site in the anterior hypothalamus (part of the fVAC, Fig. 8.2) in type I males can dramatically alter the vocal motor output of the pacemaker–motor circuit (mainly an increase in the duration of each vocalization) (Goodson and Bass, 2000). In contrast, isotocin, but not AVT, alters motor patterning in both type II males and females (mainly an increase in the number of vocalizations produced). Thus, type II males and females are convergent, but divergent from type I males, in yet another vocal neuron trait as they are in their vocal behaviors.

Neurophysiological studies have also demonstrated the influence of steroid hormones on fictive vocalization. Remage-Healey and Bass (2004) show that glucocorticoids (cortisol), 17β-estradiol, and 11-ketotestosterone (a nonaromatizable androgen found in teleosts) produce rapid (within 5 minutes) increases in the duration of fictive vocalizations in type I male midshipman fish (Fig. 8.8).

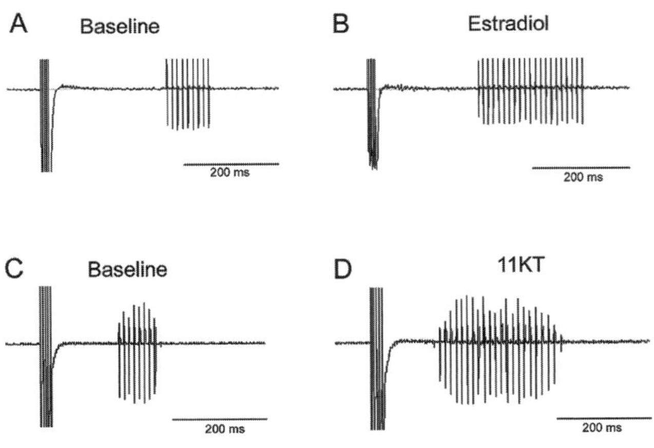

FIGURE 8.8. Oscillograms of fictive vocalizations (see Figure 8.7) evoked by midbrain electrical stimulation in type I male midshipman fish. The stimulus artifact appears on left side of each trace. Recordings shown are under baseline conditions (**A, C**) and either 15 minutes after 17-β-estradiol injection (**B**) or 120 minutes after androgen (11-ketotestosterone, 11KT) injection (**D**). There is a rapid change in burst duration in each experiment, but neither discharge frequency nor latency change following steroid administration. (Adapted from Remage Healey and Bass [2004]; copyright 2004 by the Society for Neuroscience.)

These various steroids have specific effects in terms of the longevity of the effect as well as its site specificity. Thus, the effects of 17β-estradiol, cortisol, and 11-ketotestosterone last, respectively, for 30 minutes, 60 minutes, and > 120 minutes following intramuscular (dorsal trunk) injection. Surgical isolation experiments further show that all of the neural circuitry necessary for a steroid effect on fictive vocalization is within the region containing the hindbrain–spinal pattern generator. However, steroid effects last only for 30 minutes in such a surgically isolated preparation. Thus, any steroid effects lasting beyond 30 minutes are apparently dependent on steroid action at the level of either the midbrain or the forebrain (see Remage-Healey and Bass 2004 for details). These results suggest an hierarchical pattern of descending vocal control such that the overall duration of a vocalization may depend on midbrain inputs to the hindbrain–spinal pattern generator. The most recent studies show intrasexual differences in steroid-dependent plasticity of fictive calling. While 11-ketotestosterone is the principal circulating androgen in type I males, testosterone is the main androgen in type II sneaker males and females (Brantley et al. 1993; Knapp et al. 1999a; Sisneros et al. 2004b). Consistent with these differences, testosterone but not 11-ketotestosterone induces rapid increases in fictive call duration in type II males and females (Remage-Healey and Bass 2007); the inverse pattern is observed for type I males (see above and Remage-Healey and Bass 2004).

Studies in the closely related Gulf toadfish, *O. beta*, also show steroid influences on vocal motor patterning including 11-ketotestosterone-specific effects alone in males (Remage-Healey and Bass 2006b). Behavioral field studies with Gulf toadfish also demonstrate the behavioral relevance of this rapid modulation of the vocal motor system by steroids. Male toadfish show rapid (10–20 min) and simultaneous increases in plasma levels of 11-ketotestosterone and vocal parameters (call duration and rate) when presented with playbacks of boatwhistle advertisement calls (Remage-Healey and Bass 2005). More recent studies show that feeding 11-ketotestosterone implanted scallops to nesting, calling male toadfish induces fast (within 10 min) increases in both call production and plasma 11-ketotestosterone levels (Remage-Healey and Bass 2006b). Thus, consistent with the neurophysiological studies, elevated androgens can play a causal role in shifting calling behavior. We can further expect interactions in both midshipman and toadfish between steroids and neuropeptides in the modulation of call parameters.

4. Ontogeny of Sound Production and Sound Detection

It is expected that sound production is widespread among juvenile fishes because the competition for food and space may occur regardless of reproductive stage. Immature skunk loach (*B. horae*, family Cobitidae), tigerperch (*Terapon jarbua*, family Teraponidae), and mormyrids (*Gnathonemus petersii*, family Mormyridae) all vocalize (Schneider 1964; Rigley and Marshall 1973; Valinsky and Rigley 1981). The ontogenetic development of vocalizations,

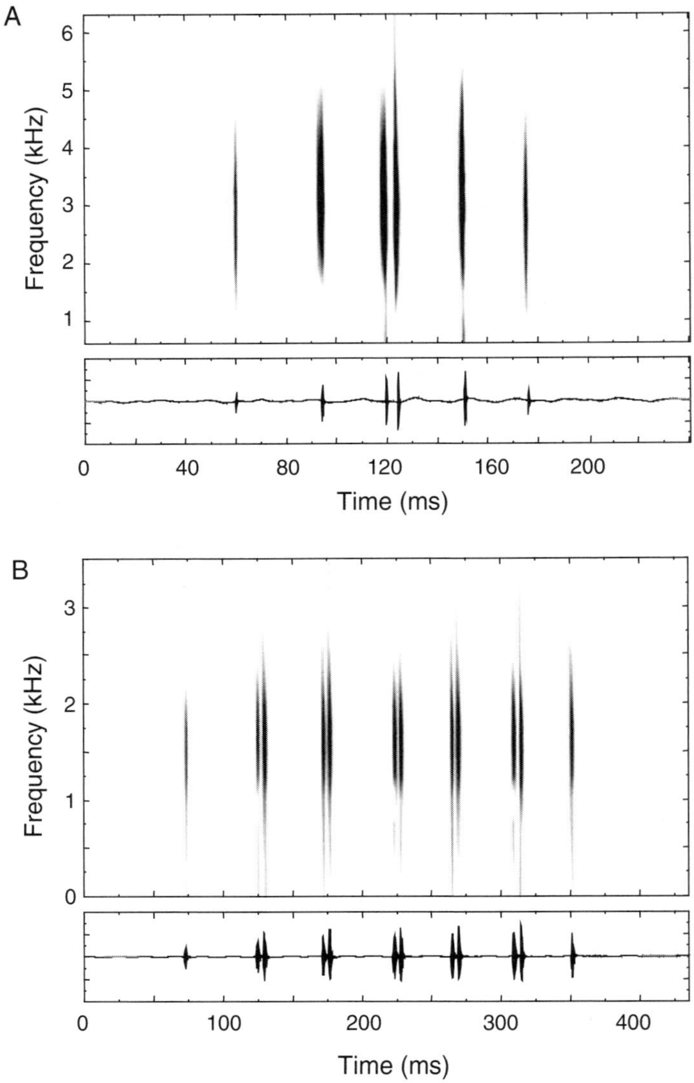

FIGURE 8.9. Development of vocalizations in *Trichopsis vittata*. Sonagrams and oscillograms of croaking sounds of a 0.11-g (**A**) and a 0.78-g (**B**) croaking gourami. Note the increase in the number of double pulses, pulse periods, and the decrease in the main energy of sounds. Also note the differences in axes-ranges in both sonograms. Sampling frequency is 16 kHz, filter bandwidth is 250 and 300 Hz for A and B respectively. (From Wysocki and Ladich [2001]; with permission of Springer Science + Business Media.)

agonistic behavior, and hearing has been investigated in only one teleost species, the croaking gourami *T. vittata* (Henglmüller and Ladich 1999; Wysocki and Ladich 2001). Agonistic behavior is first accompanied by sound emission at the age of 8 weeks. The character of sounds changes widely during ontogeny, probably due to the development of sound-generating structures and to a larger body size. Initially, croaking sounds are formed mainly of a series of single pulses, each pulse produced by one pectoral fin (Fig. 8.9A). Later, single pulses give way to an increasing number of double pulses as the pulse period and number of pulses increased (Fig. 8.9B). The dominant frequency also decreases while the sound intensity of croaks increases (Fig. 8.9A, B; note different x and y axes in A and B). Wysocki and Ladich (2001) also showed that ontogenetic changes in sound production are accompanied by changes in auditory sensitivity; the most sensitive frequency shifted from 2.5 kHz to 1.5 kHz as thresholds decreased by 14 dB. A comparison between audiograms and sound power spectra revealed that juveniles are initially unable to detect conspecific sounds; auditory sensitivity develops before the ability to vocalize while vocalizations occur before the ability to communicate (Wysocki and Ladich 2001).

Most recently, Sisneros and Bass (2005) identified ontogenetic changes in hearing mechanisms in 4- to 12-month-old juvenile plainfin midshipman fish. This was the first in vivo study of age-related changes in the encoding properties of single, peripheral auditory neurons for any fish species. They show that while the resting discharge rate and auditory threshold sensitivity of saccular afferents increased with age/size, the degree of temporal encoding of frequency was similar to that of nonreproductive females and males (McKibben and Bass 1999; Sisneros and Bass 2003). The results suggest that the saccular afferents of subadults are best adapted to encode low-frequency components (≤ 100 Hz) in vocalizations and that the shifts in temporal encoding observed among adults (Fig. 8.4) is indeed a trait of only reproductively active individuals.

5. Summary

As can be seen in this brief overview, there is a wide range of exciting new studies of the neural and behavioral mechanisms of acoustic communication among teleost fishes. Studies of neuroendocrine, vocal, and auditory mechanisms hold great promise for showing how these mechanisms interact with one another to contribute to an individual species' performance of a reproductive tactic. The remarkable diversity of reproductive and sonic behaviors among teleosts provides tremendous opportunities for those who want to work at the interface of neural and non-neural mechanisms. These studies will also likely show that such mechanisms are common to all vertebrates given the conserved pattern of the organization of the neuroendocrine, vocal, and auditory systems.

There is clearly a need for more behavioral and neural studies of the sensitivity of sonic fishes to the temporal parameters of their acoustic signals. Many of these traits are also likely to be shared with other vertebrates. Ontogenetic studies of vocal and acoustic behaviors are just beginning and they need to be coupled with developmental studies of vocal and auditory mechanisms (e.g., see Bass et al. 1996; Knapp et al. 1999b; Sisneros and Bass 2005). Last, but not least, there remains a continuing need for studies of the natural history and behavioral ecology of vocal fishes, for it is herein that we find our inspiration and our guide to behaviorally relevant, laboratory-based studies. Consider the remarkable number of sound producing species that Fish and Mowbray (1970) documented in their volume "Sounds of Western North Atlantic Fishes." There is plenty of room here for investigators just embarking on a career in mechanisms of vocal communication among teleost fishes.

Acknowledgments. The authors thank Terri Natoli for help with organizing the reference list. This work was supported in part by NIH-NIDCD (DC00092) and NSF (IBN9987241, IOB0516748) grants to AHB and the Austrian Science Fund (FWF 17263) grant to FL. Additional thanks are expressed to the editors for the invitation to write this chapter and for their helpful comments on the manuscript.

References

Amorim MCP (2006) Diversity of sound production in fish. In: Ladich F, Collin SP, Moller P, Kapoor BG (eds) Communication in Fishes. Enfield, NH: Science Publishers, pp. 69–103.

Bass AH (1996) Shaping brain sexuality. Am Sci 84:352–363.

Bass AH, Baker R (1990) Sexual dimorphisms in the vocal control system of a teleost fish: morphology of physiologically identified cells. J Neurobiol 21:1155–1168.

Bass AH, Baker R (1991) Evolution of homologous vocal control traits. Brain Behav Evol 38:240–254.

Bass AH, Baker R (1997) Phenotypic specification of hindbrain rhombomeres and the origins of rhythmic circuits in vertebrates. Brain Behav Evol 50:3–16.

Bass AH, Clark CW (2003) The physical acoustics of underwater sound communication. In: Fay RR, Popper A (eds) Acoustic Communication. New York: Springer, pp. 15–64.

Bass AH, Lu Z (2007) Neural and behavioral mechanisms of audition. In: Hara T, Zielinski B (eds) Fish Physiology, Sensory Systems Neuroscience, Vol. 25. New York: Elsevier, pp. 377–410.

Bass AH, Marchaterre MA (1989) Sound-generating (sonic) motor system in a teleost fish (*Porichthys notatus*): sexual polymorphism in the ultrastructure of myofibrils. J Comp Neurol 286:141–153.

Bass AH, McKibben JR (2003) Neural mechanisms and behaviors for acoustic communication in teleost fish. Prog Neurobiol 69:1–26.

Bass AH, Marchaterre MA, Baker R (1994) Vocal-acoustic pathways in a teleost fish. J Neurosci 14:4025–4039.

Bass AH, Horvath BJ, Brothers E (1996) Nonsequential developmental trajectories lead to dimorphic vocal circuitry for males with alternative reproductive tactics. J Neurobiol 30:493–504.

Bass AH, Bodnar DA, Marchaterre MA (1999) Complementary explanations for existing phenotypes in an acoustic communication system. In: Hauser M, Konishi M (eds) Neural Mechanisms of Communication. Cambridge: MIT Press, pp. 493–514.

Bass AH, Bodnar DA, Marchaterre MA (2000) Midbrain acoustic circuitry in a vocalizing fish. J Comp Neurol 419:505–531.

Bass AH, Bodnar DA, Marchaterre MA (2001) Acoustic nuclei in the medulla and midbrain of the vocalizing gulf toadfish, *Opsanus beta*. Brain Behav Evol 57:63–79.

Bass AH, Gilland E, Baker R (2005a) Early development of teleost vocal circuitry. Program No. 1001.5. 2005 Abstract Viewer/Itinerary Planner. Washington, DC: Society for Neuroscience, 2005. Online.

Bass AH, Rose GJ, Pritz MB (2005b) Auditory midbrain of fish, amphibians and reptiles: model systems for understanding auditory function. In: Winer JA, Schreiner CE (eds) The Inferior Colliculus. New York: Springer, pp. 459–492.

Bodnar DA, Bass AH (1997) Temporal coding of concurrent acoustic signals in auditory midbrain. J Neurosci 17:7553–7564.

Bodnar DA, Bass AH (1999) A midbrain combinatorial code for temporal and spectral information in concurrent acoustic signals. J Neurophysiol 81:552–563.

Bodnar DA, Bass AH (2001a) The coding of concurrent vocal signals by the auditory midbrain: the effects of stimulus level and depth of modulation. J Acoust Soc Am 109:809–825.

Bodnar DA, Bass AH (2001b) Coding of concurrent signals by the auditory midbrain: effects of duration. J Comp Physiol A 187:381–391.

Bradbury JW, Vehrencamp SL (1998) Principles of Animal Communication. Sunderland, MA: Sinauer Associates.

Brantley RK, Bass AH (1994) Alternative male spawning tactics and acoustic signalling in the plainfin midshipman fish, *Porichthys notatus*. Ethology 96:213–232.

Brantley RK, Wingfield JC, Bass AH (1993) Hormonal bases for male teleost dimorphisms: sex steroid levels in *Porichthys notatus*, a fish with alternative reproductive tactics. Horm Behav 27:332–347.

Capranica RR (1992) The untuning of the tuning curve: is it time? Semin Neurosci 4:401–408.

Carlson B, Bass AH (2000) Sonic/vocal motor pathways in squirrelfish (Teleostei, Holocentridae). Brain Behav Evol 56:14–28.

Cohen MJ, Winn HE (1967) Electrophysiological observations on hearing and sound production in the fish, *Porichthys notatus*. J Exp Zool 165:355–370.

Connaughton MA (2004) Sound generation in the searobin (*Prionotus carolinus*), a fish with alternate sonic muscle contraction. J Exp Biol 207:1643–1654.

Crawford JD (1997) Feature-detecting auditory neurons in the brain of a sound-producing fish. J Comp Physiol A 180:439–450.

Crawford JD, Cook AP, Heberlein AS (1997) Bioacoustic behavior of African fishes (Mormyridae): potential cues for species and individual recognition in *Pollimyrus*. J Acoust Soc Am 102:1200–1212.

Dos Santos ME, Modesto T, Matos RJ, Grober MS, Oliveira RF, Canario A (2000) Sound production by the Lusitanian toadfish, *Halobatrachus didactylus*. Bioacoustics 10:309–321.

Edds-Walton PL, Fay RR (2003) Directional selectivity and frequency tuning of midbrain cells in the oyster toadfish, *Opsanus tau*. J Comp Physiol A 189:527–543.

Fay RR (1978) Phase-locking in goldfish saccular nerve fibers accounts for frequency discrimination capacities. Nature 275:320–322.

Fay RR, Edds-Walton PL (1997a) Directional response properties of saccular afferents of the toadfish, *Opsanus tau*. Hear Res 111:1–21.

Fay RR, Edds-Walton PL (1997b) Diversity in frequency response properties of saccular afferents of the toadfish, *Opsanus tau*. Hear Res 113:235–246.

Fine ML (1997) Endocrinology of sound production in fishes. Mar Fresh Behav Physiol 29:23–45.

Fine ML, Ladich F (2003) Sound production, spine locking and related adaptations. In: Arratia G, Kapoor BG, Chardon M, Diogo R (eds) Catfishes, Vol. 1. Enfield, NH: Science Publishers, pp. 249–290.
Fine ML, Lenhardt ML (1983) Shallow-water propagation of the toadfish mating call. Comp Biochem Physiol 76A:225–231.
Fine ML, Winn HE, Olla BL (1977) Communication in fishes. In: Sebeok TA (ed) How Animals Communicate. Bloomington: Indiana University Press, pp. 472–518.
Fine ML, Freil JP, McElroy D, King CB, Loesser KE, Newton S (1997) Pectoral spine locking and sound production in the channel catfish (*Ictalurus punctatus*). Copeia 1997:777–790.
Fish JF (1972) The effect of sound playback on the toadfish. In: Winn HE, Olla BL (eds) Behavior of Marine Animals. New York: Plenum, pp. 386–434.
Fish MP, Mowbray WH (1970) Sounds of Western North Atlantic Fishes. Baltimore: The Johns Hopkins Press.
Forlano PM, Deitcher DL, Myers DA, Bass AH (2001) Neuroanatomical basis for high aromatase levels in teleost fish: aromatase enzyme and mRNA expression identify glia as source. J Neurosci 21:8943–8955.
Forlano PM, Deitcher DL, Bass AH (2005) Distribution of estrogen receptor alpha mRNA in the brain and inner ear of a vocal fish with comparisons to sites of aromatase expression. J Comp Neurol 483:91–113.
Forlano PM, Schlinger BA, Bass AH (2006) Brain aromatase: new lessons from non-mammalian vertebrates. Front Neuroendocrinol 27:247–274.
Goodson JL, Bass AH (2000) Forebrain peptide modulation of sexually polymorphic vocal motor circuitry. Nature 403:769–772.
Goodson JL, Bass AH (2002) Vocal-acoustic circuitry and descending vocal motor pathways in teleost fish: convergence with terrestrial vertebrates reveals conserved traits. J Comp Neurol 448:298–321.
Goodson JL, Evans AK, Bass AH (2003) Isotocin distributions in vocal fish: relationship to vasotocin and vocal-acoustic circuitry. J Comp Neurol 462:1–14.
Haggard M, Gaston JB (1978) Changes in auditory perception in the menstrual cycle. Br J Audiol 12:105–118.
Henglmüller SM, Ladich F (1999) Development of agonistic behavior and vocalization in croaking gouramis. J Fish Biol 54:380–395.
Heyd A, Pfeiffer W (2000) Über die Lauterzeugung der Welse (Siluroidei, Ostariophysi, Teleostei) und ihren Zusammenhang mit der Phylogenese und der Schreckreaktion. Rev Suisse Zool 107:165–211.
Kenyon TN, Ladich F, Yan HY (1998) A comparative study of hearing ability in fishes: the auditory brainstem response approach. J Comp Physiol (A) 182:307–318.
Kittelberger JM, Land BR, Bass AH (2006) The midbrain periaqueductal gray and vocal patterning in a teleost fish. J Neurophysiol 96:71–85.
Knapp R, Wingfield JC, Bass AH (1999a) Steroid hormones and paternal care in the midshipman fish (*Porichthys notatus*). Horm Behav 35:81–89.
Knapp R, Marchaterre MA, Bass AH (1999b) Early development of a sexually dimorphic vocal motor system in a vocal fish. J Neurobiol 38:475–490.
Kozloski J, Crawford JD (2000) Transformations of an auditory temporal code in the medulla of a sound-producing fish. J Neurosci 20:2400–2408.
Kratochvil H (1978) Der Bau des Lautapparates vom Knurrenden Gurami (*Trichopsis vittatus* Cuvier & Valenciennes) (Anabantidae, Belontiidae). Zoomorphologie 91: 91–99.

Ladich F (1997a) Comparative analysis of swimbladder (drumming) and pectoral (stridulation) sounds in three families of catfishes. Bioacoustics 8:185–208.

Ladich F (1997b) Agonistic behaviour and significance of sounds in vocalizing fish. Mar Fresh Behav Physiol 29:87–108.

Ladich F (2001) Sound-generating and -detecting motor system in catfish: design of swimbladder muscles in doradids and pimelodids. Anat Rec 263:297–306.

Ladich F (2004) Sound production and acoustic communication. In: von der Emde G, Mogdans J, Kapoor BG (eds) The Senses of Fishes. New Delhi: Narosa, pp. 210–230.

Ladich F, Bass AH (1996) Sonic/vocal-acousticolateralis pathways in teleost fishes: a transneuronal biocytin study in mochokid catfish. J Comp Neurol 374:493–505.

Ladich F, Bass AH (1998) Sonic/vocal motor pathways in catfishes: comparisons with other teleosts. Brain Behav Evol 51:315–330.

Ladich F, Bass AH (2003a) Underwater sound generation and acoustic reception in fishes with some notes on frogs. In: Collin SP, Marshall NJ (eds) Sensory Processing in Aquatic Environments. New York: Springer, pp. 173–193.

Ladich F, Bass AH (2003b) Audition. In: Arratia G, Kapoor BG, Chardon M, Diogo R (eds) Catfishes. Enfield, NH: Science Publishers, pp. 701–730.

Ladich F, Bass AH (2005) Sonic motor pathways in piranhas with a reassessment of phylogenetic patterns of sonic mechanisms among teleosts. Brain Behav Evol 66:167–176.

Ladich F, Fine ML (1992) Localization of pectoral fin motoneurons (sonic and hovering) in the croaking gourami *Trichopsis vittatus*. Brain Behav Evol 39:1–7.

Ladich F, Fine ML (1994) Localization of swimbladder and pectoral motoneurons involved in sound production in pimelodid catfish. Brain Behav Evol 44:86–100.

Ladich F, Fine M (2006) Sound generating mechanisms in fishes: A unique diversity in vertebrates. In: Ladich F, Collin SP, Moller P, Kapoor BG (eds) Communication in Fishes. Enfield, NH: Science Publishers, pp. 3–41.

Ladich F, Myrberg AA (2006) Agonistic behaviour and acoustic communication. In: Ladich F, Collin SP, Moller P, Kapoor BG (eds) Communication in Fishes. Enfield, NH: Science Publishers, pp. 117–145.

Ladich F, Popper AN (2004) Parallel evolution in fish hearing organs. In: Manley GA, Popper AN, Fay RR (eds) Evolution of the Vertebrate Auditory System. New York: Springer, pp. 95–127.

Lee JSF, Bass AH (2004) Does the "exaggerated" morphology preclude plasticity to cuckoldry? A test in the midshipman fish, *Porichthys notatus*. Naturwissenschaften 91:338–341.

Lee JSF, Bass AH (2006) Dimorphic male midshipman fish: reduced sexual selection or sexual selection for reduced characters? Behav Ecol 17:670–675.

Lewis MK, Nahirney PC, Chen V, Adikari BB, Wright J, Bass AH, Wang K (2003) Concentric intermediate filament lattice links to specialized Z band-junctional complexes in sonic muscle fiber of the type I male midshipman fish. J Struct Biol 143:56–71.

Lu Z (2004) Neural mechanisms of hearing in fishes. In: von der Emde G, Mogdans J, Kapoor BG (eds) The Senses of Fish: Adaptations for the Reception of Natural Stimuli. New Delhi, India: Narosa, pp. 210–230.

Mann D (2006) Propagation of fish sounds. In: Ladich F, Collin SP, Moller P, Kapoor BG (eds) Communication in Fishes. Enfield, NH: Science Publishers, pp. 103–116.

Marvit P, Crawford JD (2000) Auditory discrimination in a sound-producing electric fish (*Pollimyrus*): tone frequency and click-rate difference detection. J Acoust Soc Am 108:1819–1825.

McCormick CA (1999) Anatomy of the central auditory pathways of fish and amphibians. In: Popper A, Fay RR (eds) Comparative Hearing: Fish and Amphibians. New York: Springer, pp. 155–217.
McFadden D (1998) Sex differences in the auditory system. Dev Neuropsychol 14: 261–298.
McKibben JR, Bass AH (1998) Behavioral assessment of acoustic parameters relevant to signal recognition and preference in a vocal fish. J Acoust Soc Am 104:3520–3533.
McKibben JR, Bass AH (1999) Peripheral encoding of behaviorally relevant acoustic signals in a vocal fish: single tones. J Comp Physiol A 184:563–576.
McKibben JR, Bass AH (2001a) Effects of temporal envelope modulation on acoustic signal recognition in a vocal fish. J Acoust Soc Am 109:2934–2943.
McKibben JR, Bass AH (2001b) Peripheral encoding of behaviorally relevant acoustic signals in a vocal fish: harmonic and beat stimuli. J Comp Physiol A 187:271–285.
Myrberg AA Jr (1981) Sound communication and interception in fishes. In: Tavolga WN, Popper AN, Fay RR (eds) Hearing and Sound Communication in Fishes. New York: Springer-Verlag, pp. 395–425.
Myrberg AA, Lugli M (2006) Reproductive behavior and acoustical interactions. In: Ladich F, Collin SP, Moller P, Kapoor BG (eds) Communication in Fishes. Enfield, NH: Science Publishers, pp. 147–174.
Myrberg AA Jr, Spires JY (1972) Sound discrimination by the bicolor damselfish, *Eupomacentrus partitus* (Poey) (Pisces: Pomacentridae). J Exp Biol 57:727–735.
Nahirney PC, Fischman DA, Wang K (2006) Myosin flares and actin leptomeres as myofibril assembly/disassembly intermediates in sonic muscle fibers. Cell Tissue Res 324:127–138.
Nelson JS (2006) Fishes of the World, 4th ed. Hoboken, NJ: John Wiley & Sons.
Pappas GD, Bennett MVL (1966) Specialized junctions involved in electrical transmission between neurons. Ann NY Acad Sci 137:495–508.
Parmentier E, Vandewalle P, Lagardere JP (2003) Sound-producing mechanisms and recordings in *Carapini* species (Teleostei, Pisces). J Comp Physiol A 189:283–292.
Popper AN, Fay RR (1999) The auditory periphery in fishes. In: Popper A, Fay RR (eds) Comparative Hearing: Fish and Amphibians, . New York: Springer-Verlag, pp. 43–100.
Remage-Healey L, Bass AH (2004) Rapid, hierarchical modulation of vocal patterning by steroid hormones. J Neurosci 24:5892–5900.
Remage-Healey L, Bass AH (2005) Simultaneous, rapid, elevations in steroid hormones and vocal signaling during playback challenge: a field experiment in Gulf toadfish. Horm Behav 47:297–305.
Remage-Healey L, Bass AH (2006a) A rapid, neuromodulatory role for steroid hormones in the control of reproductive behavior. Brain Res 1126:27–35.
Remage-Healey L, Bass AH (2006b) From social behavior to neurons: rapid modulation of advertisement calling and vocal pattern generators by steroid hormones. Horm Behav 50:432–441.
Remage-Healey L, Bass AH (2007) Plasticity in brain sexuality is revealed by the rapid actions of steroid hormones. J Neurosci 27:1114–1122.
Rice AN, Lobel PS (2002) Enzyme activities of pharyngeal jaw musculature in the cichlid *Tramitichromis intermedius*: implications for sound production in cichlid fishes. J Exp Biol 205:3519–3523.
Rigley L, Marshall JA (1973) Sound production by the elephant-nose fish *Gnathonemus petersi* (Pisces, Mormyridae). Copeia 1973:134–135.

Rome LC (2006) Design and function of superfast muscles: new insights into the physiology of skeletal muscle. Annu Rev Physiol 68:193–221.

Schlinger BA, Greco C, Bass AH (1999) Aromatase activity in hindbrain vocal control region: divergence between "singing" and "sneaking" males. Proc Roy Soc Lond B 266:131–136.

Schneider H (1964) Physiologische und morphologische Untersuchungen zur Bioakustik der Tigerfische (Pisces, Theraponidae). Z Vergl Physiol 47:493–558.

Sisneros JA, Bass AH (2003) Seasonal plasticity of peripheral auditory frequency sensitivity. J Neurosci 23:1049–1058.

Sisneros JA, Bass AH (2005) Ontogenetic changes in the response properties of individual, primary auditory afferents in the vocal plainfin midshipman fish, *Porichthys notatus* Girard. J Exp Biol 208:3121–3131.

Sisneros JA, Forlano PM, Deitcher DL, Bass AH (2004a) Steroid-dependent plasticity of frequency sensitivity in the adult auditory periphery. Science 305:404–407.

Sisneros JA, Forlano PM, Knapp R, Bass AH (2004b) Seasonal variation of steroid hormone levels in an intertidal-nesting fish, the vocal plainfin midshipman. Gen Comp Endocrinol 136:101–116.

Stadler JM (2002) Evidence for a hydrodynamic mechanism of sound production by courting males of the notchtongue goby, *Bathygobius curacao* (Metzlaar). Bioacoustics 13:145–152.

Stenberg AE, Wang H, Sahlin L, Hultcranz M (1999) Mapping of estrogen receptors α and β in the inner ear of mouse and rat. Hear Res 136:29–34.

Stenberg AE, Wang H, Fish J, III, Schrott-Fischer A, Sahlin L, Hultcrantz M (2001) Estrogen receptors in the normal adult and developing human inner ear and in Turner's syndrome. Hear Res 157:87–92.

Suzuki A, Kozloski J, Crawford JD (2002) Temporal encoding for auditory computation: physiology of primary afferent neurons in sound-producing fish. J Neurosci 22: 6290–6301.

Tower R (1908) The production of sound in drumfishes, the sea-robin and the toadfish. Ann NY Acad Sci 18:149–180.

Valinsky W, Rigley L (1981) Function of sound production by the skunk loach *Botia horae* (Pisces, Cobitidae). Z Tierpsychol 55:161–172.

Vester HJ, Folkow LP, Blix AS (2004) Click sounds produced by cod (*Gadus morhua*). J Acoust Soc Am 115:914–919.

Wahlberg MH, Westerberg H (2003) Sounds produced by herring (*Clupea harengus*) bubble release. Aquat Living Resourc 16:271–275.

Walsh PW, Mommsen TP, Bass AH (1995) Biochemical and molecular aspects of singing in batrachoidid fishes. In: Hochachka PW, Mommsen TP (eds) Biochemistry and Molecular Biology of Fishes, Vol. 4: Metabolic Biochemistry. Amsterdam: Elsevier, pp. 279–289.

Weeg MS, Fay RR, Bass AH (2002) Directionality and frequency tuning of primary saccular afferents of a vocal fish, the plainfin midshipman (*Porichthys notatus*). J Comp Physiol A 188:631–641.

Wilson B, Batty RS, Dill LM (2004) Pacific and Atlantic herring produce bursts pulse sounds. Proc Roy Soc Lond B (Suppl) 271:95–97.

Winn HE (1972) Acoustic discrimination by the toadfish with comments on signal systems. In: Winn HE, Olla BL (eds) Behavior of Marine Animals. New York: Plenum, pp. 361–385.

Wysocki LE (2006) Detection of communication sounds. In: Ladich F, Collin SP, Moller P, Kapoor BG (eds) Communication in Fishes. Enfield, NH: Science Publishers, pp. 175–203.

Wysocki LE, Ladich F (2001) The ontogenetic development of auditory sensitivity, vocalization and acoustic communication in the labyrinth fish, *Trichopsis vittata*. J Comp Physiol A 187:177–187.

Wysocki LE, Ladich F (2002) Can fishes resolve temporal characteristics of sounds? New insights using auditory brainstem responses. Hear Res 169:36–46.

Wysocki LE, Ladich F (2003) The representation of conspecific sounds in the auditory brainstem of teleost fishes. J Exp Biol 206:2229–2240.

Yoshimoto M, Kikuchi K, Yamamoto N, Somiya H, Ito H (1999) Sonic motor nucleus and its connections with octaval and lateral line nuclei of the medulla in a rockfish, *Sebastiscus marmoratus*. Brain Behav Evol 54:183–204.

9 Active and Passive Acoustics to Locate and Study Fish

David A. Mann, Anthony D. Hawkins, and J. Michael Jech

1. Introduction

Two important goals in studying the biology of fishes are to detect and enumerate the fish and to define where the fish are to be found. Locating and counting fish is difficult, but defining and mapping a fish's habitat can be even more daunting. A fish's habitat is the physical, chemical, geological and biological environment in which it resides or migrates through and includes the pelagic (open water), benthic (on or in the sea floor), and demersal (on or near the sea floor) realms. With the continuing loss of estuarine and coastal habitats it is especially critical to seek out the waters and substrates that are necessary as spawning, nursery and feeding areas for fishes. In the United States, the Magnuson-Stevens Fishery Conservation and Management Act, Public Law 94-265, as amended through October 11, 1996 calls for direct action to stop or reverse the loss of fish habitat and requires the identification of "essential fish habitat" (Section 305 of the Act). In a wider context, the wish to promote conservation through the establishment of marine protected areas also requires the identification of habitats of managed, threatened, and endangered species.

Investigating the distribution of fish is especially difficult because fish can rarely be seen and counted underwater. Fisheries trawl or net surveys can provide an overall picture of fish distribution, but are destructive of the species being surveyed. One of the greatest challenges to the study of fish populations is the ability to collect data over large spatial scales and to study behavior for long periods of time, without intruding upon the lives of these animals.

Two uses of acoustics have been developed for studying fish populations and behavior. Active acoustics uses sound generated actively by transducers and the acoustic scattering properties of fish to image individual fishes and populations of fishes. Passive acoustics relies on listening to the sounds produced by fishes with a hydrophone to infer their distribution and behavior. For passive acoustics to be useful a fish must make a sound, thus this technique is limited to species that produce sounds and to the times and places where they produce them. These techniques have typically been used independently, depending on the situation and goals of the study. This chapter reviews each of these technologies and shows how they could be used synergistically to understand both the distribution of soniferous species and to define important details of the spawning habitat of fishes.

2. Active Acoustics

2.1 Introduction

The use of sound to explore the aquatic environment is a relatively young science. Although Colladon and Sturm first measured the speed of sound in water in Lake Geneva in 1826 (Medwin and Clay 1998), it was not until the sinking of the *Titanic* and the need to locate submarines during World War I that sound was used to locate objects and measure depths in the ocean. While using these early depth sounders, "false" echoes in the water column were often observed. Rallier du Baty (1927) was one of the first to attribute these echoes to Atlantic cod (*Gadus morhua*), and Kimura (1929) performed experiments that confirmed fish could be responsible for these echoes. Commercial fishers began using echo sounders to locate many species of fish (e.g., Sund 1935; Balls 1948), and the use of echo sounders revolutionized commercial fishing. The field of fisheries acoustics has its roots in this history, but it was not until improvements in computer and electronic technology took place that the quantitative use of active acoustics for scientific and management applications became possible.

Active acoustics differs from passive acoustics in that a pulse of sound is generated by a transducer, and either the same transducer (monostatic) or one or more receivers (bistatic) are used to "listen" for echoes. In fisheries acoustics, the most common configuration is to use the same transducer for transmit and receive. In fisheries for bottom dwelling species, downward-facing transducers are usually installed on the vessel's hull for permanent use. In the fisheries for pelagic species, transducers are mounted so that they can be pointed in different directions, or even scanned back and forth. In some cases, the transducers may be retractable into the hull. Because the transducers are relatively small (from a few centimeters to less than a meter in diameter) a variety of deployment configurations, such as pole mounts or towed bodies, are used. Transducers convert electrical energy, most commonly a single-frequency sinusoidal signal (the acoustic frequency (f [Hz]) of the echo sounder), to acoustical energy (this transmission is often called a "ping"), and then acoustical energy returned from objects in the water (echoes) back into electrical energy (Fig. 9.1). Transducers use multiple ceramic or piezoelectric elements to create a directional acoustic beam, which is usually conical in shape, although elliptical beam patterns are used in shallow water environments, such as rivers. Similar to a flashlight where objects in the middle of the beam appear brighter than objects on the edges, echo amplitudes are larger on the acoustic axis than on the edges (Urick 1983). This acoustic sensitivity relative to the distance off-axis is called the transducers beam pattern. The edge of the beam is defined as the angular distance between

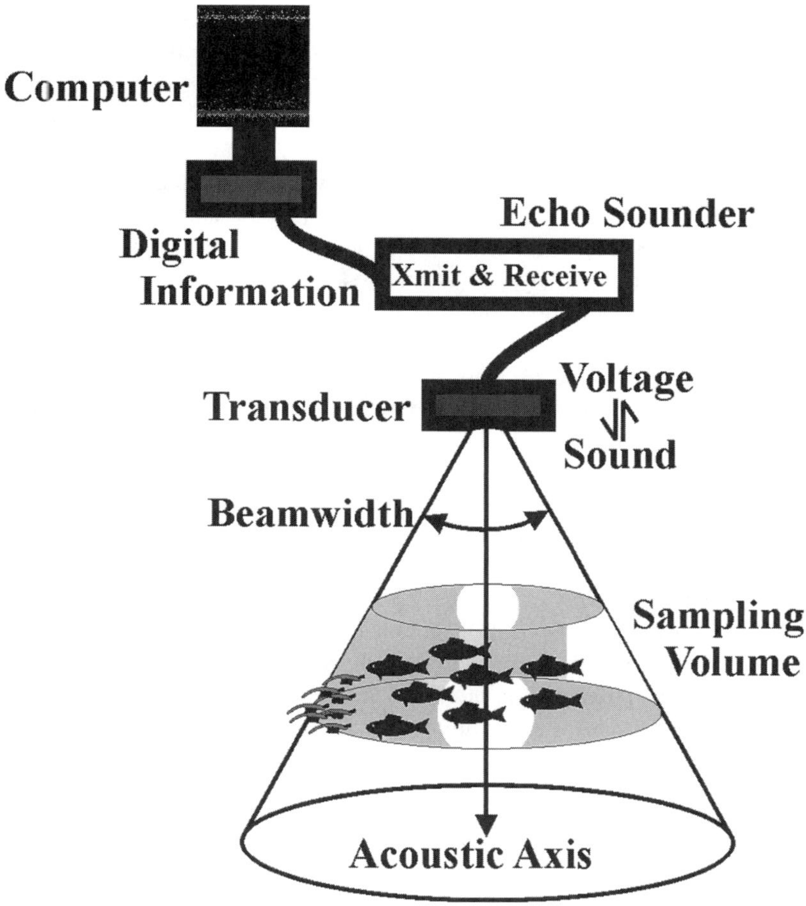

FIGURE 9.1. Schematic cartoon showing the major components of a scientific echo sounder and acoustic beam.

the axis and where the echo intensity is reduced by one-half, and twice this distance is the beam width (Medwin and Clay 1998).

Two basic measurements are derived from acoustic data: backscatter from individual targets, and volume backscatter. When organisms are dispersed, it is possible to obtain echoes from individuals. In this case, the method of echo counting can be used to derive a numeric density [no. m^{-3}] estimate (Trout et al. 1952), as well as measure the acoustic backscattering cross-sectional area (σ_{bs} [m^2]) and target strength ($TS = 10 \log_{10}(\sigma_{bs})$ [dB]) of the individuals (Medwin and Clay 1998). Abundance is estimated by multiplying the numeric density by the volume of water in the survey area. Most pelagic fish aggregate, making echo counting impossible. In this case, echo integration is used to derive density estimates (Johannesson and Mitson 1983). Echo integration is essentially the summation of the echoes within a sampling volume (volume backscatter (s_v [m^2 m^{-3}])). For most aggregations, this summation is linearly proportional to the numeric density of the organisms (Foote 1983). When the packing density [no. m^{-3}]) of the organisms is high, nonlinear effects can occur (Toresen 1991; Furusawa et al. 1992; Alvarez and Ye 1999), and other methods are required to estimate numeric density (e.g., Zhao and Ona 2003). Volume backscatter can be vertically integrated and horizontally averaged to obtain area backscatter (s_a [m^2 m^{-2}]), which is proportional to the area density [no. m^{-2}] of the organisms. Area backscatter values are those used to derive population estimates in fisheries surveys. In marine fisheries, s_a is commonly scaled to square nautical miles, s_A [m^2 nmi^{-2}] (MacLennan et al. 2002).

Selecting echo sounders and transducers and choosing operational parameters is an exercise in compromise. The choice of acoustic frequency will have the greatest effect on locating, enumerating, and studying fish and zooplankton (Horne and Clay 1998). Lower frequencies have greater transmission ranges and sampling volumes than higher frequencies, whereas higher frequencies tend to have higher resolution and are able to detect smaller targets (MacLennan and Simmonds 1992). Acoustic systems sample the pelagic environment and the sea floor very well, but have limitations near boundaries. For example, the sea floor can be efficiently mapped, but fish within 1–2 m of the bottom are not well detected. Due to the conical shape of the acoustic beam, echoes from the sea floor or air–water interface, which are much larger than echoes from fish, severely contaminate echoes from the water column. This zone is often called the acoustic “dead zone” (Ona and Mitson 1996) and limits the use of hull-mounted transducers for sampling demersal or benthic fish. This effect can be reduced by towing the transducer closer to the seabed, but at the cost of reducing the sampling volume. It is best to determine the goals of the survey or study first, then choose a system and deployment methods to accommodate the requirements.

Before computers, echoes were displayed by burning marks on long paper scrolls. This analog process was sufficient for qualitative analysis, but inadequate for the type of quantitative information required for scientific studies or fisheries management. Computers and modern electrical components have greatly increased our ability to collect, analyze and display acoustical data. The amount of data continues to increase with more sophisticated technology (e.g., multibeam, broadband, and acoustic lens systems), and computing power continues to increase our ability to process and visualize multiple data streams (Towler et al. 2003). However, this increase in information is not necessarily an improvement unless we can interpret the data and relate its importance to scientists, managers, and nonspecialists. The fundamental measurement remains a time series of voltage, and the grand challenge is interpreting and translating these signals into information useful for scientists and managers.

2.2 Surveys

Species-specific, age-based population estimates are often needed by fisheries managers, and as such, set the goal for surveying living marine resources. Traditional methods of surveying fish use nets to sample populations. Trawls are essentially point samples, where the spatial scale of a trawl haul is on the order of 0.1 km^2, and many trawls are required to estimate a population's abundance. However, net catches provide species composition and biological information, such as age, diet, maturity, and sex. In contrast, fisheries acoustics methods continuously sample in the vertical and horizontal dimensions along the ship track, but require other methods, such as trawls or underwater optics, to identify the source of the backscatter. The largest sources of error in fisheries acoustic surveys are incomplete coverage of the entire population, inaccuracy in proportioning backscatter to taxa, and error in estimating the target strengths of individual fish (MacLennan and Simmonds 1992). Minimizing the first source can be accomplished using historical knowledge, commercial catch information, and fisheries-independent surveys. The other two sources of error are discussed later.

Trawl surveys commonly employ a stratified-random design, where strata are based on bathymetry and multiple trawls are conducted in each stratum, whereas acoustic surveys commonly use a standardized design where the entire survey area is sampled systematically. The most common designs are systematic-parallel (Fig. 9.2) and systematic zigzag, where the transects are oriented perpendicular

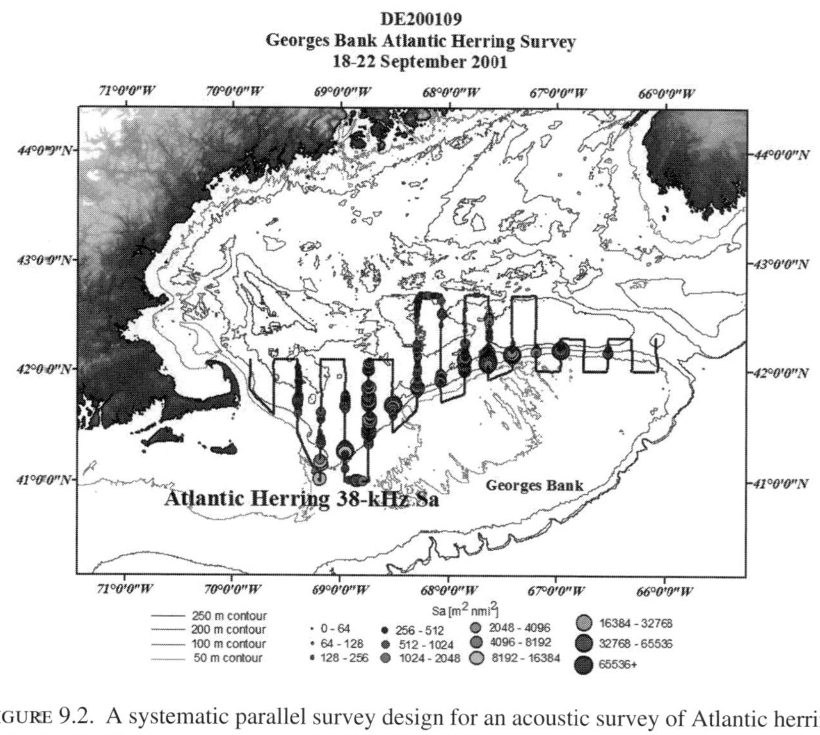

FIGURE 9.2. A systematic parallel survey design for an acoustic survey of Atlantic herring in the Gulf of Maine and Georges Bank regions. This survey was conducted in September 2001. Area backscatter values (s_A) show the spatial distribution of prespawning Atlantic herring.

to the major bathymetric features. Zigzag patterns can be advantageous because they provide greater coverage per unit time than parallel transects and are often used along coastlines. However, zigzag patterns have the disadvantage of transects being correlated at the nodes (i.e., samples between transects are more correlated at the nodes than in the middle of the transects), which can be difficult to deal with statistically (MacLennan and Simmonds 1992). "Star" patterns are often used to survey small areas, such as seamounts (Doonan et al. 2003), but as with zigzag patterns, consideration must be given to the correlation between transects. Systematic-parallel designs are commonly used because they are logistically easy to conduct, and are amenable to design-based analyses and model-based analyses. Design-based analyses treat each transect as a sample by averaging backscatter data along transects and then scale these transect estimates to the survey area. Model-based analyses, such as geostatistical methods, are gaining in popularity for acoustical surveys. Geostatistical methods were originally developed for land-based mining applications, and are designed to incorporate the spatial autocorrelation between samples (e.g., between pings) and the spatial structure inherent in fish distributions (Petitgas 1993).

Long-term time series of abundance are desired in fisheries management to track age classes and cohorts, and to monitor population trends in relation to fishing activities and the environment. In almost every case, populations are monitored using relative indices of abundance or biomass, rather than "absolute" estimates. Relative indices require consistent measurement methodology and the assumption that spatial and temporal distributions are constant over time, or if not, that the survey design accounts for these changes. Acoustical systems achieve measurement consistency through calibration and continuous monitoring of system performance. Fisheries acoustics echo sounders are calibrated before, during, and after surveys to ensure high-quality data. There are many different methods to calibrate acoustic systems (Urick 1983), but the standard method for scientific echo sounders is the standard-target method (Foote et al. 1987). The standard-target method uses a calibration sphere whose material properties and size have been optimized and for which the echo amplitude is known for each frequency (Foote 1982). The sphere is positioned below the transducer and the measured echo amplitudes, beam pattern, and beam widths are related to standard values.

Volume backscatter (s_v) and area backscatter values (s_A) can be used as relative indices of two- and three-dimensional spatial distributions of density and abundance. While conceptually simple, converting these relative indices to absolute estimates is, in practice, difficult. Numeric density is calculated from volume backscatter (s_v) and acoustic backscattering cross-sectional area (σ_{bs}) by:

$$N(f) = \frac{s_v(f)}{\hat{\sigma}_{bs}(f)} \; [\text{no. m}^{-3}], \tag{9.1}$$

where s_v and σ_{bs} are dependent on the acoustic frequency (f) and the "∧" refers to a value that is estimated empirically or theoretically. This equation is valid when packing densities are not high (i.e., nonlinear effects are not significant), all organisms within the sampling volume are "similar," and a single value for σ_{bs} can be used for all the organisms in the sampling volume. Observations of fish aggregations have shown that species tend to aggregate with their own (Pitcher and Parish 1993) and length–frequency distributions are similar within aggregations (Ranta et al. 1992). However, if multiple types of organisms (e.g., swimbladder-bearing fish and non-swimbladder–bearing animals) are present, then s_v and $\hat{\sigma}_{bs}$ are partitioned to the different types of scatterers and multiple frequency methods are essential to solve for N. If the organisms are dispersed, $\hat{\sigma}_{bs}$ can be estimated directly from in situ measures (Jech and Horne 2001). When organisms are aggregated, echoes tend to overlap, making in situ measures difficult, if not impossible and $\hat{\sigma}_{bs}$ must be estimated in other ways. The most common approach for estimating $\hat{\sigma}_{bs}$ is to use an empirical target strength to length regression to convert length-frequency distributions from trawl catches and then calculate the mean backscattering cross section ($\bar{\sigma}_{bs}$). This assumes the trawl catch is representative of the backscatter and the mean best represents the distribution. This is true in many cases but may not be when distributions are multimodal or highly skewed (Jech and Horne 2001).

Empirical models are statistical relationships between observed and measured variables. In fisheries acoustics, the most common relationship, target strength to length regression, is used to convert echo amplitudes to fish lengths. Other common relationships are volume backscatter to abundance or biomass regressions. These regressions are parameterized using in situ and/or ex situ measurements (see reviews in Love 1971; Foote 1991; MacLennan and Simmonds 1992). In situ measurements are optimal because they incorporate the natural behavior of the fish. However, obtaining these measurements can be difficult or prohibitive due to finding suitable densities and length distributions that are representative of the population, the ability to observe and monitor a fish's behavior without disturbing the fish, and minimizing backscatter from unwanted targets. Ex situ measurements include laboratory or controlled field experiments where the fish can be tethered or free swimming. Advantages of ex situ measurements are that factors affecting backscatter can be controlled and behavior can be monitored. The primary disadvantages of ex situ measures are the assumption that ex situ behavior is equivalent to that encountered during surveys and measuring all possible factors affecting echo amplitude can be prohibitive.

The relationships among numeric density, volume backscatter, and backscattering by individuals highlight the critical need for accurate classification of backscatter to species, and accurate estimates of $\hat{\sigma}_{bs}$ to relate echo amplitude to the length or biomass of the fish. Because the acoustic system is calibrated, we have confidence in the validity of these values to the accuracy and precision of the calibration. However, the backscatter must be correctly identified to the appropriate species. The dependence of abundance estimates on σ_{bs} is significant. An error of 3 dB in target strength will result in a factor of two difference in abundance estimates.

2.3 Acoustic Backscatter

Fish are complicated scatterers of sound due to their size, shape, anatomy, and behavior. The single anatomical attribute that is the most significant to sound scattering is the presence or absence of a gas-filled swimbladder. All fish reflect sound, but those with a gas-filled swimbladder will scatter more sound than an identically sized fish without a swimbladder. Reflected sound from the swimbladder has been shown to comprise up to 90% of the total energy backscattered by the fish (Haslett 1962; Foote 1980a). The gas-filled swimbladder is shaped like an irregular, prolate spheroid. The dorsal surface tends to be relatively straight and is constrained by the vertebral column and musculature. The ventral surface tends to be more rounded and the shape is more prone to deformation due to feeding or gonad location than the dorsal surface. Swimbladders have a variety of shapes and chambers (e.g., Jones and Marshall 1953; Whitehead and Blaxter 1964; Foote 1985) that may be lipid-filled (Hayashi and Takagi 1980) or reduced (Neighbors 1992) swimbladders. Two important categories of gas-filled swimbladders are physostome and physoclist. A physostomous fish has ducts connecting the swimbladder to the aquatic environment and is required to gulp air at the surface to inflate the swimbladder. The swimbladder is compressed as the fish migrates downward and gets larger as it shoals; i.e., the swimbladder shape and volume are strongly dependent on the ambient pressure. A physoclistic fish has a closed swimbladder and is able to inflate its swimbladder at depth via the rete mirabile. These fish tend to have small vertical migrations; thus the swimbladder shape and volume tend to remain constant. Whether the fish is a physoclist or a physostome will significantly influence the swimbladder volume and shape when exposed to a change in pressure, and hence echo amplitude (e.g., Mukai and Iida 1996; Gauthier and Rose 2002; Gorska and Ona 2003).

An echo is generated when the acoustic wave encounters an object with an acoustic impedance that is different than that of the surrounding water. The acoustic impedance depends on the material properties of the object: density (ρ [kg m^{-3}]) and the speed of sound (c [m s^{-1}]). The proportion of sound backscattered at the interface of two media (e.g., water and fish body, or swimbladder and flesh) depends on the difference between the acoustic impedances of the two objects; the greater the impedance difference, the greater the backscattered amplitude. In addition to the acoustic impedance, backscatter amplitude is dependent on the size of the object relative to the acoustic wavelength (λ [m], $\lambda = c/f$). For fish, sound scattering can be characterized three different ways depending on the length of the fish. If the ratio of fish length (L) to wavelength is less than 1, fish will reflect sound as an omnidirectional point source. Sound intensity in this Rayleigh region is proportional to L/λ. As the L/λ ratio approaches 1, sound resonates within the gas-filled swimbladder and the sound intensity is proportional to the volume of the swimbladder. When fish length exceeds the insonifying acoustic wavelength (i.e., $L/\lambda > 1$) sound reflection is specular, and is referred to as the "interference" (Love 1971) or geometric scattering region. In this scattering region, the sound intensity is proportional to the insonified surface area of the fish body or swimbladder.

Any anatomical/morphological, physiological, or behavioral attribute that will modify the shape, size, or orientation of the swimbladder will influence echo energy. While density [kg m^{-3}] values are fairly well known for fish anatomical features, sound speeds of different anatomical features or species-specific differences are not well documented. Direct measurements of sound speed have been done for zooplankton (Chu et al. 2000) or for specific components of fish such as lipids (McClatchie and Ye 2000), but sound speeds are inferred for most fish species. Gut fullness and gonad development are two physiological factors that may influence echo amplitude. Gut contents and gonad material can directly influence echo amplitude by modifying the material properties of the fish or indirectly by modifying the shape and volume of the swimbladder (Ona 1990). Other than the general rule that gas-filled swimbladders are about 3–5% of the fish body volume (Sand and Hawkins 1973), relationships of other metrics such as fish body and swimbladder length, or how these relationships change with ontogeny (e.g., isometric or allometric) are not well known. Allometric changes in swimbladder volumes have been recorded in teleost species (Butler and Pearcy 1972) and among individuals within a species (Gee 1968), but generalized relationships have not been established.

Behavioral factors that influence acoustic backscatter can be divided into the orientation and the activity of the fish. Orientation is the tilt, roll, and yaw of the animal relative to the incident wave front. Activity is the swimming motion of a fish. The influence of activity on backscatter has been observed for many years and is well known. However, this influence has only been empirically quantified (Huang and Clay 1980), and predicting echo amplitude as a function of activity with theoretical models remains elusive. The effect of orientation on echo amplitude has been well studied (e.g., Haslett 1965; Foote 1980b; Horne and Jech 1999) with ex situ measurements. Incorporating behavior in estimates of the acoustic backscattering cross section will improve the accuracy of abundance estimates. One method includes incorporating tilt angles in the target strength to length regression (Foote 1980c). This requires information on tilt angle distributions encountered during surveys, which can be difficult to obtain without disturbing the fish. Observations of fish swimming naturally or in cages (Foote and Ona 1987) are quantified using underwater video or photographic images. Because visual ranges in water are limited, optical instruments must be close to the fish and light sources may be required. The presence of a towed body, camera, and artificial light at depth may cause avoidance reactions (e.g., Koslow et al. 1995) or alter "natural" tilt angles (e.g., Kloser and Horne 2003). Techniques used to minimize behavioral responses include the use of low-light cameras and choosing wavelengths of supplementary lights that are not visible to fish.

Early active acoustic measurements were conducted at frequencies typically less than 10 kHz (and often less than a few kilohertz) due to the interest in military and commercial application and backscatter models were based on approximating the swimbladder as a gas-filled spherical bubble (Anderson 1950; Andreeva 1964). Spherical models with corrections for elongation (Weston 1967) and viscosity (Love 1978) are appropriate for modeling acoustic backscatter near resonance frequencies. An advantage of resonance scattering by fish is that echo amplitude is less affected by fish orientation than at higher frequencies (Feuillade and Werby 1994) and echo amplitude is dependent on the volume of the swimbladder, not the shape. Early model results and measurements suggested that resonance peaks could be used to separate fish sizes, and if species were segregated by size, then species could be classified (Holliday 1972). The primary disadvantages to these systems are that they can be quite large and difficult to deploy for survey operations, although recent interest in resonance scattering has generated convenient instrumentation (e.g., Nero et al. 2004) for fisheries surveys, and they tend to be less directional than higher frequency systems.

The majority of fisheries echo sounders operate in the 12–200 kHz frequency range. For locating and enumerating fish, the combination of high resolution and a directional transducer is advantageous. For nearly all juvenile and adult fish, backscatter is in the geometric scattering region at these frequencies. In the geometric scattering region simple geometric shapes are inadequate representations of irregularly shaped swimbladders. Realistic models of backscatter at geometric scattering frequencies must incorporate the anatomy and morphology of all major scattering components. Owing to the complexity in developing analytical and/or numerical scattering models for irregular shapes, most modeling efforts approximate fish bodies and swimbladders as objects that have analytical or numerical solutions. In essence, efforts to predict acoustic scattering by fish have approximated and/or simplified anatomical characteristics, and used scattering models that are themselves mathematical approximations.

Early efforts to predict acoustic backscatter attempted to create a "standard" body form (Haslett 1965) or approximate the fish as arrays of point scatterers (Huang and Clay 1980; Clay and Heist 1984). Foote (1985) was the first to use actual swimbladder morphometry to predict fish backscatter. Clay and Horne (1994) utilized finite cylinder (Stanton 1989) and Kirchhoff approximations to model backscatter by fish. This Kirchhoff-ray mode (KRM) model utilizes digital morphometry of the fish body and swimbladder, usually obtained from radiograph images (Jech and Horne 2002) or Computer Tomography imaging of live, anaesthetized fish (JF Webb, personal communication), to predict backscatter as a function of fish length, orientation, and acoustic frequency for a variety of species (Jech et al. 1995; Horne et al. 2000). The KRM model can be used to predict echo amplitude from the fish body, different anatomical features (e.g., swimbladder), and by coherently adding the backscatter from the different anatomical components, backscatter for the whole fish (Fig. 9.3). The dominant characteristic of all echo amplitude response curves in the geometric scattering region is the undulating pattern of peaks and nulls (Fig. 9.3). These peaks and nulls are due to the constructive and destructive interference between the incident acoustic signal and the echoes from the interfaces between anatomical features. Because the acoustic impedance contrast between the water and body is less than that of the swimbladder, the target strength of the fish body is lower than that of the swimbladder for this L/λ range. At higher L/λ ratios, target strength of the fish body becomes comparable to the swimbladder

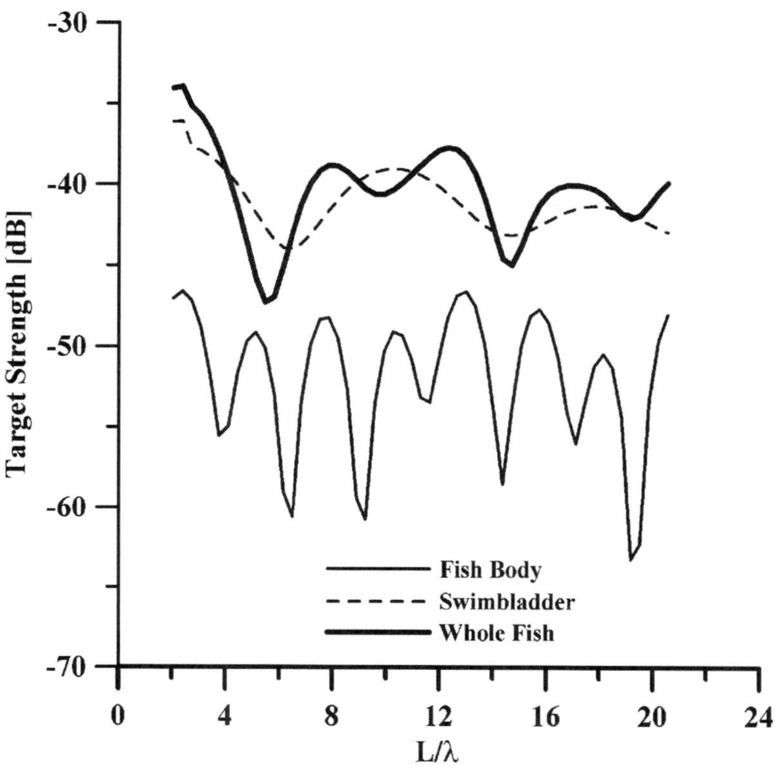

FIGURE 9.3. Target strength as a function of fish length (L) and acoustic wavelength (λ) at dorsal incidence (tilt angle = 90°) for a 25-cm (total length) alewife (*Alosa pseudoharengus*) as predicted by the KRM model. KRM model parameters: $c_{water} = 1460\,m\,s^{-1}$, $\rho_{water} = 1000\,kg\,m^{-3}$, $c_{fishbody} = 1570\,m\,s^{-1}$, $\rho_{fishbody} = 1080\,kg\,m^{-3}$, $c_{swimbladder} = 240\,m\,s^{-1}$, $\rho_{swimbladder} = 2.64\,kg\,m^{-3}$. Fish length was kept constant, so acoustic frequency ranged from 12 kHz ($L/\lambda = 2.05$) to 120 kHz ($L/\lambda = 20.55$).

(not shown). Echo amplitude from the "whole" fish (combining the fish body and swimbladder) is comparable to that from the swimbladder, demonstrating that gas-filled swimbladder dominates the echo.

Changes in fish orientation will significantly influence backscatter amplitude and variability, which directly affects accuracy of echo amplitude to fish length conversions and density, abundance, or biomass estimates (Eq. 9.1). Similar to a transducer's beam pattern, backscatter from a fish is directional (Stanton et al. 2003). A fish's beam pattern is dependent on the insonifying frequency and the shape and orientation of the fish body and swimbladder (Fig. 9.4). As the L/λ ratio increases, smaller tilt angles result in greater changes in target strength and the number of side lobes increases (i.e., variability in target strength increases with increasing frequency). Another characteristic of gas-filled swimbladders is that the maximum echo amplitude corresponds to the angle of the swimbladder within the fish body. Swimbladders of pelagic species are often oriented with

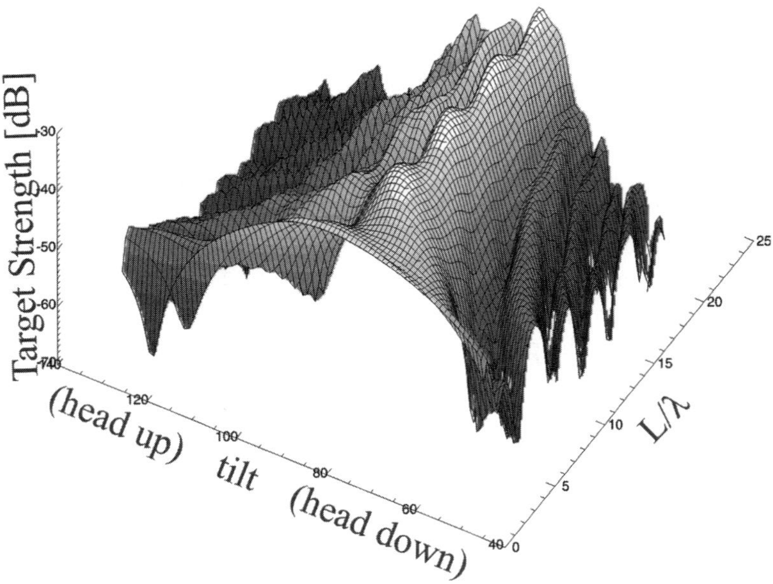

FIGURE 9.4. Target strength as a function of fish length (*L*), acoustic wavelength (λ), and tilt angle as predicted by the KRM model for a 25-cm alewife. KRM parameters are equivalent to those in Figure 9.3.

the posterior end downward relative to the rostrocaudal axis (i.e., along the vertebrae) of the fish. For alewife, the swimbladder is angled about 10°, which corresponds to a maximum target strength at 80° tilt angle.

The fish's backscattering beam pattern has significant implications for absolute estimates of density and abundance. For example, at 38 kHz, a horizontally swimming alewife (*Alosa pseudoharengus*) would have a target strength of approximately –43 dB. If that same fish swims toward the bottom at a 10° angle, the target strength would increase to –32 dB. Using a generic target strength to length equation (Love 1971), the estimated length of the horizontally swimming fish would be 13 cm while the downward swimming fish would be 48 cm. Assuming fish mass is proportional to the cube of the length, this 400% change in estimated length potentially alters biomass estimates by a factor of 64.

2.4 Acoustic Identification

Relating and identifying acoustic backscatter to taxa (e.g., species) is one of the great challenges of active acoustics (MacLennan and Holliday 1996). Acoustically determining fish length and identifying species is complicated by nonlinear and nonmonotonic relationships among fish structure, size, behavior, and acoustic frequency. This is especially true when using a single frequency. The de facto method for classifying and apportioning backscatter is to sample the backscatter with nets or optics, and over time build a visual library of the relationship between backscatter patterns and species composition. This method has been successful when the species of interest is amassed in monospecific aggregations. However, this method requires significant initial effort to discriminate backscatter and continuous monitoring using nets or optics, and it is subjective. Visual interpretation of echograms is based on the experience of the personnel scrutinizing the echograms. Objective methods to classify backscatter are advantageous because they are not dependent on personal experience, should provide consistent results, and the method can be quantified.

Ultimately, remote species identification will require more information, which acoustically means increasing the number of frequencies. Multifrequency data are obtained from multiple, discrete frequencies or broadband sources. Multifrequency data require multiple, single-frequency echo sounders, which require additional laboratory and hull space. Broadband transducers transmit a continuous, wideband signal (e.g., a frequency modulated "chirp") over a wide frequency range (>10% of the center frequency). In addition to increasing frequency information, broadband systems can provide greater spatial resolution (Chu and Stanton 1998) than single-frequency systems and may be able to measure fish orientation directly (Stanton et al. 2003; Coombs and Barr 2004). Because broadband systems transmit over a wide bandwidth, the energy per frequency is lower than that transmitted in single-frequency transducers, which limits the useable range of these systems. Ideally, measurements should span the transition from resonance (hundreds of hertz) to geometric scattering (tens of kilohertz).

The "ideal" acoustic system has yet to be developed, but significant effort has been applied to classification. Initial efforts focused on plankton (McNaught 1969; Holliday 1972) and later fish (e.g., Miyanohana et al. 1990; McClatchie et al. 1996). Holliday (1977, 1980) introduced an inverse method to estimate size-based density using multiple frequencies. The inverse approach has been used to estimate densities and size distributions of fish (Johnson 1977; Kalish et al. 1986), to separate fish from plankton (Saetersdal et al. 1984; Cochrane et al. 1991), and to classify and discriminate species (Simmonds and Armstrong 1990; Zakharia et al. 1996; Barr 2001). However, in the geometric scattering region, nonmonotonic scattering characteristics of fish limit the use of the inverse approach (Horne and Jech 1999), and the inverse approach tends to be more effective with the availability of more frequencies.

While 38 kHz is the standard frequency used for abundance estimates of marine fish, it is common to have echo sounders collecting data at two or three frequencies during acoustic surveys. A common problem when surveying fish is to separate the backscatter from zooplankton such as krill or shrimp from the backscatter by fish (Korneliussen and Ona 2003). Because most zooplankton do not have a gas-bearing organ, the combination of a lower frequency system (e.g., 12, 18, or 38 kHz) with a higher frequency system (e.g., 120 or 200 kHz) can be used to separate backscatter from these two types of scatters (McKelvey 2000; Kloser et al. 2002).

A recent technological innovation (DIDSON) and an innovative application of long-range sonar are at the opposite ends of the resolution spectrum, but give us new perspectives on spatial and temporal distributions of fish. The dual-frequency identification sonar (DIDSON) is an acoustic "imaging" system that provides very high resolution (order of millimeters) images, analogous to underwater video (e.g., Tiffan et al. 2004). The sampling range of the DIDSON is tens of meters, so while it will have limited capability to survey populations, it has significant potential for monitoring organisms in and around reefs, benthic habitat classification, and observing behavior. Long-range sonar systems have been in use for decades, but a recent set of measurements on the US continental shelf (Makris et al. 2006) may renew interest for fisheries applications. The Makris et al. system can detect fish aggregations over tens of kilometers, but has a resolution of about 500 m. This system shows significant promise for monitoring the behavior of aggregations (e.g., vertical and horizontal migration, dispersion, and coalescence) and may be useful for estimating local density and abundance.

2.5 Future of Active Acoustics

Fisheries acoustics continues to evolve at a rapid pace. Availability of new, digital hardware and development of backscatter models has initiated research in many areas. Integration of experimental measures with modeling efforts is becoming a standard approach and will continue to guide the use and interpretation of acoustic data. The amount of information and data are increasing through wider frequency bandwidth, larger sample volumes from multibeam systems or long-range sonars, and integration of multiple environmental sensors with acoustic data. Multibeam systems were originally developed for mapping bathymetry, but recent interest in applying this instrumentation to fisheries (e.g., Gerlotto and Paramo 2003) has resulted in the development of new systems. Quantifying multibeam data is still in its infancy, and will require development of calibration methods (Cochrane et al. 2003; Foote et al. 2005), volume backscattering and target detection algorithms (Mayer et al. 2002), and echo amplitude prediction at all angles of orientation (e.g., Jech and Horne 2002). Numeric and analytic models of acoustic backscatter continue to progress through improved measures of fish material properties and corroboration with ex situ and in situ measures (Reeder et al. 2004). Integration of biological and environmental sensors will advance the development of multi-species and ecosystems approaches in resource conservation and management. Combining broadband and multibeam acoustic technologies with environmental sensors in a single instrument should provide adequate bandwidth and frequency range to acoustically survey, discriminate, and track all organisms in the water column. New deployment methods, such as remotely operative vehicles (ROVs), autonomously operated vehicles (AUVs), and stationary buoys need to be developed to house multiple sensors to enhance our survey capabilities. The grand challenge remains to intelligently and efficiently utilize this information.

3. Passive Acoustics

3.1 Introduction

Passive acoustics takes advantage of the sounds produced by fishes to eavesdrop on their behavior. Most fish sounds are associated with aggression, courtship, and spawning. The courtship and spawning sounds have been most intensively studied and can provide high resolution time series of reproductive behavior (e.g., Connaughton and Taylor 1995). The ultimate goal of using passive acoustics is to provide data on where and when spawning takes place, and how many fish are involved.

The study of fish sounds began in earnest after World War II with the work of Fish and Mowbray (1970), who sought to catalog the sounds made by many different species of fish for the US Navy. Their work showed that a great number of fishes produced species-specific sounds. However, most of their work was taken out of ecological context, where the fish were prodded to make sounds. The next major advances were the studies of Winn (1964) and Tavolga (1977), who investigated the sounds produced by naturally behaving fish as well as the mechanisms of sound production. A high point of this period was the monograph by Breder (1967) on the diel and seasonal production of sounds by several fish species in southwest Florida. As most of these sounds were associated with courtship and spawning, they were taken as proxies of the reproductive behaviors of these fishes.

From these early studies passive hydrophone surveys have been used as powerful, nondestructive tools for locating sound-producing males and to document the time and place of spawning over large spatial and temporal scales (Mok and Gilmore 1983; Saucier et al. 1992; Saucier and Baltz 1993; Mann and Lobel 1995; Luczkovich et al. 1999; Gilmore 2003). While active sonar has been widely adopted in the past 20 years for estimating fish abundance, passive acoustics has not seen widespread adoption. Part of the reason for this is that the technology (both hardware and software) for conducting these studies has not been readily available commercially. Another major reason is that with the exception of a few studies (notably Luczkovich et al. 1999b) passive acoustics has not been developed to the point of ground-truthing data sets so that measurements of sound production can be converted into estimates of fish numbers or egg production, which are often sought by fisheries agencies. This is analogous to active acoustics where the target strengths of fish being studied need to be measured before an echo sounder can be used to estimate fish abundance. A further reason is that listening for fish takes place against the noisy background of the sea. Ambient noise in the sea can vary widely for natural reasons, and levels increase as wind and weather conditions deteriorate (Wenz 1962; Urick 1983). There are also many human-made sources of noise including shipping, harbors, seismic surveys and other sources that can mask the sounds from fish. In particular, noise from a listening vessel itself may make it difficult to detect the sounds, especially if the vessel is under way.

The field of passive acoustics will likely see increasingly rapid development because of the recent interest in ecosystem-level management. It is increasingly important to understand the spawning habitats of fishes and the patterns of fish spawning over large spatial and temporal scales.

3.2 Instrumentation

Passive acoustic instrumentation uses a hydrophone to convert the sounds produced by fishes into a voltage that can be recorded and analyzed. In its simplest form, the sounds can be recorded, played back in real time, and the sounds characteristic of particular species identified by an experienced listener. Sounds can also be identified by converting them into sound spectrograms, which display changes in their frequency and temporal characteristics with time (e.g., Fig. 9.5). However, fish sounds also lend themselves to automated sampling and analysis. Most fish sounds are relatively low in frequency (< 3 kHz), so that

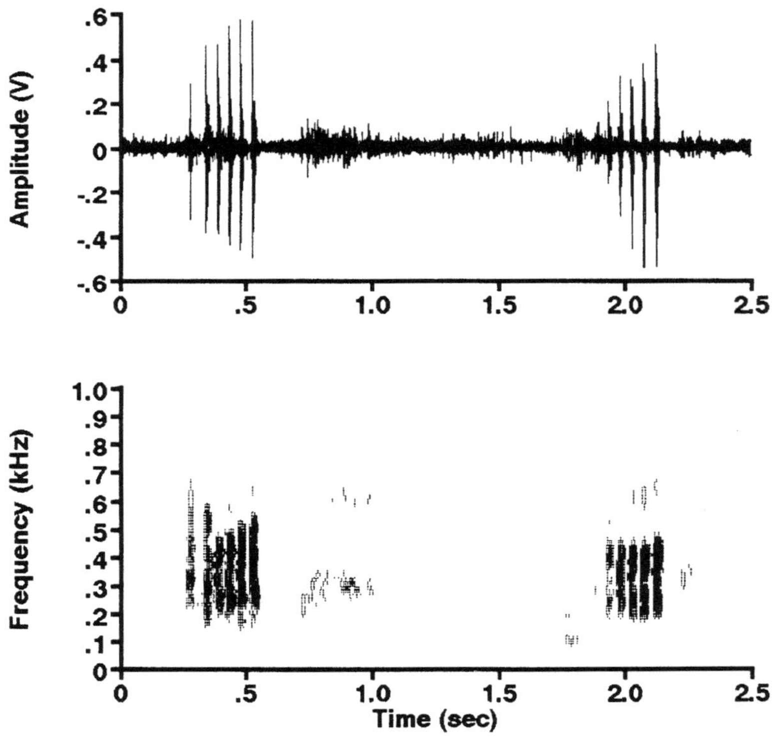

FIGURE 9.5. Courtship sounds of the male domino damselfish *Dascyllus albisella*. (**A**) Oscillogram of sounds produced during two signal jumps, the courtship behavior in which the male rises in the water and then swims down rapidly while making this sound. (**B**) Spectrogram of the same call in (**A**) showing that the domino damselfish sounds have frequencies between 200 and 600 Hz. (From Lobel and Mann [1995].)

systems digitizing at low sampling rates (e.g., 6 kHz) can capture the frequencies found in fish sounds. The computerized data acquisition system could be directly connected to the hydrophone or connected through a wireless system (e.g., Mann and Lobel 1995). Alternatively, a self-contained battery-operated data recording computer can be used to record the fish sounds to a hard drive or flash memory (e.g., Locascio and Mann 2005). Each type of system has its advantages. A hard-wired or wireless system allows the user to verify that it is working on a regular basis. But, there are many situations, such as recording in the open ocean or in remote locations, where it is not feasible to implement such a system. A data-logging system can be used anywhere, but there is usually no way of knowing that it is working between deployment and retrieval.

Passive acoustic systems generate large amounts of raw data. Listening to the raw recordings and looking at their spectrograms is useful for exploratory studies, for identifying species that are calling, and for identifying unknown calls. The current challenge of passive listening is to develop automated signal processing routines for identifying species and measuring the level of sound production. It is simply not possible to listen to all of the data that will be collected. But, it is important to remember that the best signal processing routine may still not be as good as the human brain in being able to distinguish sounds.

3.3 *Examples*

In this section we provide three examples of passive acoustics to study reproductive behavior in three fish families: damselfish (Pomacentridae), croakers and drums (Sciaenidae), and haddock (Gadidae). We then consider future advances in passive acoustics needed to meet the challenges of identifying fish habitat, enumerating the number of fish present and investigating the relationship between sound production and egg production.

3.3.1 Damselfishes (Pomacentridae)

Damselfishes (Pomacentridae) are small fishes that typically inhabit coral reefs. Male damselfish produce a behavior known as the signal jump in which a male rises in the water column, and then swims down rapidly while producing a pulsed sound (Myrberg 1972; Spanier et al. 1979; Lobel and Mann 1995). The sounds produced by different species are typically unique in the number of pulses and/or pulse rate (Spanier et al. 1979). However, there appears to be overlap between some closely related species, such as the three-spot damselfish (*Dasyllus trimaculatus*) and the domino damselfish (*Dascyllus albisella*) (*D. trimaculatus*: Spanier 1970; Luh and Mok 1986; *D. albisella*: Lobel and Mann 1995).

Because damselfish sounds are stereotypical, it is possible to devise relatively simple detection schemes to automatically detect and count sounds produced by individual damselfish. This approach is also made easier because males are territorial and establish territories that may be limited to individual coral heads. Thus, a hydrophone can be placed in the center of a male's territory in order to

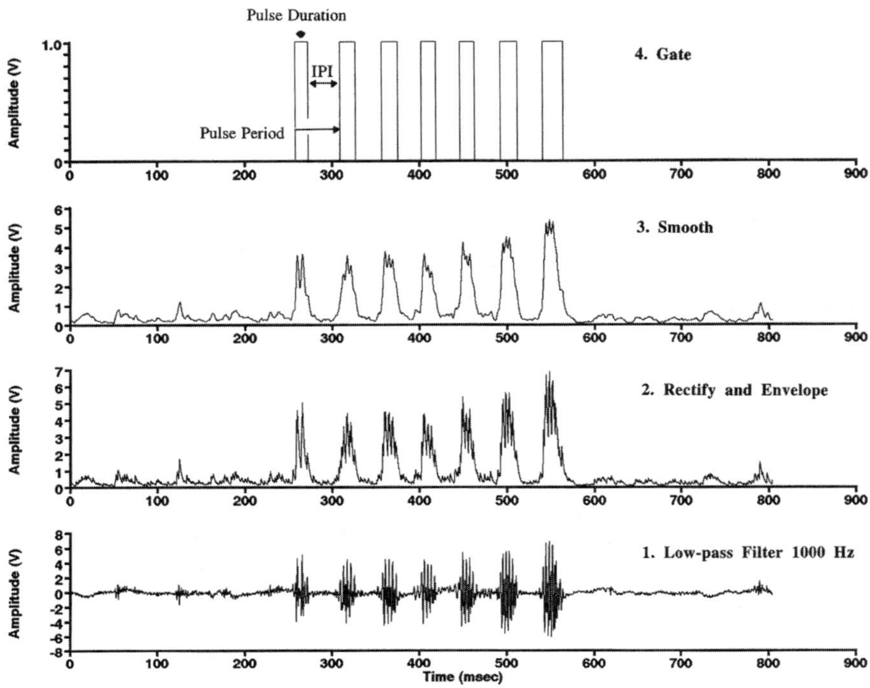

FIGURE 9.6. Automated detection of *Dascyllus albisella* sounds. Processing steps are shown from the bottom panel toward the top panel. The first step is a 1000-Hz low-pass filter, followed by rectification (i.e., taking absolute value) and enveloping of the signal. In step 3 the signal is then smoothed before gating in step 4 with a threshold at 1.5 V.

record sounds from that individual. Damselfishes do not chorus as do sciaenids, which makes detection of individual sounds simple because there is usually little overlap in sounds of the same level.

A wireless system was developed to broadcast sounds from the field to a real-time detection system (Mann and Lobel 1995). This system used a basic detection algorithm that involved the following steps (Fig. 9.6):

1. Filtering: Bandpass filtering limits the signal to the frequencies containing the most energy of the damselfish call.
2. Rectify and envelope: Rectifying is simply taking the absolute value of the hydrophone signal. The envelope effectively "traces" the outlines of the signal so that individual pulses can be detected.
3. Smoothing: This further smooths the signal just leaving behind a clean envelope of the pulses within a signal.
4. Gate. The gate is a simple threshold detection to identify the times when the processed signal went above a certain voltage. This threshold was set by hand, but not adjusted during data collection, so that sounds that were heard were detected by the system.

The system recorded the time of each pulse detected and its duration and amplitude. It did not save actual recordings of the sounds owing to the limitations of hard disks available at the time. Today, it would make sense to detect and record each pulse or call, which would allow other types of postprocessing to be implemented after the data had been collected. However, the same detection scheme could be implemented in software to process large amounts of data.

Female domino damselfish lay eggs in nests that are guarded by males for about 5 days before they hatch at dusk, and the larvae enter the water column. Nesting in this species is synchronized so that males typically receive eggs over the same 1- or 2-day period. There is no additional egg laying on subsequent days while a male guards a nest (Mann and Lobel 1995). In other damselfishes, such as the sergeant major (*Abudefduf* sp.), males may guard nests containing eggs of many different ages. The nesting behavior of the domino damselfish allows simple documentation of the timing and number of eggs received by a given male. This makes an ideal test case to compare the level of sound production to the number of eggs received by a male. Figure 9.7 shows a time series of sound production for an individual male along with the status of egg presence in his nest. This time series shows that sound production peaks the day before or day of egg laying.

Detailed analysis of such a time series shows that the courtship rates of an individual are highly variable over time. Thus, simply measuring sound production for a few minutes would not be adequate for detecting the periods

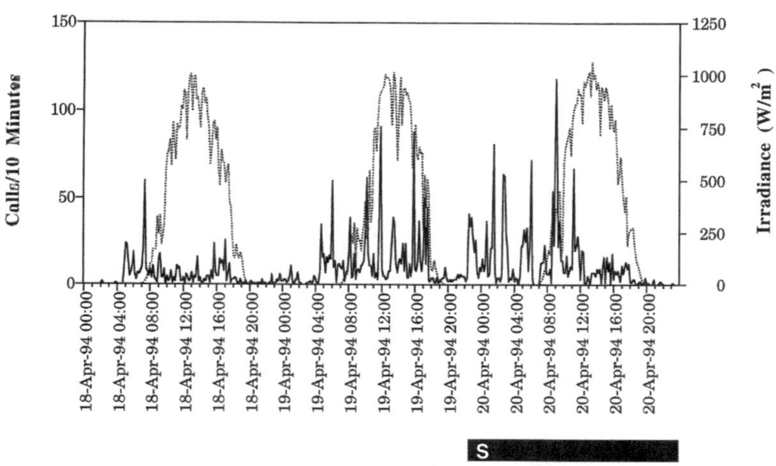

FIGURE 9.7. Time series of *Dascyllus albisella* courtship and spawning behavior. Solid line indicates number of calls detected per 10-minute period. Dotted line indicates surface irradiance (i.e., night is the period with no irradiance). S indicates day when spawning occurred, and the black bar indicates when eggs present in nest. (From Mann and Lobel [1995].)

of highest courtship levels, which often occurred at night. Thus, determining the variability of sound production in time is important for understanding what sampling schedule is adequate for characterizing sound production. In the domino damselfish, there are many periods with no sound production and periods of intense sound production, which coincide when a female is moving between nests and deciding where to lay her eggs.

3.3.2 Croakers and Drum (Sciaenidae)

Members of the family Sciaenidae, the croakers and drums, are likely the most soniferous fishes in the temperate and subtropical coastal oceans of the world. Many sciaenid species form chorusing groups where the sounds of many individuals overlap. Indeed, the sounds can be so loud that they can be heard above water. It can often be difficult to distinguish an individual's call when a hydrophone is lowered into a chorus of sciaenids.

Like damselfishes, sciaenids tend to have species-specific sounds (Fig. 9.8). Although, some closely related species, such as sand seatrout (*Cynoscion arenarius*) and weakfish (*Cynoscion regalis*) have calls that are indistinguishable. Sciaenids have been the subject of numerous studies on the distribution of spawning fishes (Mok and Gilmore 1983; Saucier et al. 1992; Saucier and Baltz 1993; Luczkovich et al. 1999; Gilmore 2003).

Since sciaenids typically form choruses, it is not possible (with a single hydrophone) to count the number of sounds produced by an individual in a chorus. Thus, measurements of the overall sound level in different frequency bands have been recently used to give a relative estimate of the level of chorusing activity (e.g., Locascio and Mann 2005). The received sound level is a function of several factors including the number of fish calling, the call source level, the call rate, the distribution of fish, and the sound propagation characteristics of the environment. There are surprisingly few estimates of many of these parameters. For example, there is only one paper with published source levels for a sciaenid, silver perch (Sprague and Luczkovich 2004). There are currently no published papers on the propagation of sciaenid sounds, although there are sound propagation studies on other species (oyster toadfish [Fine and Lenhardt 1983]; domino damselfish [Mann Lobel 1997]; midshipman [Bass and Clark 2003]). Finally, most of these species inhabit murky waters and call at night, so it is not possible to see the distribution of calling fish. This is one area where a combination of active and passive acoustics is especially promising.

Despite these current limitations, sciaenids typically produce a strong signal that is easily recorded and quantified. A composite spectrogram of sounds recorded from a datalogger in Charlotte Harbor, FL overnight is shown in Figure 9.9. The datalogger was programmed to record 10 s of sound every 10 minutes. The results show that sound production starts in the early evening with sand seatrout calls, and then progresses to include both sand seatrout and spotted seatrout. In this chorus, it is difficult to distinguish sand seatrout calls from silver perch calls.

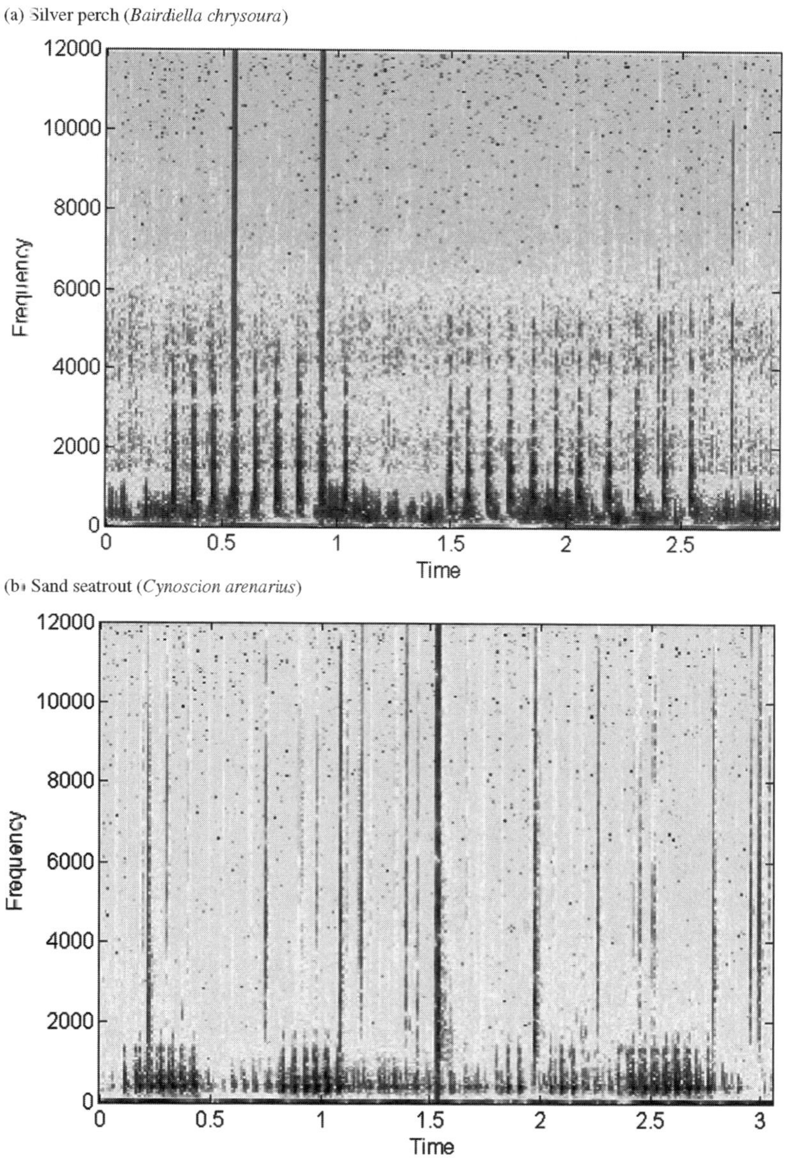

FIGURE 9.8. Spectrograms of sound production by (**a**) silver perch and (**b**) spotted seatrout. Time is in seconds and frequency is in hertz. The fish sounds are pulsed, with silver perch having a larger bandwidth (about 300–6000 Hz) than spotted seatrout (about 100–1200 Hz). Irregularly spaced vertical lines are clicks from invertebrates.

3.3.3 Cod and Haddock (Gadidae)

The Gadidae, or cod family, forms the basis of important fisheries in northern waters. Many species within the family, including the cod *Gadus morhua* and the

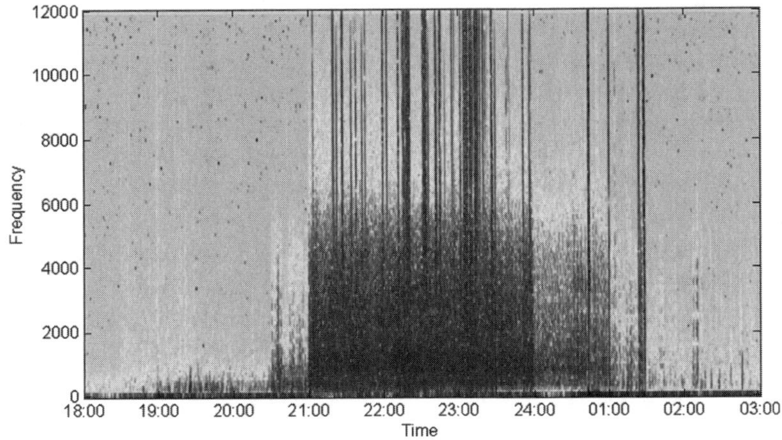

FIGURE 9.9. Composite spectrogram showing time series of sciaenid sound production in Charlotte Harbor, FL. Sound production begins around 1900 hours with sand seatrout sound production. Silver perch sound production starts around 2030 hours, showing the increased frequency range produced by their calls. Silver perch sound levels increase dramatically at 2100 hours continuing through approximately 0100 hours, with indications of clipping of recordings where sounds cover the recording bandwidth.

haddock *Melanogrammus aeglefinus*, are active sound producers (Hawkins and Rasmussen 1978). The sounds are low in frequency, with most of their energy below 1 kHz, but the calls of different species can readily be identified via their temporal characteristics.

Figure 9.10 shows the waveforms of sounds from four species of gadoid fish, cod, haddock, pollack (*Pollachius pollachius*), and tadpole fish (*Raniceps raninus*). During aggressive activities and at spawning, the male cod produces short grunts, made up of a series of rapidly repeated pulses (Finstad and Nordeide 2004). The male haddock produces longer calls, made up of a series of more widely spaced pulses, the repetition rate varying with the context of the call (Hawkins and Amorim 2000). Thus, a solitary male haddock may engage in a territorial display, producing long sequences of sound with the pulses repeated at a slow rate. As a female approaches and courtship proceeds the pulses are repeated more and more rapidly, culminating in a continuous humming from the male fish (Fig. 9.11). In this instance it is possible not only to locate fish that are ready to spawn but also to determine whether active courtship is taking place.

In the management of cod and haddock stocks more attention is being paid to the adoption of measures to protect spawning fish. It is becoming increasingly evident that spawning in these fish is a structured activity, with elaborate courtship behavior and strong mate selection. Spawning may take place at particular locations, especially when population sizes are reduced. In the past, fishing often concentrated on accumulations of spawning fish, with unknown effects upon spawning success. With declining stocks, especially of cod, it would seem prudent to close spawning areas to fishing. However, it will be a major

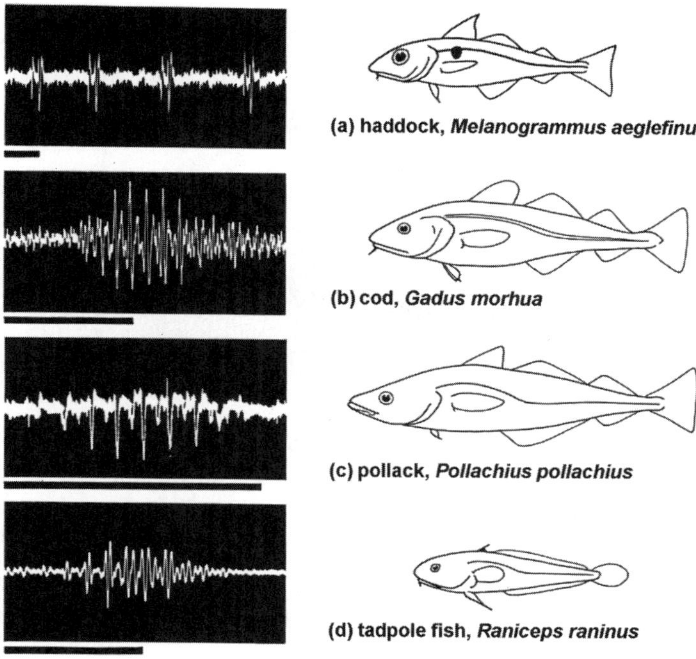

FIGURE 9.10. Calls produced by four different members of the family Gadidae from the northern North Sea. The time base (black bar) in each case is 100 ms. All of these calls were produced during aggressive behavior outside the spawning season. (**a**) A short series of repeated knocks from the haddock. (**b**) A grunt from the cod. (**c**) A grunt from the lythe; these are repeated at short intervals. (**d**) A grunt from the tadpole fish. All the sounds are made up of rapidly repeated pulses (each associated with the sharp contraction of paired muscles attached to the swimbladder). The variations in the calls result from differences in the patterns of contraction of the muscles in the different species.

task to identify the spawning areas of these widely distributed fishes, living as they do on the continental shelf of the North Atlantic.

Sounds have been recorded from cod on the well known cod spawning grounds close to the Lofoten Islands (Nordeide and Kjellsby 1999). It has also proved possible to identify a discrete spawning area for haddock within a Norwegian fjord simply by listening for sounds from the spawning fish (Hawkins et al. 2002). It is now necessary to extend such listening surveys over much wider areas, including those open areas of the sea where the majority of cod and haddock spawn. Such surveys pose a number of challenges.

3.4 Challenges

Although some species of fish produce relatively loud sounds (reaching 160 dB re: 1 μPa at 1 m in the case of some sciaenids [J. Locascio, personal communication]), the majority of sounds from fish are relatively low in amplitude.

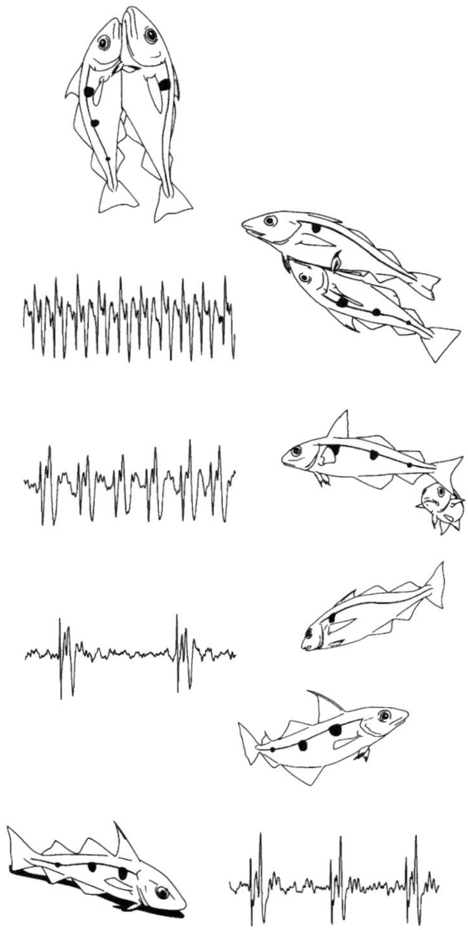

FIGURE 9.11. Spawning behavior of the haddock *Melanogrammus aeglefinus* and accompanying sounds. (**a**) Solitary display by male, moving in tight circles on the substrate, emitting a long sequence of regularly repeated low frequency sound pulses. (**b**) Male moves up to meet approaching female, presenting his flank, emitting a short sequence of sound pulses. (**c**) Male leading female, swinging body from side to side and emitting a continuous sequence of fast sound pulses. (**d**) Male attempting to mount female from below, emitting very rapid sound pulses. (**e**) Male pressed firmly to female, with synchronous release of eggs and sperm, in silence. Each sound sequence is 280 ms long. Male fish shows three distinctive marks on the flank, the female only one.

Moreover, they tend to be composed mainly of low frequencies (below 1–2 kHz), which fall within the noise spectrum from ships. For this reason, passive listening is usually conducted using fixed hydrophones, or from ships that are moored and have their engines switched off. It would not be possible to conduct a passive listening survey using a conventional omnidirectional hydrophone from a large moving research vessel—even the exceptionally quiet fisheries research ship *Scotia* (described by Fernandes et al. 2000).

Coverage of an area can be achieved by deploying a hydrophone from a ship that stops and starts, but the process is inevitably slow. There may be some scope in the future for the deployment of large towed hydrophone arrays, which can be directed to listen in the vertical plane and reject the sound from the ship itself. Here the limitation may be set by the self noise from the array itself as it is towed through the water. Coverage of inshore areas may also be achieved through the siting of a large number of discrete listening stations.

Despite the many years that passive acoustics has been used to study fishes, the field is still in its infancy. New techniques for detecting and characterizing populations exist including hydrophone arrays and wavelet analysis. Hydrophone arrays have been used for many years for locating sound-producing animals, perhaps with the greatest success in tracking cetaceans.

4. Summary

Because fish scatter sounds, especially from their swimbladders, active acoustics can be used to map and quantify fish abundance. Active acoustics uses a pulse of sound generated by a transducer, and either the same transducer (monostatic) or one or more receivers (bistatic) are used to "listen" for echoes. In fisheries acoustics, the most common configuration is to use the same transducer to both transmit and receive acoustic signals. The challenge of active acoustics has been to develop models of fish sound scattering and to use these to quantify numbers and identify species of scatterers.

Passive acoustics uses the sounds made by fishes to understand their distribution and, because most sounds are produced in relation to courtship and spawning, to understand the dynamics of spawning. In passive acoustics, it is usually clear which species are being studied, as most species make a species-specific sound. It has been more difficult to quantify fish abundance from the sounds they produce.

The future of each of these fields depends on developing algorithms to process large data sets and to automatically classify the species under study. Finally, there is great potential for combining passive and active acoustic systems to study fish populations. Many of the issues related to understanding fish numbers and distributions that are difficult to study with passive acoustics could be answered with active acoustic systems.

Acknowledgments. This work was supported by the Office of Naval Research and the Northeast Fisheries Science Center by awards to JMJ, and also by Mote Marine Laboratory, Florida Sea Grant, and the National Undersea Research Program (NOAA/NURP) by awards to DAM.

References

Alvarez A, Ye Z (1999) Effects of fish school structures on acoustic scattering. ICES J Mar Sci 56:361–369.

Anderson VC (1950) Sound scattering from a fluid sphere. J Acoust Soc Am 22:426–431.

Andreeva IB (1964) Scattering of sound by air bladders of fish in deep sound scattering ocean layers. Sov Phys-Acoust 10:17–21.

Balls R (1948) Herring fishing with the echometer. J Cons Int Explor Mer 15:193–206.

Barr R (2001) A design study of an acoustic system suitable for differentiating between orange roughy and other New Zealand deep-water species. J Acoust Soc Am 109: 164–178.

Bass AH, Clark CW (2003) The physical acoustics of underwater sound communication. In: Simmons AM, Popper AN, Fay RR (eds) Acoustic Communication. New York: Springer-Verlag, pp. 15–64.

Brawn VM (1961) Sound production by the cod (*Gadus callarias* L). Behaviour 18: 239–255.

Breder CM (1967) Seasonal and diurnal occurrences of fish sounds in a small Florida Bay. Bull Am Mus Nat Hist 138:329–278.

Butler JL, Pearcy WG (1972) Swimbladder morphology and specific gravity of myctophids off Oregon. J Fish Res B Can 29:1145–1150.

Chu D, Stanton TK (1998) Application of pulse compression techniques to broadband acoustic scattering by live individual zooplankton. J Acoust Soc Am 104:39–55.

Chu D, Wiebe P, Copley N (2000) Inference of material properties of zooplankton from acoustic and resistivity measurements. ICES J Mar Sci 57:1128–1142.

Clay CS, Heist BG (1984) Acoustic scattering by fish- acoustic models and a 2–parameter fit. J Acoust Soc Am 75:1077–1083.

Clay CS, Horne JK (1994) Acoustic models of fish: the Atlantic cod (*Gadus morhua*). J Acoust Soc Am 96:1661–1668.

Cochrane NA, Sameoto D, Herman AW, Neilson J (1991) Multiple-frequency acoustic backscattering and zooplankton aggregations in the inner Scotian shelf basins. Can J Fish Aquat Sci 48:340–355.

Cochrane NA, Li Y, Melvin GD (2003) Extraction of calibrated volume backscattering strength from multi-beam sonar.ICES J Mar Sci 60:669–677.

Connaughton MA, Taylor MH (1995) Seasonal and daily cycles in sound production associated with spawning in weakfish, *Cynoscion regalis*. Environ Biol Fish 42: 233–240.

Coombs RF, Barr R (2004) Acoustic remote sensing of swimbladder orientation and species mix in the oreo population on the Chatham Rise. J Acoust Soc Am 115: 1516–1524.

Doonan IJ, Bull B, Coombs RF (2003) Star acoustic surveys of localized fish aggregations. ICES J Mar Sci 60:132–146.

Dragesund O, Olsen S (1965) On the possibility of estimating year-class strength by measuring echo-abundance of 0–group fish. Fiskeridir Skr Ser Havunders 13:47–75.

Fernandes PG, Brierley A, Simmonds EJ, Millard NW, McPhail SD, Armstrong F, Stevenson P, Squires M (2000) Fish do not avoid survey vessels. Nature 404:35–36.

Feuillade C, Werby MF (1994) Resonances of deformed gas bubbles in liquids. J Acoust Soc Am 96:3684–3692.

Fine ML, Lenhardt ML (1983) Shallow-water propagation of the toadfish mating call. Comp Biochem Physiol 76A:225–231.

Finstad JL, Nordeide JT (2004) Acoustic repertoire of spawning cod, *Gadus morhua*. Environ Biol Fish 70:427–433.

Fish MP, Mowbray WH (1970) Sounds of Western North Atlantic fishes. A Reference File of Biological Underwater Sounds. Baltimore: The John Hopkins University Press.

Foote KG (1980a) Importance of the swimbladder in acoustic scattering by fish: a comparison of gadoid and mackerel target strengths. J Acoust Soc Am 67:2084–2089.

Foote KG (1980b) Effect of fish behaviour on echo energy: the need for measurement of orientation distributions. J Cons Int Explor Mer 39:193–201.

Foote KG (1980c) Averaging of fish target strength functions. J Acoust Soc Am 67: 504–515.

Foote KG (1982) Optimizing copper spheres for precision calibration of hydroacoustic equipment. J Acoust Soc Am 71:742–747.

Foote KG (1983) Linearity of fisheries acoustics, with addition theorems. J Acoust Soc Am 73:1932–1940.

Foote KG (1985) Rather-high-frequency sound scattering by swimbladdered fish. J Acoust Soc Am 78:688–700.

Foote KG (1991) Summary of methods for determining fish target strength at ultrasonic frequencies. J Cons Int Explor Mer 48:211–217.

Foote KG, Ona E (1987) Tilt angles of schooling penned saithe. J Cons Int Explor Mer 43:118–121.

Foote KG, Knudsen HP, Vestnes G, MacLennan DN, Simmonds EJ (1987) Calibration of acoustic instruments for fish density estimation: a practical guide ICES Coop Res Rep 44.

Foote KG, Chu D, Hammar TR, Baldwin KC, Mayer LA, Hufnagle, LC Jr, Jech JM (2005) Protocols for calibrating multibeam sonar. J Acoust Soc Am 117:2013–2027.

Furusawa M, Ishii K, Miynohana Y (1992) Attenuation of sound by schooling fish. J Acoust Soc Am 92:987–994.

Gauthier S, Rose GA (2002) *In situ* target strength studies on Atlantic redfish (*Sebastes* spp). ICES J Mar Sci 59:805–815.

Gee JH (1968) Adjustment of buoyancy by longnose dace (*Rhinichthys cataractae*) in relation to velocity of water. J Fish Res B Can 25:1485–1496.

Gerlotto F, Paramo J (2003) The three dimensional morphology and internal structure of Clupeid schools as observed using vertical scanning multibeam sonar.Aquatic Living Resources 16:113–122.

Gilmore RG (2003) Sound production and communication in the spotted seatrout. In: Bortone SA (ed) Biology of the Spotted Seatrout. New York: CRC Marine Biology Series, pp. 177–195.

Gorska N, Ona E (2003) Modelling the effect of swimbladder compression on the acoustic backscattering from herring at normal or near-normal dorsal incidences. ICES J Mar Sci 60:1381–1391.

Haslett RWG (1962) Determination of the acoustic scatter patterns and cross sections of fish models and ellipsoids. Br J Appl Phys 13:611–620.

Haslett RWG (1965) Acoustic backscattering cross sections of fish at three frequencies and their representation on a universal graph. Br J Appl Phys 16:1143–1150.

Hawkins AD, Amorim MCP (2000) Spawning sounds of the male haddock, *Melanogrammus aeglefinus*. Environ Biol Fish 59:29–41.

Hawkins AD, Rasmussen KJ (1978) The calls of gadoid fish. J Mar Biol Assoc UK 58:891–911.

Hawkins AD, Casaretto L, Picciulin M, Olsen K (2002) Locating spawning haddock by means of sound. Bioacoustics 12:284–285.

Hayashi K, Takagi T (1980) Occurrence of unusually high levels of wax esters in deep-sea teleost fish muscle, *Hoplostethus atlanticus*. Bull Jpn Soc Sci Fish 46:459–463.

Holliday DV (1972) Resonance structure in echoes from schooled pelagic fish. J Acoust Soc Am 51:1322–1333.

Holliday DV (1977) Extracting bio-physical information from the acoustic signature of marine organisms. In: Anderson NR, Zahuranec BJ (eds) Oceanic Sound Scattering Prediction. New York: Plenum Press.

Holliday DV (1980) Use of acoustic frequency diversity for marine biological measurements. In: Diemer FP, Vernberg FJ, Mirkes DZ (eds) Advanced Concepts in Ocean Measurements for Marine Biology. BW Baruch Library in Marine Science no. 10, pp. 423–460.

Horne JK, Clay CS (1998) Sonar systems and aquatic organisms: matching equipment and model parameters. Can J Fish Aquat Sci 55:1296–1306.

Horne JK, Jech JM (1999) Multi-frequency estimates of fish abundance: constraints of rather high frequencies. ICES J Mar Sci 56:184–199.

Horne JK, Walline PD, Jech JM (2000) Comparing acoustic model predictions to in situ backscatter measurements of fish with dual-chambered swimbladders. J Fish Biol 57:1105–1121.

Huang K, Clay CS (1980) Backscattering cross-sections of live fish: PDF and aspect. J Acoust Soc Am 67:795–802.

Jech JM, Horne JK (2001) Effects of in situ target spatial distributions on acoustic density estimates. ICES J Mar Sci 58:123–136.

Jech JM, Horne JK (2002) Three-dimensional visualization of fish morphometry and acoustic backscatter. Acoust Res Lett Online 3:35–40. http://wwwojpsaiporg/ARLO.

Jech JM, Schael DM, Clay CS (1995) Application of three sound scattering models to threadfin shad (*Dorosoma petenense*). J Acoust Soc Am 98:2262–2269.

Johannesson KA, Mitson RA (1983) Fisheries acoustics. FAO Fisheries Technical Paper 240:1–249.

Johnson RK (1977) Acoustic estimation of scattering-layer composition. J Acoust Soc Am 61:1636–1639.

Jones FRH, Marshall NB (1953) The structure and function of the teleostean swimbladder. Biol Rev 28:16–83.

Kalish JH, Greenlaw CF, Pearcy WG, Holliday DV (1986) The biological and acoustical structure of sound scattering layers off Oregon. Deep-Sea Res 33:631–653.

Kimura K (1929) On the detection of fish-groups by an acoustic method. J Imp Fish Inst Tokyo 24:41–45.

Kloser RJ, Ryan T, Sakov P, Williams A, Koslow JA (2002) Species identification in deep water using multiple acoustic frequencies. Can J Fish Aquat Sci 59:1065–1077.

Kloser RK, Horne JK (2003) Characterizing uncertainty in target-strength measurements of a deepwater fish: orange roughy (Hoplostethus atlanticus). ICES J Mar Res 60: 516–523.

Korneliussen RJ, Ona E (2003) Synthetic echograms generated from the relative frequency response. ICES J Mar Sci 60:636–640.

Koslow AJ, Kloser R, Stanley CA (1995) Avoidance of a camera system by a deepwater fish, the orange roughy (*Hoplostethus atlanticus*). Deep-Sea Res (Pt 1) 42:233–244.

Lobel PS, Mann DA (1995) Spawning sounds of the domino damselfish, *Dascyllus albisella* (Pomacentridae), and the relationship to male size. Bioacoustics 6:187–198.

Locascio JV, Mann DA (2005) Effects of Hurricane Charley on fish chorusing. Biol. Lett. 1: 362–365.

Love RH (1971) Measurements of fish target strength: a review. Fish Bull 69:703–715.

Love RH (1978) Resonant scattering by swimbladder bearing fish. J Acoust Soc Am 64:571–580.

Luczkovich JJ, Sprague MW, Johnson SE, Pullinger RC (1999a) Delimiting spawning areas of weakfish, *Cynoscion regalis* (family Sciaenidae), in Pamlico Sound, North Carolina using passive hydroacoustic surveys. Bioacoustics 10:143–160.

Luczkovich JJ, Daniel HJ III, Sprague MW, Johnson SE, Pullinger RC, Jenkins T, Hutchinson M (1999b) Characterization of critical spawning habitats of weakfish, spotted seatrout and red drum in Pamlico Sound using hydrophone surveys. North Carolina Dept of Environ and Nat Resour, Div Mar Fish, Morehead City, NC.

Luh HK, Mok HK (1986) Sound production in the domino damselfish *Dascyllus trimaculatus* (Pomacentridae) under laboratory conditions. Jpn J Ichthyol 33: 70–74.

MacLennan DN, Holliday DV (1996) Fisheries and plankton acoustics: past present, and future. ICES J Mar Sci 53:513–516.

MacLennan DN, Simmonds EJ (1992) Fisheries Acoustics. London: Chapman and Hall.

MacLennan DN, Fernandes PG, Dalen J (2002) A consistent approach to definitions and symbols in fisheries acoustics. ICES J Mar Sci 59: 365–369.

Makris NC, Ratilal P, Symonds DT, Jagannathan S, Lee S, Nero RW (2006) Fish population and behavior revealed by instantaneous continental shelf-scale imaging. Science 311:660–663.

Mann DA, Jarvis SM (2004) Potential sound production by a deep-sea fish. J Acoust Soc Am 115:2331–2333.

Mann DA, Lobel PS (1995) Passive acoustic detection of sounds produced by the damselfish, *Dascyllus albisella* (Pomacentridae). Bioacoustics 6:199–213.

Mann DA, Lobel PS (1997) Propagation of damselfish (Pomacentridae) courtship sounds. J Acoust Soc Am 101:3783–3791.

Mann DA, Bowers-Altman J, Rountree RA (1997) Sounds produced by the striped cusk eel, *Ophidion marginatum*, during courtship and spawning. Copeia 610–612.

Mayer L, Li Y, Melvin G (2002) 3D visualization for pelagic fisheries research and assessment. ICES J Mar Sci 59:216–225.

McClatchie S, Ye Z (2000) Target strength of an oily deep-water fish, orange roughy (*Hoplostethus atlanticus*) II. Modeling. J Acoust Soc Am 107:1280–1285.

McClatchie S, Alsop J, Coombs RF (1996) A re-evaluation of relationships between fish size, acoustic frequency, and target strength. ICES J Mar Sci 53:780–791.

McKelvey DR (2000) The use of two frequencies to interpret acoustic scattering layers. MS Thesis, University of Washington, Seattle, WA.

McNaught DC (1969) Developments in acoustic plankton sampling. In: Proceedings of the 12th Conference on Great Lakes Research, Ann Arbor, Michigan, USA: 61–68.

Medwin H, Clay CS (1998) Fundamentals of Acoustical Oceanography. New York: Academic Press.

Miyanohana Y, Ishii K, Furusawa M (1990) Measurements and analyses of dorsal-aspect target strength of six species of fish at four frequencies. Rapp P-V Réun Cons Int Explor Mer 189:317–324.

Mok HK, Gilmore RG (1983) Analysis of sound production in estuarine spawning aggregations of *Pogonia cromis, Bairdiella chrysoura*, and *Cynoscion nebulosus* (Sciaenidae). Bull Acad Zool Acad Sinica 22:157–186.

Mukai T, Iida K (1996) Depth dependence of target strength of live kokanee salmon in accordance with Boyle's Law. ICES J Mar Sci 53:245–248.

Myrberg AA Jr (1972) Ethology of the bicolor damselfish, *Eupomacentrus partitus* (Pisces:Pomacentridae): a comparative analysis of laboratory and field behavior. Anim Behav Mon 5:197–283.

Neighbors MA (1992) Occurrence of inflated swimbladders in five species of lanternfishes (family Myctophidae) from waters off southern California. Mar Biol 114:355–363.

Nero RW, Thompson CH, Jech JM (2004) In situ acoustic estimates of the swimbladder volume of Atlantic herring (*Clupea harengus*). ICES J Mar Sci 61:323–337.

Nordeide JT, Kjellsby E (1999) Sound from spawning cod at their spawning grounds. ICES J Mar Sci 56:326–332.

Ona E (1990) Physiological factors causing natural variations in acoustic target strength of fish. J Mar Biol Assoc UK 70:107–127.

Ona E, Mitson R (1996) Acoustic sampling and signal processing near the seabed: the deadzone revisited. ICES J Mar Sci 53:677–690.

Petitgas P (1993) Geostatistics for fish stock assessments: a review and an acoustic application. ICES J Mar Sci 50:285–298.

Pitcher TJ, Parish JK (1993) Functions of shoaling behaviour in teleosts. In: Pitcher TJ (ed) Behaviour of Teleost Fishes. London: Chapman & Hall, pp. 363–439.

Rallier du Baty R (1927) La pêche sur le banc de Terre-Neuve et autour del îles. Saint-Pierre et Miquelon Office Scientifique et Technique des Pêches Maritimes Mémoires (Série Spécial), 7.

Ranta E, Lindström K, Peuhkuri N (1992) Size matters when three-spined sticklebacks go to school. Anim Behav 43:160–162.

Reeder DB, Jech JM, Stanton TK (2004) Broadband acoustic backscatter and high-resolution morphology of fish: measurement and modeling. J Acoust Soc Am 116: 729–746.

Saetersdal G, Stromme T, Bakken B, Piekutowski L (1984) Some observations on frequency-dependent backscattering strength. FAO Fish Rep 300:150–156.

Sand O, Hawkins AD (1973) Acoustic properties of the cod swimbladder. J Exp Biol 58:797–820.

Saucier MH, Baltz DM (1993) Spawning site selection by spotted seatrout, *Cynoscion nebulosus*, and black drum, *Pogonias cromis*, in Louisiana. Environ Biol Fishes 36: 257–272.

Saucier MH, Baltz DM, Roumillat WA (1992) Hydrophone identification of spawning sites of spotted seatrout *Cynoscion nebulosus* (Osteichthys: Sciaenidae) near Charleston, South Carolina. NE Gulf Sci 12:141–145.

Simmonds EJ, Armstrong F (1990) A wideband echo sounder:measurements on cod, saithe and herring, and mackerel from 27 to 54 kHz. Rapp P-V Réun Cons Int Explor Mer 189:381–387.

Spanier E (1970) Analysis of sounds and associated behavior of the domino damselfish *Dascyllus trimaculatus* (Ruppell, 1828) (Pomacentridae). MSc Thesis, Tel-Aviv University, Tel-Aviv.

Spanier E (1979) Aspects of species recognition by sound in four species of damselfishes, genus *Eupomacentrus* (Pisces: Pomacentridae). Z Tierpsychol 51:301–316.

Sprague M, Luczkovich J (2004) Measurement of an individual silver perch *Bairdiella chrysoura* sound pressure level in a field recording. J Acoust Soc Am 116:3186–3191.

Stanton TK (1989) Sound scattering by cylinders of finite length. III. Deformed cylinders. J Acoust Soc Am 86:691–705.

Stanton TK, Reeder DB, Jech JM (2003) Inferring fish orientation from broadband-acoustic echoes. ICES J Mar Sci 60:524–531.
Sund O (1935) Echo sounding in fishery research. Nature 135:953.
Tavolga WN (1977) Mechanism for directional hearing in the sea catfish (*Arius felis*). J Exp Biol 67:97–115.
Tiffan KF, Rondorf DW, Skalicky JJ (2004) Imaging fall Chinook salmon redds in the Columbia River with a dual-frequency identification sonar. N Am J Fish Manag 24:1421–1426.
Toresen R (1991) Absorption of acoustic energy in dense herring schools studied by the attenuation in the bottom echo signal. Fish Res 10:317–327.
Towler RH, Jech JM, Horne JK (2003) Visualizing fish movement, behavior, and acoustic backscatter. Aquat Liv Res 16:277–282.
Trout GC, Lee AJ, Richardson ID, Harden Jones RR (1952) Recent echosounder studies. Nature 170:71–72.
Urick RJ (1983) Principles of Underwater Sound. New York: McGraw-Hill.
Wenz GM (1962) Acoustic ambient noise in the ocean: spectra and sources. J Acoust Soc Am 34:1936–1956.
Weston DE (1967) Sound propagation in the presence of bladder fish. In Albers VM (ed) Underwater Acoustics. New York: Plenum Press, pp. 55–88.
Whitehead PJP, Blaxter JHS (1964) Swimbladder form in clupeoid fishes. Zool J Linn Soc 97:299–372.
Winn HE (1964) The biological significance of fish sounds. In: Tavolga WN (ed) Marine Bio-Acoustics. Oxford: Pergamon Press, pp. 213–231.
Zakharia ME, Magand F, Hetroit F, Diner N (1996) Wideband sounder for fish species identification at sea. ICES J Mar Sci 53:203–209.
Zhao X, Ona E (2003) Estimation and compensation models for the shadowing effect in dense fish aggregations. ICES J Mar Sci 60:155–163.

Index

SPRINGER HANDBOOK OF AUDITORY RESEARCH *(continued from page ii)*

Volume 22: Evolution of the Vertebrate Auditory System
Edited by Geoffrey A. Manley, Arthur N. Popper, and Richard R. Fay

Volume 23: Plasticity of the Auditory System
Edited by Thomas N. Parks, Edwin W. Rubel, Arthur N. Popper, and Richard R. Fay

Volume 24: Pitch: Neural Coding and Perception
Edited by Christopher J. Plack, Andrew J. Oxenham, Richard R. Fay, and Arthur N. Popper

Volume 25: Sound Source Localization
Edited by Arthur N. Popper and Richard R. Fay

Volume 26: Development of the Inner Ear
Edited by Matthew W. Kelley, Doris K. Wu, Arthur N. Popper, and Richard R. Fay

Volume 27: Vertebrate Hair Cells
Edited by Ruth Anne Eatock, Richard R. Fay, and Arthur N. Popper

Volume 28: Hearing and Sound Communication in Amphibians
Edited by Peter M. Narins, Albert S. Feng, Richard R. Fay, and Arthur N. Popper

Volume 29: Auditory Perception of Sound Sources
Edited by William A. Yost, Arthur N. Popper and Richard R. Fay

Volume 30: Active Processes and Otoacoustic Emissions in Hearing
Edited by Geoffrey A. Manely, Arthur N. Popper and Richard R. Fay

Volume 31: Auditory Trauma, Protection, and Repair
Edited by Jochen Schacht, Arthur N. Popper and Richard R. Fay

Volume 32: Fish Bioacoustics
Edited by Jacqueline F. Webb, Richard R. Fay, Arthur N. Popper

Volume 33: Hair Cell Regeneration, Repair, and Protection
Edited by Richard J. Salvi, Arthur N. Popper and Richard R. Fay

For more information about the series, please visit www.springer-ny.com/shar.

Printed In The United States Of America